MICROBIOLOGY FOR SURGICAL INFECTIONS

MICROBIOLOGY FOR SURGICAL INFECTIONS

DIAGNOSIS, PROGNOSIS AND TREATMENT

Edited by

KATERYNA KON MD, PHD
Kharkiv National Medical University, Kharkiv, Ukraine

MAHENDRA RAI MSC, PHD
Sant Gadge Baba Amravati University, Amravati, Maharashtra, India
Department of Chemical Biology, Institute of Chemistry, University of Campinas, Campinas SP, Brazil

ELSEVIER

AMSTERDAM • BOSTON • HEIDELBERG • LONDON
NEW YORK • OXFORD • PARIS • SAN DIEGO
SAN FRANCISCO • SINGAPORE • SYDNEY • TOKYO
Academic Press is an imprint of Elsevier

Academic Press is an imprint of Elsevier
32 Jamestown Road, London NW1 7BY, UK
225 Wyman Street, Waltham, MA 02451, USA
525 B Street, Suite 1800, San Diego, CA 92101-4495, USA

British Library Cataloguing-in-Publication Data
A catalogue record for this book is available from the British Library

Library of Congress Cataloging-in-Publication Data
A catalog record for this book is available from the Library of Congress

ISBN: 978-0-12-411629-0

For information on all Academic Press publications
visit our website at elsevierdirect.com

Printed and bound by CPI Group (UK) Ltd, Croydon, CR0 4YY

14 15 16 17 18 10 9 8 7 6 5 4 3 2 1

Contents

5. Anastomotic Leakage after Colonic and Rectal Surgery: Operative and Conservative Management

ELIA POIASINA, MICHELA GIULII CAPPONI AND LUCA ANSALONI

6. Acute Appendicitis: An Open Issue. Current Trends in Diagnostic and Therapeutic Options

MICHELE PISANO, MICHELA GIULII CAPPONI AND LUCA ANSALONI

7. Assessment of Severity and Prognosis in Surgical Patients with Secondary Peritonitis

RODOLFO LEONEL BRACHO-RIQUELME

8. Pathogenesis and Management of Dialysis Access Infections

HARIHARAN REGUNATH, VENKATESH KUMAR ARIYAMUTHU AND KUNAL CHAUDHARY

9. Microbiological Diagnosis and Antimicrobial Therapy in Acute Pancreatitis

LIPIKA SINGHAL, AMBER PRASAD AND VIKAS GAUTAM

III

CARDIOVASCULAR INFECTIONS

10. Infective Endocarditis

ALI ALIZZI, JOHN-PAUL TANTIONGCO AND SARAH SHEPARD

IV

SKIN AND SOFT TISSUE INFECTIONS

11. Natural Remedies for the Treatment of Wounds and Wound Infection

KATERYNA KON AND MAHENDRA RAI

12. Bioactive Molecules of Herbal Extracts with Anti-Infective and Wound Healing Properties

ALI PARSAEIMEHR, YI-FENG CHEN AND ELMIRA SARGSYAN

13. The Role of Honey and Propolis in the Treatment of Infected Wounds

ANANIAS PASCOAL, XESUS FEÁS, TERESA DIAS, LUÍS G. DIAS AND LETICIA M. ESTEVINHO

14. Necrotizing Soft Tissue Infections

DONALD E. FRY

15. Significance of Surgical Intervention in the Management of Diabetic Foot Infections

SHALBHA TIWARI, DALIPARTHY DEVI PRATYUSH, SANJEEV KUMAR GUPTA AND SURYA KUMAR SINGH

V

SKELETAL INFECTIONS

16. Spinal Epidural Abscesses

ANDREAS K. DEMETRIADES AND FLORIAN RINGEL

VI

SURGICAL INFECTIONS OF NON-BACTERIAL ORIGIN

Preface

Surgical infections represent a diverse group of diseases, which despite advances in techniques of surgery and anesthesia, the presence of modern equipment and improving perioperative health care in hospitals, still lead to significant morbidity and mortality. The rapid increase in levels of antibiotic resistance and the appearance of new multidrug-resistant pathogens makes it necessary to constantly update recommendations on the management of infections, and the present book reviews the most recent guidelines on the prevention, diagnosis and treatment of surgical infections of different locations, with particular emphasis on intra-abdominal, cardiovascular and skin and soft tissue infections.

The scope of the chapters encompasses reviews of *in vitro* studies of the principles of prevention of surgical infections, such as the evaluation of bacterial adherence to surgical materials, as well as clinical studies on the management of a broad spectrum of surgical infections, including anastomotic leakage after colorectal surgery, infectious complications of dialysis access, infective endocarditis, necrotizing soft tissue infections, diabetic foot infections and others. In addition, alternative methods of antimicrobial treatment of surgical infections are also discussed in several chapters, such as *in vitro* and *in vivo* studies on wound healing and anti-infectious properties of plant extracts, essential oils, and zootherapeutics methods.

Although the most common cause of surgical infections is bacteria, the role of other microorganisms should not be disregarded. In consideration of this fact, one chapter is also devoted to understanding diagnostic approaches for invasive mycoses in surgical patients, as this pathology has attracted much attention in recent years.

Selection of the optimal treatment strategy is impossible without predicting a probable outcome of the infection based on a patient's laboratory and clinical parameters. Nowadays, there are a large number of studies dedicated to the development of scoring systems using modern statistical methods for assessment of the severity of a patient's state and for predicting the course and outcome in different surgical infections. Some such scoring systems have gained great popularity in the medical community, such as APACHE II (Acute Physiology and Chronic Health Evaluation II) system, SAPS (Simplified Acute Physiology Score), MPI (Mannheim Peritonitis Index), etc., and this book also summarizes studies of the efficacy of these and other scoring systems in the prognosis of surgical infections, particularly in secondary peritonitis.

The book explores current trends in the etiology and antibiotic resistance of pathogens causing different types of surgical infections; it discusses recent advances in diagnostic approaches in bacterial and non-bacterial surgical infections; it reviews methods of prognosis of the course and outcome of surgical infections; and it also summarizes recent guidelines for prophylaxis of infectious complications in surgery, and for improvement of diagnosis and treatment of surgical infections.

The book will be very useful to microbiologists, surgeons, infectious diseases specialists, researchers in surgery, clinical microbiologists, pharmacologists, and those who are interested in tackling the problem of antibiotic resistance.

The editors would like to thank Elizabeth Gibson, Editorial Project Manager, Academic Press/Elsevier S&T Books, Waltham, MA, USA for her help and valuable suggestions, the contributors for devoting their time and efforts to this book, and the reviewers for their comments for improving the chapters. Prof. Rai thankfully acknowledges FAPESP, Brazil for providing financial support to visit the Institute of Chemistry, Biological Chemistry Laboratory, Universidade Estadual de Campinas, Campinas, SP, Brazil.

Kateryna Kon
Mahendra Rai

List of Contributors

Ali Alizzi Cardiothoracic Surgical Unit, Ashford Hospital, Adelaide, Australia.

Luca Ansaloni General Surgery Department, Papa Giovanni XXIII Hospital, Bergamo, Italy.

Venkatesh Kumar Ariyamuthu Department of Medicine, University of Missouri Columbia, Missouri.

Ramkrishna Bhalchandra Tata Medical Center, Kolkata, India.

Sanjay Bhattacharya Tata Medical Center, Kolkata, India.

Rodolfo Leonel Bracho-Riquelme Instituto de Investigación Científica de la Universidad Juárez del Estado de Durango, Durango, México.

Michela Giulii Capponi General Surgery Department, Papa Giovanni XXIII Hospital, Bergamo, Italy.

Kunal Chaudhary Department of Medicine, University of Missouri Columbia, Missouri.

Yi-Feng Chen Institute of Biotechnology, Jiangsu Academy of Agricultural Sciences, Nanjing, Jiangsu, China.

Andreas K. Demetriades Department of Neurosurgery, Western General Hospital, Edinburgh, UK.

Daliparthy Devi Pratyush Department of Endocrinology and Metabolism, Institute of Medical Sciences, Banaras Hindu University, Varanasi, India.

Luís G. Dias CIMO-Mountain Research Center, Department of Biology and Biotechnology, Agricultural College of Bragança, Polytechnic Institute of Bragança, Bragança, Portugal.

Teresa Dias CIMO-Mountain Research Center, Department of Biology and Biotechnology, Agricultural College of Bragança, Polytechnic Institute of Bragança, Bragança, Portugal.

Elena Eraso Departamento de Inmunología, Microbiología y Parasitología, Universidad del País Vasco/Euskal Herriko Unibertsitatea (UPV/EHU), Bilbao, Spain.

Jaime Esteban Department of Clinical Microbiology, IIS-Fundación Jiménez Díaz, Madrid, Spain.

Leticia M. Estevinho CIMO-Mountain Research Center, Department of Biology and Biotechnology, Agricultural College of Bragança, Polytechnic Institute of Bragança, Bragança, Portugal.

Guillermo Ezpeleta Departamento de Medicina Preventiva y Salud Pública, Facultad de Medicina y Odontología, Universidad del País Vasco/Euskal Herriko Unibertsitatea (UPV/EHU), Bilbao, Spain.

Xesus Feás Department of Organic Chemistry, Faculty of Sciences, University of Santiago de Compostela, Lugo, Spain.

Donald E. Fry Department of Surgery, Northwestern University Feinberg School of Medicine, Chicago, Illinois, and the Department of Surgery, University of New Mexico School of Medicine, Albuquerque, New Mexico.

Vikas Gautam Department of Medical Microbiology, Postgraduate Institute of Medical Education and Research, Chandigarh, India.

Gaurav Goel Tata Medical Center, Kolkata, India.

Enrique Gómez-Barrena Department of Orthopedics, IdiPAZ-Hospital Universitario La Paz, Madrid, Spain.

Karolin Graf Institute for Medical Microbiology and Hospital Epidemiology, Hannover Medical School, Hannover, Germany.

Sanjeev Kumar Gupta Department of General Surgery, Institute of Medical Sciences, Banaras Hindu University, Varanasi, India.

Torsten Herzog Department of Surgery, St. Josef Hospital, Ruhr-University Bochum, Germany.

Lalawmpuia Hmar Tata Medical Center, Kolkata, India.

Kateryna Kon Department of Microbiology, Virology and Immunology, Kharkiv National Medical University, Kharkiv, Ukraine.

Surya Kumar Singh Department of Endocrinology and Metabolism, Institute of Medical Sciences, Banaras Hindu University, Varanasi, India.

Ali Parsaeimehr G. S. Davtyan Institute of Hydroponics Problems, National Academy of Sciences, Yerevan, Republic of Armenia.

Ananias Pascoal CIMO-Mountain Research Center, Department of Biology and Biotechnology, Agricultural College of Bragança, Polytechnic Institute of Bragança, Bragança, Portugal.

Javier Pemán Servicio de Microbiología, Hospital Universitario La Fe, Valencia, Spain.

Concepción Pérez-orge-eremarch Department of Clinical Microbiology, IIS-Fundación Jiménez Díaz, Madrid, Spain.

Ramón Pérez-Tanoira Department of Clinical Microbiology, IIS-Fundación Jiménez Díaz, Madrid, Spain.

Michele Pisano General Surgery Department, Papa Giovanni XXIII Hospital, Bergamo, Italy.

Elia Poiasina General Surgery Department, Papa Giovanni XXIII Hospital, Bergamo, Italy.

Amber Prasad Department of Medical Microbiology, Postgraduate Institute of Medical Education and Research, Chandigarh, India.

Guillermo Quindós Departamento de Inmunología, Microbiología y Parasitología, Universidad del País Vasco/Euskal Herriko Unibertsitatea (UPV/EHU), Bilbao, Spain.

Mahendra Rai Department of Biotechnology, Sant Gadge Baba Amravati University, Amravati, Maharashtra, India.

Hariharan Regunath Department of Medicine, University of Missouri Columbia, Missouri.

Florian Ringel Department of Neurosurgery, Klinikum rechts der Isar, Technische Universität München, München, Germany.

Ferran Sanchez-Reus Servicio de Microbiología, Hospital de la Santa Creu i Sant Pau, Barcelona, Spain.

Elmira Sargsyan G. S. Davtyan Institute of Hydroponics Problems, National Academy of Sciences, Yerevan, Republic of Armenia.

Sarah Shepard SA Pathology, Adelaide, Australia.

Lipika Singhal Department of Medical Microbiology, Postgraduate Institute of Medical Education and Research, Chandigarh, India.

John-Paul Tantiongco Cardiothoracic Surgical Unit, Flinders Medical Center, Adelaide, Australia.

Shalbha Tiwari Department of Endocrinology and Metabolism, Institute of Medical Sciences, Banaras Hindu University, Varanasi, India.

Waldemar Uhl Department of Surgery, St. Josef Hospital, Ruhr-University Bochum, Germany.

Ralf-Peter Vonberg Institute for Medical Microbiology and Hospital Epidemiology, Hannover Medical School, Hannover, Germany.

SECTION I

INFECTION CONTROL MEASURES FOR THE PREVENTION OF SURGICAL INFECTIONS

CHAPTER

1

Infection Control Measures for the Prevention of Surgical Site Infections

Karolin Graf and Ralf-Peter Vonberg

Institute for Medical Microbiology and Hospital Epidemiology, Hannover Medical School, Hannover, Germany

CONSEQUENCES OF SURGICAL SITE INFECTIONS

It is as simple as this: The best you can do about surgical site infections (SSI) is not to let them happen in the first place. Today the occurrence of SSI represents one of the most severe complications in all types of surgical procedures, and it will have an enormous impact on the ongoing course of recovery of the affected patient. It leads to significantly increased morbidity and mortality of patients in various medical disciplines.[1–3] Besides their importance with respect to such clinical consequences, SSI should also concern hospitals and health care systems for economic reasons.[4,5] Depending upon the type of surgery, the prolonged length of stay (LOS) in the hospital may vary from 3 to 21 days.[1] According to a review of the literature on SSI in general[6] and to a recently published large estimation on SSI in orthopedics (412,356 total hip and 784,335 total knee arthroplasties) in particular,[7] SSI will roughly double the magnitude of the cost for a case patient. Based on the 2005 Healthcare Cost and Utilization Project National Inpatient Sample (HCUP NIS), which included 723,490 surgical patients with 6,891 SSI, the average LOS was 9.7 days and costs increased by $20,842 per admission. This leads to nearly 1,000,000 additional inpatient-days and $1,600,000,000 excess costs for the US alone.[8]

Evidence-based infection control guidelines for the prevention of SSI have been published, for example by the Hospital Infection Control Practices Advisory Committee (HICPAC) from the Centers for Disease Control and Prevention (CDC)[9] and by the Society for Healthcare Epidemiology of America (SHEA) in collaboration with the Infectious Diseases Society of America (IDSA) (Table 1.1).[10] This chapter provides an overview of patient-derived (endogenous) risk factors for SSI development, and summarizes the most

Microbiology for Surgical Infections
DOI: http://dx.doi.org/10.1016/B978-0-12-411629-0.00001-5

TABLE 1.1 Summary of Evidence-based Recommendations of the Society for Healthcare Epidemiology of America (SHEA) for SSI Prevention

Recommendation	Grade
Use evidence-based standards when implementing prevention measures.	A-II
Control serum blood glucose levels; reduce glycosylated hemoglobin A1c levels to <7% before surgery, if possible.	A-II
Increase dosing of prophylactic antimicrobial agent for morbidly obese patients.	A-II
Encourage smoking cessation within 30 days before procedure.	A-II
No formal recommendation with respect to immunosuppressive medications; avoid them in the perioperative period, if possible.	C-II
Do not routinely delay surgery to provide parenteral nutrition.	A-I
Do not remove unless hair will interfere with the operation; if removal is necessary, remove by clipping and do not use razors.	A-II
Identify and treat infections remote to the surgical site before elective surgery.	A-II
Wash and clean skin around incision site; use an appropriate antiseptic agent.	A-II
Use appropriate antiseptic agent (e.g., an alcohol-based surgical hand antisepsis product) for a 2–5 minute preoperative surgical scrub.	A-II
Administer antimicrobial prophylaxis only when indicated in accordance with evidence-based standards and guidelines.	A-I
Administer antimicrobial prophylaxis within 1 hour before incision to maximize tissue concentration.	A-I
Select appropriate agents on the basis of surgical procedure, most common pathogens, and published recommendations.	A-I
Do not routinely use vancomycin for antimicrobial prophylaxis.	B-II
Stop prophylaxis within 24 hours after the procedure for all procedures except cardiac surgery (stop within 48 hours here).	A-I
Handle tissue carefully and eradicate dead space.	A-III
Control blood glucose level during the immediate postoperative period for patients undergoing cardiac surgery.	A-I
Adhere to standard principles of operating room asepsis.	A-III
No formal recommendation with respect to operative time; minimize as much as possible.	A-III
Adhere to ventilation, follow American Institute of Architects' recommendations for ventilation.	C-I
Minimize operating room traffic.	B-II
Use an approved hospital disinfectant to clean surfaces and equipment.	B-III
Sterilize all surgical equipment according to published guidelines; minimize the use of flash sterilization.	B-I
Perform surveillance for SSI.	A-II

(Continued)

TABLE 1.1 (Continued)

Recommendation	Grade
Provide ongoing feedback on SSI surveillance and process measures to surgical and perioperative personnel and leadership.	A-II
Perform expanded SSI surveillance to determine the source and extent of the problem and to identify possible targets for intervention.	B-II
Measure and provide feedback of SSI rates, antimicrobial prophylaxis, proper hair removal, and glucose control (for cardiac surgery)	A-III
Increase the efficiency of surveillance through the use of automated data.	A-II
Educate surgeons and perioperative personnel about SSI prevention.	A-III
Educate patients and their families about SSI prevention, as appropriate.	A-III

important infection control measures according to the time point of their implementation (before, during or after the operation on the patient).

INFECTION CONTROL MEASURES BEFORE THE START OF THE SURGICAL PROCEDURE

Patient-Derived (Endogenous) Risk Factors for Surgical Site Infections

Diseases. Some patients present with certain characteristics that will increase their risk of subsequent SSI acquisition. Unfortunately, a number of those endogenous risk facts cannot be influenced by proper infection control measures. This includes (but does not exclusively apply to) underlying diseases which increase the overall comorbiditiy burden such as cancer, cirrhosis and other liver diseases, congestive heart failure, coagulopathies, and chronic obstructive lung diseases,[7,11,12] patients with any kind of severe immune-suppression,[13] infants aged less than 1 year,[14,15] older patients aged more than 65 years,[16] a higher score on the American Society of Anaesthesiology (ASA) classification,[16,17] and patients with a need for surgical procedures in microbiologically contaminated areas as expressed by the primary wound contamination class.[18–20] Whenever staff recognizes one or more of the above mentioned risk factors, an increased awareness of potential SSI is necessary. In addition, there are several patient-derived risk factors that can, in fact, be positively influenced in principle. This may often be difficult, but is still worthwhile in terms of the patient's outcome. These modifiable characteristics will be discussed in more detail in the following sections.

Overweight. Overweight patients with an increased body mass index (BMI) show higher SSI rates than do patients with a normal body weight.[13,17,21,22] In a randomized prospective study in a group of 1,032 patients, Beldi et al.[23] showed that a BMI $> 30\,\text{kg/m}^2$ doubled the risk of an SSI. Tran et al. checked for SSI risk factors in 969 women after cesarean section: every five-unit increment in the BMI increased the odds ratio (OR) for an SSI by a factor of two.[24] Yeung et al. recruited a consecutive cohort of 210 patients from varicose vein surgery. Nine of 53 patients (17.0%) with a BMI $\geq 30\,\text{kg/m}^2$ suffered from an

SSI, compared to 0.7of the remainder.[25] Obese patients should therefore be encouraged to lose weight before elective surgical procedures are performed on them. It is noteworthy that cachexia also worsens the postoperative outcome. Thus, one should aim for a BMI within normal limits.[26]

Diabetes. The risk of an SSI in patients who suffer from diabetes mellitus (DM) may be dramatically increased due to high intra-operative blood glucose levels and poor microcirculation.[27] Bykowski et al. calculated the OR for the risk of SSI in diabetic patients to be as high as 2.8.[28] Similarly Davies found a SSI infection rate after mastectomy in patients with DM of 19.0% vs. 11.2% in patients without DM[17] and, even more striking, 18.75% vs. 1.97% in adult spinal trauma patients.[16] Ensure that blood sugar levels remain within the physiological range.[29]

Smoking. Smokers, too, are more likely to acquire SSI.[22] In a large single-center retrospective review of 8,850 cases, smoking status even tripled the risk of an SSI (OR = 3.0).[28] Sorensen and coworkers performed a systematic review and meta-analysis of 479,150 patients in 140 cohort studies. The pooled adjusted OR for a SSI in smokers was 1.79.[30] Thus the prevention of SSI is just one more good reason among many others to encourage patients to quit smoking.

Infection. Patients who already suffer from a previous SSI infection or an infection at another body site are at higher risk of SSI thereafter. Webster et al. showed in 827 patients undergoing several kinds of elective and emergency surgery that the presence of any previous SSI was associated with an adjusted OR of 2.5.[21] Similar findings are reported by Xing and coworkers in a systematic review of evidence-based independent risk factors for SSI in 226 patients after spinal surgery.[31] Thus, it is strongly recommended that any existing SSI or other infection is cured before performing an elective surgical procedure, if possible.

Colonization. Nasal colonization of the patient by *Staphylococcus aureus*, both methicillin-resistant (MRSA) and methicillin-sensitive (MSSA), increases the risk of subsequent SSI.[32–34] Ramirez et al. found that 6.4% of the 1,137 screened patients were MRSA positive. In 70% of all previously colonized patients, MRSA was also cultured from the wound after major gastrointestinal surgery.[35] Donker et al. calculated a 10-fold increased risk for the development of SSI in patients with *S. aureus* nasal carriage compared to non-carriers in vascular surgery.[36] Thus patients should be screened before elective surgical procedures in order to detect any *S. aureus* colonization – preferably before hospital admission. In colonized patients, appropriate decolonization measures are recommended – such as application of nasal mupirocin ointment, antiseptic mouth wash and whole-body antiseptic scrubs.[37–40] These decolonization measures should be continued after the surgical procedure if necessary.

Hospital-Derived (Exogenous) Risk Factors for Surgical Site Infections

A well-known independent risk factor for SSI occurrence is a prolonged stay by the patient within the hospital prior to surgery; this increases the risk of subsequent SSI occurrence.[21,41] A recently published systematic review on risk-adjusted models for SSI by Gibbons et al.[42] identified the duration of preoperative stay as the most common risk factor in coronary artery bypass graft surgery, large and small bowel surgery, hip and knee prosthesis, and vascular surgery. As a consequence, the time frame between admission of the patient and start of the surgical procedure should be as short as possible.

Expertise. Of course the rate of SSI also depends upon the expertise of staff.[43] A total of 117 hospitals were compared with respect to their trainee-to-bed ratio in the latest American College of Surgeons-National Surgical Quality Improvement (ACS-NSQIP): outlier hospitals with extraordinary high SSI rates also had a significantly increased proportion of trainees among their staff compared to outlier hospitals with extremely low SSI rates.[44] In addition, Meyer et al. recently showed in a comprehensive multivariate analysis of 120,564 procedures in 206 departments that the larger the annual number of specific procedures in a hospital, the lower the resulting SSI rates will be; departments that performed less than 50 knee replacements per year ended up with a SSI rate of 1.81% compared 0.79% in hospitals that performed more than 100 operations of that kind per year. Corresponding data for hip replacement and arthroscopy were 1.11% vs. 0.84% and 2.16% vs. 0.24%, respectively.[45]

Devices. Urinary tract catheters (UTC)[12] and other medical devices[15] may be associated with higher SSI rates. Lonjon et al. report an SSI rate of 2/99 (2.0%) in patients after spine surgery when an UTC was in place for less than five days compared to an infection rate of 4/31 (12.9%) in patients with an UTC for more than five days.[16] Similar results are reported by Bucher et al. in 159 children; usage of an UTC (OR = 3.56) and implanted medical devices (OR = 3.05), respectively, significantly increased the risk for a SSI.[46] Note that the application of total parenteral nutrition may also significantly increase the risk of SSI.[15] It seems reasonable to check for the correct indication of existing medical devices carefully, and to remove them if they are no longer required.

Shaving. Hair should only be removed if it interferes with the site of incision. At present the use of clippers (instead of shaving with razors) seems to be the most appropriate way of hair removal for SSI prevention. If razors are used, keep the time frame short between shaving and incision. Ng et al. used posters and enhanced prenatal education for obstetric patients before cesarean section. The rate of hair self-removal decreased significantly from 41% in 2008 to 27% in 2011. Concurrently, a 51% reduction was seen in the SSI rate following cesarean section.[47] Thus it is strongly recommended that patients are discouraged from shaving themselves on the day before admission.

Antiseptic shower. Preoperative antiseptic showering of the patient e.g., using chlorhexidine gluconate, may be helpful in the prevention of SSI, as it reduces the number of bacteria in the residual skin flora. Studies exist that show a benefit of a total body shower of the patient shortly before the surgical procedure (preferably pre-admission or at the evening before surgery, using an antiseptic lotion).[48] A systematic review of this topic by Kamel et al. suggests that skin antiseptic preparations may be effective in preventing SSI.[49] However, up to now overall data on this measure are sparse, and a meta-analysis by Chlebicki et al.[50] and a recently updated Cochrane review including 7,791 patients failed to show a statistically significant reduction in SSI after bathing with chlorhexidine compared with placebo (risk ratio (RR) = 0.91).[51]

INFECTION CONTROL MEASURES DURING THE SURGICAL PROCEDURE

Antibiotic prophylaxis. Due to conflicting results, the use of antimicrobial prophylaxis (AP) is currently controversial. There are numerous studies that show a drop in the SSI rate when AP is administered properly and multiple Cochrane reviews have been carried out on this topic. The pooled results of a total of nine studies including 2,260 patients who underwent surgery for breast cancer demonstrated that AP significantly reduced the incidence for a SSI with a RR of 0.71.[52] Single dose antibiotic prophylaxis also significantly reduced deep SSI (RR = 0.40) after surgery for closed fracture fixation in 8,447 participants in 23 studies.[53] Recently Ott et al. showed that application of perioperative AP independently reduces the risk for a SSI after arterial vascular surgery by as much as 80%.[13] Darouiche showed that systemic AP significantly reduced the incidence of SSI (RR = 0.14) when performing cardiac implantable electronic device implantation.[54] On the other hand, there are just as many studies that failed to show a benefit of AP, including analysis of patients who have undergone plastic surgery, elective laparoscopic cholecystectomy, neurosurgery, and hernia repair.[55–58] A single-center retrospective review by Bykowski et al. calculated a SSI rate of 0.54% in 2,755 patients who received AP compared to a SSI rate of 0.26% in 6,095 patients who did not.[28] The choice to use AP, and which antimicrobial substance to employ, depends upon the kind of surgical procedure, nasal screening results and existing infections elsewhere,[12,59] but vancomycin is generally not an appropriate substance, and is thus explicitly not recommended for this purpose.[60] No recommendation can be made regarding the use of antimicrobial substances that cover a broad spectrum of Gram-negative and/or anaerobe bacteria as there is still a lack of clinical data. Timely application may also be important.[61–63] It is recommended to aim for highest concentrations of the antimicrobial substance in the tissue at the time point of incision. Brown et al. showed remarkable decline in the SSI rates from 10.8% in 2010 to 2.8% in 2011 after a change of timing of AP application in cesarean section.[64] Once again, contradictory results have also been published. Hawn et al. compared timely application to untimely application of AP in 9,195 elective procedures (orthopedic, colon, and vascular) performed in 95 hospitals. Corresponding SSI rates were 4.6% and 5.8% respectively in a bivariable unadjusted analysis. Use of multiple dosages may be necessary in some cases, especially if the duration of the surgical procedure is prolonged.[53,65,66] AP is discontinued (usually within 24 hours) after the end of surgery, as prolonged AP will not further decrease the risk of an SSI.[67] It might become essential to increase the dosage in order to adapt AP to overweight patients.[68–70] Forse et al. reported a drop in the SSI rate from 16.5% to 5.6% after the dosage of cefazolin PA was changed to 2 g in morbidly obese patients undergoing gastroplasty, compared to 1 g in normal weight patients.[71]

Air flow systems. Laminar air flow (LAF) is often used as a measure for SSI prevention, on the assumption that driving out potentially contaminated room air by directed, filtered air from the ceiling would lower the number of pathogens in the operation area, and thus infections of the patient. However, a recently published study by Breier et al. failed to detect any advantage of LAF use in patients undergoing 20,554 knee, 33,463 elective and 7,749 urgent hip prostheses regardless of the size of the LAF ceiling.[72] Diab-Elschahawi and coworkers showed that a LAF ventilation system does not provide bacteria-free conditions at the surgical site and on the instrument table.[73] A meta-analysis by Gastmeier et al. even questions LAF use in principle, as SSI rates turned out to be higher when LAF systems were used.[74]

Instruments. All medical equipment that comes into contact with primary sterile body sites must previously be sterilized itself.[75] This includes but is not limited to surgical instruments, rinse liquids, suture material, and wound dressings. There are hardly any randomized control trials that address this question for ethical reasons, but several nosocomial outbreaks are reported to have been due to insufficient sterilization processes. For example Dancer et al. experienced such an outbreak involving 15 orthopedic patients following metal insertion, and five ophthalmology patients who developed endophthalmitis. SSIs caused by coagulase-negative staphylococci and *Bacillus* spp. were traced to post-sterilization contamination of sets containing surgical instruments.[76] Rutala et al. published a study of an outbreak of podiatric infections due to *Proteus mirabilis* following outpatient surgery. Bone drills served as the reservoir for the organism, because the gas sterilization procedure that was used to sterilize the drills was found to be deficient.[77]

Clothing. It is more or less common sense that people who participate in a surgical procedure directly at the patient's site need to wear a sterile gown,[78] sterile gloves,[79] and a surgical face mask[80,81] in order to maintain sterile barrier precautions and also to protect themselves from infectious agents that may derive from the patient. It is also recommended to keep the overall number of people in the surgical theater to a minimum, because the quality of the operating room environment is mainly affected by the number of persons in it.[82–84] Talking by the participating staff during the operation should also be minimized, as it is well known that this will promote pathogen spread via the lateral spaces between face mask and the face,[85,86] especially when worn by bearded surgeons.[87] In addition, a thorough surgical hand scrub is highly recommended before each surgical procedure. Parienti et al. could not show a significant difference between the groups using alcoholic hand rub or antiseptic soaps including povidone-iodine or chlorhexidine gluconate. The corresponding SSI rates did not differ at all (2.44% vs. 2.48%), but the alcohol-based hand rub was better tolerated, producing less skin dryness and irritation.[88] A potential influence of nail polish on the number of bacteria on the skin post-scrub – and thus on the SSI rate – cannot be excluded, as the amount of data available is still scarce.[89]

Scrub. Antiseptic is usually administered to the patient's skin at the incision site immediately before the cut is performed. In a prospective observational study in 1,014 patients, Tschudin-Sutter et al. showed that there was no risk of SSI from residual bacteria after disinfection of the preoperative site with povidone-iodine-alcohol.[90] Similar findings are reported by Boston et al.[91]: in their study povidone-iodine for preoperative skin antisepsis was found to be protective against SSI acquisition (OR = 0.16). A systematic review and meta-analysis by Lee et al. on nine randomized controlled trials with a total of 3,614 patients revealed that chlorhexidine/(+/− isopropyl alcohol) antisepsis was associated with significantly fewer SSI incidences (RR = 0.64: CI95%:0.51–0.80) than was iodine/(+/− isopropyl alcohol) antisepsis.[92] Thus proper skin disinfection at the site of incision by an appropriate skin disinfectant is strongly recommended. Skin areas with a higher amount of sebum may require a prolonged time for accurate antisepsis.

Surgical technique. Prolonging the surgical procedure increases the risk for SSI acquisition in most types of surgical procedures, including coronary artery bypass, gastric surgery, total hip replacement, knee prosthesis, large bowel surgery, and vascular surgery.[93] SSI rates after 2,644 procedures of femoral-popliteal bypass with autogenous vein were compared by Tan et al. on the basis of operative duration quartiles. Corresponding SSI rates were 6.3%, 9.0%, 10.1%, and 13.9%.[94] Similar findings have been reported in many studies by other authors.[14,44] Proper surgical technique is itself of great importance.[23,95] Minimize the amount of necrotic tissue,[96–98] avoid excessive bleeding during the surgical procedure,[16,99] and aim for an intra-operative blood glucose level of < 8 mmol/L.[29]

Implants. Foreign material such as implants may be inserted for different reasons during a variety of surgical procedures. Whenever this takes place, the risk of SSI increases dramatically.[83,100,101] Thus, the use of implants should be limited insofar as possible and maximum effort should be employed in terms of infection control measures. In addition, the use of material that has been impregnated with antimicrobial or antiseptic substances may be helpful and should therefore be considered.[102–105]

INFECTION CONTROL MEASURES AFTER THE END OF THE SURGICAL PROCEDURE

Wound dressing. Wounds are often dressed after the end of the surgical procedure. The decision on whether or not to do this, and on what type of wound dressing is used should be made individually. A recently published Cochrane review concludes that:

> "at present, there is no evidence to suggest that covering surgical wounds healing by primary intention with wound dressings reduces the risk of SSI or that any particular wound dressing is more effective than others in reducing the rates of SSI".[106]

A systematic review by Walter et al. of 16 controlled trials with 2,594 participants also found no evidence that dressing wounds reduces rates of SSI compared to uncovered wounds.[107] The frequency at which the dressings are changed should also be determined individually and will depend on the type of surgery, on the kind of wound dressing used, and on host factors of the patient.[108,109]

Drainage. Wound drainage systems are often used when wound secretion is expected. Akinyoola and coworkers observed no benefit in 65 patients from the routine use of wound drains following open reduction and internal fixation of femoral shaft fractures,[110] and a Cochrane review on this topic concluded that there is insufficient evidence from randomized trials to support or refute the routine use of closed suction drainage in orthopedic surgery.[111] So the decision on the use of wound drains should be made individually.[112,113] However, if inserted after surgery the duration of drain usage should be kept short. Arabshahi et al. followed 918 patients for 30 days postoperatively. A total of 77 cases of SSI were observed, and the presence of a wound drain was a significant risk factor for SSI development (OR = 2.2).[114] Rao et al. showed in a retrospective, case-control study on 57 deep SSIs out of 1,587 procedures that the OR increased by 1.6 for every day in which a drain was present.[115] Continuously check whether drainages that are currently in place are still needed or may be removed, and do this as soon as possible.[13,116]

Surveillance. Surveillance of nosocomial infections in general and of SSI in particular may significantly reduce infection rates. Data from the German National Nosocomial Infections Surveillance System (KISS) demonstrates the effect of active surveillance on SSI. Fourteen hospitals participated in KISS continuously for three years and performed 15,457 hip prostheses during this time. A multiple logistic regression analysis confirmed that the

SSI rate was significantly lower – almost cut by half – in the third year compared to the first year of surveillance (OR = 0.57).[117] Brandt et al. showed that lower SSI rates were independently associated with increasing duration of SSI surveillance in 14 out of 19 operative procedure categories. After two years of surveillance, the SSI rate dropped by 16% (OR = 0.84; CI 95%: 0.77–0.93) and decreased further after one more year of surveillance (OR = 0.75; CI95%: 0.68–0.82).[118] It is therefore strongly recommended to perform active and prospective surveillance of SSI with a standardized protocol; for example according to CDC-definitions.[9] If an increase in the local SSI rate is recognized, one should immediately check for any breaches in the adherence to the above mentioned infection control measures.

ADHERENCE TO INFECTION CONTROL RECOMMENDATIONS

Despite the existence and knowledge of evidence-based guidelines for the prevention of SSI, there may still be a remarkable lack of adherence to them.[119] In a descriptive study, Castella et al. showed that in 799 operations, only 29% of the patients showered the day before surgery, 12% of the surgical team members did not wear a cap/hood and mask correctly, and insufficient or incorrect preoperative hand and forearm scrubbing technique were observed in 22% of cases.[120] Young et al. estimated that as few as two errors in infection control procedures will increase the risk of subsequent SSI by a factor of four.[121] Despite the accessibility of national and international standards for SSI prevention, Durando et al.[122] reported that shavers were used for hair removal in 92% of 717 elective interventions performed on 703 patients, 83.5% of which on the day before the operation. While compliance in giving a preoperative shower was 100% and antisepsis of the incision area was done in at least 97.4% of cases, antimicrobial prophylaxis was inappropriately provided to more than half (55.2%) of the patients. In addition, the doors of the operating theaters remained open during the duration of the operation in more than one third (36.3%) of the cases.[122]

There are several ways to improve compliance. These comprise standard operation procedure (SOP) manuals for the handling of patients in surgery.[123,124] Furthermore, education of staff (both nurses and physicians) on the causes and consequences of SSI should be performed on a regular basis.[125–128] In addition, an easy-to-use checklist may be a very valuable tool to assure adherence to infection control guidelines.[129] Of course, a number those items, for example the timely administration of preoperative antibiotics and the maintenance of perioperative normothermia, will be the responsibility of other groups of staff such as anesthesiologists.[130] Finally, infection control protocols of hospitals should be continuously reviewed, modified and adapted to evidence-based guidelines if appropriate.[131]

References

1. Coello R, Charlett A, Wilson J, Ward V, Pearson A, Borriello P. Adverse impact of surgical site infections in English hospitals. *J Hosp Infect* 2005;**60**:93–103.
2. Vogel TR, Dombrovskiy VY, Carson JL, Graham AM, Lowry SF. Postoperative sepsis in the United States. *Ann Surg* 2010;**252**:1065–71.

3. Vogel TR, Dombrovskiy VY, Lowry SF. Impact of infectious complications after elective surgery on hospital readmission and late deaths in the US Medicare population. *Surg Infect (Larchmt)* 2012;**13**:307–11.
4. Ho D, Lynch RJ, Ranney DN, Magar A, Kubus J, Englesbe MJ. Financial impact of surgical site infection after kidney transplantation: implications for quality improvement initiative design. *J Am Coll Surg* 2010;**211**:99–104.
5. Graf K, Ott E, Vonberg RP, Kuehn C, Schilling T, Haverich A, et al. Surgical site infections – economic consequences for the health care system. *Langenbecks Arch Surg* 2011;**396**:453–9.
6. Broex EC, van Asselt AD, Bruggeman CA, van Tiel FH. Surgical site infections: how high are the costs? *J Hosp Infect* 2009;**72**:193–201.
7. Poultsides LA, Ma Y, Della Valle AG, Chiu YL, Sculco TP, Memtsoudis SG. In-hospital surgical site infections after primary hip and knee arthroplasty – incidence and risk factors. *J Arthroplasty* 2012;**28**:385–9.
8. de Lissovoy G, Fraeman K, Hutchins V, Murphy D, Song D, Vaughn B. Surgical site infection: incidence and impact on hospital utilization and treatment costs. *Am J Infect Control* 2009;**37**:387–97.
9. Mangram AJ, Horan TC, Pearson ML, Silver LC, Jarvis WR. Guidelines for prevention of surgical site infection, 1999. Centers for disease control and prevention (CDC) hospital infection control practices advisory committee. *Am J Infect Control* 1999;**27**:97–132.
10. Anderson DJ, Kaye KS, Classen D, Arias KM, Podgorny K, Burstin H, et al. Strategies to prevent surgical site infections in acute care hospitals. *Infect Control Hosp Epidemiol* 2008;**29**(**Suppl 1**):S51–61.
11. Haridas M, Malangoni MA. Predictive factors for surgical site infection in general surgery. *Surgery* 2008;**144**:496–501.
12. Pessaux P, Atallah D, Lermite E, Msika S, Hay JM, Flamant Y. Risk factors for prediction of surgical site infections in "clean surgery". *Am J Infect Control* 2005;**33**:292–8.
13. Ott E, Bange FC, Sohr D, Teebken O, Mattner F. Risk factors associated with surgical site infections following vascular surgery at a German university hospital. *Epidemiol Infect* 2003;**141**:1207–13.
14. Costello JM, Graham DA, Morrow DF, Morrow J, Potter-Bynoe G, Sandora TJ, et al. Risk factors for surgical site infection after cardiac surgery in children. *Ann Thorac Surg* 2010;**89**:1833–41.
15. Sohn AH, Schwartz JM, Yang KY, Jarvis WR, Guglielmo BJ, Weintrub PS. Risk factors and risk adjustment for surgical site infections in pediatric cardiothoracic surgery patients. *Am J Infect Control* 2010;**38**:706–10.
16. Lonjon G, Dauzac C, Fourniols E, Guigui P, Bonnomet F, Bonnevialle P, et al. Early surgical site infections in adult spinal trauma: A prospective, multicenter study of infection rates and risk factors. *Orthop Traumatol Surg Res* 2012;**98**:788–94.
17. Davis GB, Peric M, Chan LS, Wong AK, Sener SF. Identifying risk factors for surgical site infections in mastectomy patients using the national surgical quality improvement program database. *Am J Surg* 2012;**205**:194–9.
18. Mannien J, van den HS, Brandt C, Behnke M, Wille JC, Gastmeier P. Comparison of the national surgical site infection surveillance data between the netherlands and germany: PREZIES versus KISS. *J Hosp Infect* 2007;**66**:224–31.
19. Suljagic V, Jevtic M, Djordjevic B, Jovelic A. Surgical site infections in a tertiary health care center: prospective cohort study. *Surg Today* 2010;**40**:763–71.
20. Hubner M, Diana M, Zanetti G, Eisenring MC, Demartines N, Troillet N. Surgical site infections in colon surgery: the patient, the procedure, the hospital, and the surgeon. *Arch Surg* 2011;**146**:1240–5.
21. Webster J, Croger S, Lister C, Doidge M, Terry MJ, Jones I. Use of face masks by non-scrubbed operating room staff: a randomized controlled trial. *ANZ J Surg* 2010;**80**:169–73.
22. Haas JP, Evans AM, Preston KE, Larson EL. Risk factors for surgical site infection after cardiac surgery: the role of endogenous flora. *Heart Lung* 2005;**34**:108–14.
23. Beldi G, Bisch-Knaden S, Banz V, Mühlemann K, Candinas D. Impact of intraoperative behavior on surgical site infections. *Am J Surg* 2009;**198**:157–62.
24. Tran TS, Jamulitrat S, Chongsuvivatwong V, Geater A. Risk factors for postcesarean surgical site infection. *Obstet Gynecol* 2000;**95**:367–71.
25. Yeung TM, Cope AC, Whittome N, Lintott P. Raised BMI is an independent risk factor for groin surgical site infections in patients undergoing varicose vein surgery. *Am J Infect Control* 2008;**36**:152–3.
26. Pausch T, Hartwig W, Hinz U, Swolana T, Bundy BD, Hackert T, et al. Cachexia but not obesity worsens the postoperative outcome after pancreatoduodenectomy in pancreatic cancer. *Surgery* 2012;**152**:S81–8.
27. Malone DL, Genuit T, Tracy JK, Gannon C, Napolitano LM. Surgical site infections: reanalysis of risk factors. *J Surg Res* 2002;**103**:89–95.

28. Bykowski MR, Sivak WN, Cray J, Buterbaugh G, Imbriglia JE, Lee WP. Assessing the impact of antibiotic prophylaxis in outpatient elective hand surgery: a single-center, retrospective review of 8,850 cases. *J Hand Surg Am* 2011;**36**:1741–7.
29. Graf K, Sohr D, Haverich A, Kühn C, Gastmeier P, Chaberny IF. Decrease of deep sternal surgical site infection rates after cardiac surgery by a comprehensive infection control program. *Interact Cardiovasc Thorac Surg* 2009;**9**:282–6.
30. Sorensen LT. Wound healing and infection in surgery. The clinical impact of smoking and smoking cessation: a systematic review and meta-analysis. *Arch Surg* 2012;**147**:373–83.
31. Xing D, Ma JX, Ma XL, Song DH, Wang J, et al. A methodological, systematic review of evidence-based independent risk factors for surgical site infections after spinal surgery. *Eur Spine J* 2012;**22**:605–15.
32. Kluytmans JA, Mouton JW, Ijzerman EP, Vandenbroucke-Grauls CM, Maat AW. Nasal carriage of *Staphylococcus aureus* as a major risk factor for wound infections after cardiac surgery. *J Infect Dis* 1995;**171**: 216–9.
33. Munoz P, Hortal J, Giannella M, Barrio JM, Rodríguez-Créixems M, Pérez MJ, et al. Nasal carriage of *S. aureus* increases the risk of surgical site infection after major heart surgery. *J Hosp Infect* 2008;**68**:25–31.
34. Humphreys H. *Staphylococcus aureus*: The enduring pathogen in surgery. *Surgeon* 2012;**10**:357–60.
35. Ramirez MC, Marchessault M, Govednik-Horny C, Jupiter D, Papaconstantinou HT. The impact of MRSA colonization on surgical site infection following major gastrointestinal surgery. *J Gastrointest Surg* 2012;**17**: 144–52.
36. Donker JM, van der Laan L, Hendriks YJ, Kluytmans JA. Evaluation of *Staphylococcus aureus* nasal carriage screening before vascular surgery. *PLoS One* 2012;**7**:e38127.
37. Kluytmans JA, Mouton JW, VandenBergh MF, Manders MJ, Maat AP, Wagenvoort JH, et al. Reduction of surgical-site infections in cardiothoracic surgery by elimination of nasal carriage of *Staphylococcus aureus*. *Infect Control Hosp Epidemiol* 1996;**17**:780–5.
38. van Rijen M, Bonten M, Wenzel R, Kluytmans J. Mupirocin ointment for preventing *Staphylococcus aureus* infections in nasal carriers. *Cochrane Database Syst Rev* 2008. CD006216.
39. van Rijen MM, Bonten M, Wenzel RP, Kluytmans JA. Intranasal mupirocin for reduction of *Staphylococcus aureus* infections in surgical patients with nasal carriage: a systematic review. *J Antimicrob Chemother* 2008;**61**:254–61.
40. Goyal N, Aggarwal V, Parvizi J. Methicillin-resistant *Staphylococcus aureus* screening in total joint arthroplasty: a worthwhile endeavor. *J Knee Surg* 2012;**25**:37–43.
41. Song KH, Kim ES, Kim YK, Jin HY, Jeong SY, Kwak YG, et al. Differences in the risk factors for surgical site infection between total hip arthroplasty and total knee arthroplasty in the Korean Nosocomial Infections Surveillance System (KONIS). *Infect Control Hosp Epidemiol* 2012;**33**:1086–93.
42. Gibbons C, Bruce J, Carpenter J, Wilson AP, Wilson J, Pearson A, et al. Identification of risk factors by systematic review and development of risk-adjusted models for surgical site infection. *Health Technol Assess* 2011;**15**:1–156.
43. Geubbels EL, Wille JC, Nagelkerke NJ, Vandenbroucke-Grauls CM, Grobbee DE, de Boer AS. Hospital-related determinants for surgical-site infection following hip arthroplasty. *Infect Control Hosp Epidemiol* 2005;**26**:435–41.
44. Campbell Jr DA, Henderson WG, Englesbe MJ, Hall BL, O'Reilly M, Bratzler D, et al. Surgical site infection prevention: the importance of operative duration and blood transfusion – results of the first American College of Surgeons National Surgical Quality Improvement Program Best Practices Initiative. *J Am Coll Surg* 2008;**207**:810–20.
45. Meyer E, Weitzel-Kage D, Sohr D, Gastmeier P. Impact of department volume on surgical site infections following arthroscopy, knee replacement or hip replacement. *BMJ Qual Saf* 2011;**20**:1069–74.
46. Bucher BT, Guth RM, Elward AM, Hamilton NA, Dillon PA, Warner BW, et al. Risk factors and outcomes of surgical site infection in children. *J Am Coll Surg* 2011;**212**:1033–8.
47. Ng W, Alexander D, Kerr B, Ho MF, Amato M, Katz K. A hairy tale: successful patient education strategies to reduce pre-hospital hair removal by patients undergoing elective caesarean section. *J Hosp Infect* 2013;**83**:64–7.
48. Edmiston Jr CE, Okoli O, Graham MB, Sinski S, Seabrook GR. Evidence for using chlorhexidine gluconate preoperative cleansing to reduce the risk of surgical site infection. *AORN J* 2010;**92**:509–18.
49. Kamel C, McGahan L, Polisena J, Mierzwinski-Urban M, Embil JM. Preoperative skin antiseptic preparations for preventing surgical site infections: a systematic review. *Infect Control Hosp Epidemiol* 2012;**33**:608–17.

50. Chlebicki MP, Safdar N, O'Horo JC, Maki DG. Preoperative chlorhexidine shower or bath for prevention of surgical site infection: A meta-analysis. *Am J Infect Control* 2012;**41**:167–73.
51. Webster J, Osborne S. Preoperative bathing or showering with skin antiseptics to prevent surgical site infection. *Cochrane Database Syst Rev* 2012;**9**:CD004985.
52. Bunn F, Jones DJ, Bell-Syer S. Prophylactic antibiotics to prevent surgical site infection after breast cancer surgery. *Cochrane Database Syst Rev* 2012;**1**:CD005360.
53. Gillespie WJ, Walenkamp GH. Antibiotic prophylaxis for surgery for proximal femoral and other closed long bone fractures. *Cochrane Database Syst Rev* 2000;**3**:CD000244.
54. Darouiche R, Mosier M, Voigt J. Antibiotics and antiseptics to prevent infection in cardiac rhythm management device implantation surgery. *Pacing Clin Electrophysiol* 2012;**35**:1348–60.
55. Landes G, Harris PG, Lemaine V, Perreault I, Sampalis JS, Brutus JP. Prevention of surgical site infection and appropriateness of antibiotic prescribing habits in plastic surgery. *J Plast Reconstr Aesthet Surg* 2008;**61**:1347–56.
56. Sanabria A, Dominguez LC, Valdivieso E, Gomez G. Antibiotic prophylaxis for patients undergoing elective laparoscopic cholecystectomy. *Cochrane Database Syst Rev* 2010;**12**:CD005265.
57. Ratilal BO, Costa J, Sampaio C, Pappamikail L. Antibiotic prophylaxis for preventing meningitis in patients with basilar skull fractures. *Cochrane Database Syst Rev* 2011;**8**:CD004884.
58. Sanchez-Manuel FJ, Lozano-Garcia J, Seco-Gil JL. Antibiotic prophylaxis for hernia repair. *Cochrane Database Syst Rev* 2012;**2**:CD003769.
59. Morrill MY, Schimpf MO, Abed H, Carberry C, Margulies RU, White AB, et al. Antibiotic prophylaxis for selected gynecologic surgeries. *Int J Gynaecol Obstet* 2013;**120**:10–15.
60. Freeman JT, Anderson DJ, Hartwig MG, Sexton DJ. Surgical site infections following bariatric surgery in community hospitals: a weighty concern? *Obes Surg* 2011;**21**:836–40.
61. Classen DC, Evans RS, Pestotnik SL, Horn SD, Menlove RL, Burke JP. The timing of prophylactic administration of antibiotics and the risk of surgical-wound infection. *N Engl J Med* 1992;**326**:281–6.
62. van Kasteren ME, Mannien J, Ott A, et al. Antibiotic prophylaxis and the risk of surgical site infections following total hip arthroplasty: timely administration is the most important factor. *Clin Infect Dis* 2007;**44**:921–7.
63. Junker T, Mujagic E, Hoffmann H, Kullberg B-J, de Boer AS, Gyssens IC. Prevention and control of surgical site infections: review of the Basel Cohort Study. *Swiss Med Wkly* 2012;**142**:w13616.
64. Brown J, Thompson M, Sinnya S, Jeffery A, de Costa C, Woods C, et al. Pre-incision antibiotic prophylaxis reduces the incidence of post-caesarean surgical site infection. *J Hosp Infect* 2013;**83**:68–70.
65. Zani EL, Clark OA, Rodrigues Jr. NN. Antibiotic prophylaxis for transrectal prostate biopsy. *Cochrane Database Syst Rev* 2011;**5**:CD006576.
66. Morrison S, White N, Asadollahi S, Lade J. Single versus multiple doses of antibiotic prophylaxis in limb fracture surgery. *ANZ J Surg* 2012;**82**:902–7.
67. De Chiara S, Chiumello D, Nicolini R, Vigorelli B, Cesana N, Bottino G, et al. Prolongation of antibiotic prophylaxis after clean and clean-contaminated surgery and surgical site infection. *Minerva Anestesiol* 2010;**76**:413–9.
68. Barbour A, Schmidt S, Rout WR, Ben-David K, Burkhardt O, Derendorf H. Soft tissue penetration of cefuroxime determined by clinical microdialysis in morbidly obese patients undergoing abdominal surgery. *Int J Antimicrob Agents* 2009;**34**:231–5.
69. El Nashar SA, Diehl CL, Swanson CL, Thompson RL, Cliby WA, Famuyide AO, et al. Extended antibiotic prophylaxis for prevention of surgical-site infections in morbidly obese women who undergo combined hysterectomy and medically indicated panniculectomy: a cohort study. *Am J Obstet Gynecol* 2010;**202**:306–9.
70. Pevzner L, Swank M, Krepel C, Wing DA, Chan K, Edmiston CE Jr. Effects of maternal obesity on tissue concentrations of prophylactic cefazolin during cesarean delivery. *Obstet Gynecol* 2011;**117**:877–82.
71. Forse RA, Karam B, MacLean LD, Christou NV. Antibiotic prophylaxis for surgery in morbidly obese patients. *Surgery* 1989;**106**:750–6.
72. Breier AC, Brandt C, Sohr D, Geffers C, Gastmeier P. Laminar airflow ceiling size: no impact on infection rates following hip and knee prosthesis. *Infect Control Hosp Epidemiol* 2011;**32**:1097–102.
73. Diab-Elschahawi M, Berger J, Blacky A, Kimberger O, Oguz R, Kuelpmann R, et al. Impact of different-sized laminar air flow versus no laminar air flow on bacterial counts in the operating room during orthopedic surgery. *Am J Infect Control* 2011;**39**:e25–9.

74. Gastmeier P, Breier AC, Brandt C. Influence of laminar airflow on prosthetic joint infections: a systematic review. *J Hosp Infect* 2012;**81**:73–8.
75. Thompson SM, Middleton M, Farook M, Cameron-Smith A, Bone S, Hassan A, et al. The effect of sterile versus non-sterile tourniquets on microbiological colonisation in lower limb surgery. *Ann. R. Coll. Surg. Engl* 2011;**93**:589–90.
76. Dancer SJ, Stewart M, Coulombe C, Gregori A, Virdi M. Surgical site infections linked to contaminated surgical instruments. *J Hosp Infect* 2012;**81**:231–8.
77. Rutala WA, Weber DJ, Thomann CA. Outbreak of wound infections following outpatient podiatric surgery due to contaminated bone drills. *Foot Ankle* 1987;**7**:350–4.
78. Bible JE, Biswas D, Whang PG, Simpson AK, Grauer JN. Which regions of the operating gown should be considered most sterile? *Clin Orthop Relat Res* 2009;**467**:825–30.
79. van den Broek PJ. Sterile gloves are necessary in minor surgery. *Ned Tijdschr Geneeskd* 2011;**155**:A3341.
80. Tunevall TG. Postoperative wound infections and surgical face masks: a controlled study. *World J Surg* 1991;**15**:383–7.
81. Romney MG. Surgical face masks in the operating theatre: re-examining the evidence. *J Hosp Infect* 2001;**47**: 251–6.
82. Ritter MA, Eitzen H, French ML, Hart JB. The operating room environment as affected by people and the surgical face mask. *Clin Orthop Relat Res* 1975;**111**:147–50.
83. Madu KA, Enweani UN, Katchy AU, Madu AJ, Aguwa EN. Implant associated surgical site infection in orthopaedics: a regional hospital experience. *Niger J Med* 2011;**20**:435–40.
84. Panahi P, Stroh M, Casper DS, Parvizi J, Austin MS. Operating room traffic is a major concern during total joint arthroplasty. *Clin Orthop Relat Res* 2012;**470**:2690–4.
85. Letts RM, Doermer E. Conversation in the operating theater as a cause of airborne bacterial contamination. *J Bone Joint Surg Am* 1983;**65**:357–62.
86. Doshi RR, Leng T, Fung AE. Reducing oral flora contamination of intravitreal injections with face mask or silence. *Retina* 2012;**32**:473–6.
87. McLure HA, Mannam M, Talboys CA, Azadian BS, Yentis SM. The effect of facial hair and sex on the dispersal of bacteria below a masked subject. *Anaesthesia* 2000;**55**:173–6.
88. Parienti JJ, Thibon P, Heller R, Le Roux Y, von Theobald P, Bensadoun H, et al. Hand-rubbing with an aqueous alcoholic solution vs. traditional surgical hand-scrubbing and 30-day surgical site infection rates: a randomized equivalence study. *JAMA* 2002;**288**:722–7.
89. Arrowsmith VA, Taylor R. Removal of nail polish and finger rings to prevent surgical infection. *Cochrane Database Syst Rev* 2012;**5**:CD003325.
90. Tschudin-Sutter S, Frei R, Egli-Gany D, Eckstein F, Valderrabano V, Dangel M, et al. No risk of surgical site infections from residual bacteria after disinfection with povidone-iodine-alcohol in 1,014 cases: a prospective observational study. *Ann Surg* 2012;**255**:565–9.
91. Boston KM, Baraniuk S, O'Heron S, Murray KO. Risk factors for spinal surgical site infection, Houston, Texas. *Infect Control Hosp Epidemiol* 2009;**30**:884–9.
92. Lee I, Agarwal RK, Lee BY, Fishman NO, Umscheid CA. Systematic review and cost analysis comparing use of chlorhexidine with use of iodine for preoperative skin antisepsis to prevent surgical site infection. *Infect. Control Hosp Epidemiol* 2010;**31**:1219–29.
93. Leong G, Wilson J, Charlett A. Duration of operation as a risk factor for surgical site infection: comparison of English and US data. *J Hosp Infect* 2006;**63**:255–62.
94. Tan TW, Kalish JA, Hamburg NM, Rybin D, Doros G, Eberhardt RT, et al. Shorter duration of femoral-popliteal bypass is associated with decreased surgical site infection and shorter hospital length of stay. *J Am Coll Surg* 2012;**215**:512–8.
95. Sherertz RJ, Karchmer TB. Surgical site infection as a surrogate marker of physician impairment. *Infect Control Hosp Epidemiol* 2009;**30**:1120–2.
96. Delaney A, Diamantis S, Marks VJ. Complications of tissue ischemia in dermatologic surgery. *Dermatol Ther* 2011;**24**:551–7.
97. Kobayashi L, Konstantinidis A, Shackelford S, Chan LS, Talving P, Inaba K, et al. Necrotizing soft tissue infections: delayed surgical treatment is associated with increased number of surgical debridements and morbidity. *J Trauma* 2011;**71**:1400–5.

98. Meredith DS, Kepler CK, Huang RC, Brause BD, Boachie-Adjei O. Postoperative infections of the lumbar spine: presentation and management. *Int Orthop* 2012;**36**:439–44.
99. Xue DQ, Qian C, Yang L, Wang XF. Risk factors for surgical site infections after breast surgery: a systematic review and meta-analysis. *Eur J Surg Oncol* 2012;**38**:375–81.
100. Hetrick EM, Schoenfisch MH. Reducing implant-related infections: active release strategies. *Chem Soc Rev* 2006;**35**:780–9.
101. Khan MS, ur Rehman S, Ali MA, Sultan B, Sultan S. Infection in orthopedic implant surgery, its risk factors and outcome. *J Ayub Med Coll Abbottabad* 2008;**20**:23–5.
102. Allababidi S, Shah JC. Efficacy and pharmacokinetics of site-specific cefazolin delivery using biodegradable implants in the prevention of post-operative wound infections. *Pharm Res* 1998;**15**:325–33.
103. de Bruin AF, Gosselink MP, van der HE, Rutten HJ. Local application of gentamicin collagen implants in the prophylaxis of surgical site infections following gastrointestinal surgery: a review of clinical experience. *Tech Coloproctol* 2010;**14**:301–10.
104. Jawa A, Shi L, O'Brien T, Wells J, Higgins L, Macy J, et al. Prosthesis of antibiotic-loaded acrylic cement (PROSTALAC) use for the treatment of infection after shoulder arthroplasty. *J Bone Joint Surg Am* 2011;**93**:2001–9.
105. Knaepler H. Local application of gentamicin-containing collagen implant in the prophylaxis and treatment of surgical site infection in orthopaedic surgery. *Int J Surg* 2012;**10(Suppl 1)**:S15–20.
106. Dumville JC, Walter CJ, Sharp CA, Page T. Dressings for the prevention of surgical site infection. *Cochrane Database Syst Rev* 2011;**7**:CD003091.
107. Walter CJ, Dumville JC, Sharp CA, Page T. Systematic review and meta-analysis of wound dressings in the prevention of surgical-site infections in surgical wounds healing by primary intention. *Br J Surg* 2012;**99**:1185–94.
108. Veiga-Filho J, Veiga DF, Sabino-Neto M, Damasceno CA, Sales EM, Garcia ES, et al. Dressing wear time after reduction mammaplasty: a randomized controlled trial. *Plast Reconstr Surg* 2012;**129**:1e–7e.
109. Ritting AW, Leger R, O'Malley MP, Mogielnicki H, Tucker R, Rodner CM. Duration of postoperative dressing after mini-open carpal tunnel release: a prospective, randomized trial. *J Hand Surg Am* 2012;**37**:3–8.
110. Akinyoola AL, Odunsi A, Yusu MB. Use of wound drains following open reduction and internal fixation of femoral shaft fractures. *J Wound Care* 2012;**21**:279–84.
111. Parker MJ, Roberts C. Closed suction surgical wound drainage after orthopaedic surgery. *Cochrane Database Syst Rev* 2007;**3**:CD001825.
112. Minnema B, Vearncombe M, Augustin A, Gollish J, Simor AE. Risk factors for surgical-site infection following primary total knee arthroplasty. *Infect Control Hosp Epidemiol* 2004;**25**:477–80.
113. Shukla RM, Tripathy BB, Mukhopadhyay B, Chattopadhyay A, Saha K, Basu KS. Outcome of colostomy closure with different skin closure techniques in a developing country. *Afr J Paediatr Surg* 2010;**7**:156–8.
114. Arabshahi KS, Koohpayezade J. Investigation of risk factors for surgical wound infection among teaching hospitals in Tehran. *Int Wound J* 2006;**3**:59–62.
115. Rao SB, Vasquez G, Harrop J, Maltenfort M, Stein N, Kaliyadan G, et al. Risk factors for surgical site infections following spinal fusion procedures: a case-control study. *Clin Infect Dis* 2011;**53**:686–92.
116. Drinkwater CJ, Neil MJ. Optimal timing of wound drain removal following total joint arthroplasty. *J Arthroplasty* 1995;**10**:185–9.
117. Gastmeier P, Geffers C, Brandt C, Zuschneid I, Sohr D, Schwab F, et al. Effectiveness of a nationwide nosocomial infection surveillance system for reducing nosocomial infections. *J Hosp Infect* 2006;**64**:16–22.
118. Brandt C, Sohr D, Behnke M, Daschner F, Rüden H, Gastmeier P. Reduction of surgical site infection rates associated with active surveillance. *Infect Control Hosp Epidemiol* 2006;**27**:1347–51.
119. Beckmann A, Doebler K, Schaefer E, Koetting J, Gastmeier P, Graf K. Sternal surgical site infection prevention – is there any room for improvement? *Eur J Cardiothorac Surg* 2011;**40**:347–51.
120. Castella A, Charrier L, Di Legami V, Pastorino F, Farina EC, Argentero PA, et al. Surgical site infection surveillance: analysis of adherence to recommendations for routine infection control practices. *Infect Control Hosp Epidemiol* 2006;**27**:835–40.
121. Young B, Ng TM, Teng C, Ang B, Tai HY, Lye DC. Nonconcordance with surgical site infection prevention guidelines and rates of surgical site infections for general surgical, neurological, and orthopedic procedures. *Antimicrob Agents Chemother* 2011;**55**:4659–63.

122. Durando P, Bassetti M, Orengo G, Crimi P, Battistini A, Bellina D, et al. Adherence to international and national recommendations for the prevention of surgical site infections in Italy: Results from an observational prospective study in elective surgery. *Am J Infect Control* 2012;**40**:969–72.
123. McBride T, Beamer J. Pre-operative patient preparation in the prevention of surgical site infections. *Can Oper Room Nurs J* 2007;**25**:34 26–32.
124. Rosenberger LH, Politano AD, Sawyer RG. The surgical care improvement project and prevention of post-operative infection, including surgical site infection. *Surg Infect (Larchmt)* 2011;**12**:163–8.
125. Galvin P. Cultivating quality: reducing surgical site infections in children undergoing cardiac surgery. *Am J Nurs* 2009;**109**:49–55.
126. Labeau SO, Witdouck SS, Vandijck DM, Claes B, Vandewoude KH, Lizy CM, et al. Nurses' knowledge of evidence-based guidelines for the prevention of surgical site infection. *Worldviews Evid Based Nurs* 2010;**7**:16–24.
127. Rauk PN. Educational intervention, revised instrument sterilization methods, and comprehensive preoperative skin preparation protocol reduce cesarean section surgical site infections. *Am J Infect Control* 2010;**38**:319–23.
128. Eskicioglu C, Gagliardi AR, Fenech DS, Forbes SS, McKenzie M, McLeod RS, et al. Surgical site infection prevention: a survey to identify the gap between evidence and practice in University of Toronto teaching hospitals. *Can J Surg* 2012;**55**:233–8.
129. Duff P. A simple checklist for preventing major complications associated with cesarean delivery. *Obstet Gynecol* 2010;**116**:1393–6.
130. Forbes SS, McLean RF. Review article: The anesthesiologist's role in the prevention of surgical site infections. *Can J Anaesth* 2012;**60**:176–83.
131. Reyes GE, Chang PS. Prevention of surgical site infections: Being a winner. *Oper Tech Sports Med* 2011;**19**:238–44.

CHAPTER

2

Microbiological Assessment before Surgical Intervention

Sanjay Bhattacharya, Ramkrishna Bhalchandra, Gaurav Goel and Lalawmpuia Hmar

Tata Medical Center, Kolkata, India

INTRODUCTION

Infection remains an important determinant in the outcome of patients undergoing surgery.[1] Surgical site infections (SSI) are classified as: superficial, deep incisional and organ space, based on the anatomical extent of the infection (Table 2.1). SSI rates vary from center to center, between developing and developed countries, and between anatomical locations.[1] Various studies of SSI have quoted figures between 0 to 30% depending on the infection related risk factors involved.[2] Data from the International Nosocomial Infection Control Consortium collected from 30 predominantly developing and middle-income countries between 2005 to 2010 show that there are significantly more SSIs in these countries compared to the USA-based figures from the Centers for Disease Control and Prevention National Healthcare Safety Network.[3] However, reliable surgical infection surveillance with robust validation is required as there is a significant possibility of underreporting SSI rates, either because of inadequate follow-up post-surgery, subjectivity in clinical assessment of SSI, and potential conflict of interests.[4] SSI rates have been used to rank hospitals, and there are also reports of underreporting of SSI rates even in the developed world.[5] However, the issue of ranking hospitals based on SSI rates is difficult, complex and controversial because of the diversity of cases (case mix). A study from The Netherlands on "rankability" (the reliability of ranking) of hospitals based on SSI rates showed that this varied between 0 to 80% based on surgical types (with an overall "rankability" of only 8% when all surgery types were combined).[6] Moreover, SSI rates become statistically reliable only when there is adequate number of cases.[2] Risk stratification

Microbiology for Surgical Infections
DOI: http://dx.doi.org/10.1016/B978-0-12-411629-0.00002-7

TABLE 2.1 Classification of Surgical Wounds

	Clean	Clean contaminated	Contaminated	Dirty
Surgical wound infection risk	<2%	2–10%	10–20%	>30%
Surgery	Elective	Emergency or elective	Emergency or elective	Emergency or elective
Transection or soiling	No transection of tracts	Controlled transection of tracts	Soiling of the operative field	Pre-operative perforation, abscess, purulence
Breakage in aseptic techniques	No break in aseptic techniques	Minor aseptic break, minimal spillage	Major break in technique or significant spillage	Penetrating lesion or trauma > 4 hours
Examples	Neuro-, orthopedic, cardiothoracic, vascular	Head & neck, biliary, urology, gynaecology	Colorectal, biliary/ urologic with infection	Perforated appendix with abscess
Antibiotics	1 dose or nil (maximum 1 day for long surgeries)	1 day maximum duration	2 days maximum duration (continue treatment if required)	Treat the infection

between surgical cases is essential to make intra-hospital and inter-hospital comparisons meaningful.[8] A study from the Duke Medical Center in the USA showed that procedure-specific rates of complex SSI were significantly correlated with their increasing National Nosocomial Infection Surveillance System risk index score.[7]

The economic burden of surgical site infections is huge. According to one European estimate, SSI forms a significant proportion of surgical costs, amounting to 1.47 to 19.1 billion euros.[9] Various SSI prevention strategies have been employed in attempts to reduce SSI rates, with variable success.[10–15] These have included the World Health Organization Surgical Safety Checklist (SSC), the Surgical Care Improvement Project (SCIP) and SSI reduction bundles.[10] In a study in Texas, USA it was shown that implementation of an integrated SSC can improve compliance with SSI reduction strategies such as SCIP, SSI reduction bundle performance and maintenance of normothermia, but this did not correlate with an improvement in overall SSI rates.[10] The nature of the bundles may not have been uniform and they have mostly concentrated on perioperative factors such as appropriate antibiotic prophylaxis, antibiotic timing, nature of hair removal, skin disinfection, regulation of body temperature, etc. Various studies have outlined patient related risk factors as important determinants in the outcome of patients undergoing surgery. However, there is inadequate information or practice about a comprehensive microbiological assessment of surgical patients, taking into account not only patient related risk factors, but also other factors (anesthetic, surgical, environmental, etc.) which may contribute to the genesis of infection.

For decades, antibiotics have dominated the discourse in the prevention of SSI. This has been the case despite the presence of other diverse measures for preventing infection in patients undergoing surgery, namely clean operating rooms, use of sterile instruments, use of skin disinfectants and aseptic techniques. Despite the obvious importance of antibiotics, in many cases the importance of other infrastructural measures and perhaps more important patient factors have not been addressed adequately in the SSI prevention strategies. As a result, the following situations seem to have emerged:

a) An over-reliance on antibiotics for preventing SSI,
b) Lack of individualization of antibiotic regimens for prophylaxis,
c) Over-prescription of antibiotics, together with increased cost and development of antibiotic resistance,
d) A relative neglect of basic infection control measures such as hand hygiene, housekeeping, patient isolation etc.,
e) An underestimation of the infrastructural and engineering requirements for infection prevention, such as quality controlled central sterile supply department (CSSD) services, microbiologically safe water supplies and clean air in operating rooms.

The current chapter emphasizes the point that a comprehensive and individualized microbiological assessment (both from a patient and institutional perspective) is an essential prerequisite to prevent infections in surgical patients. That assessment should look at both common organizational factors and individual patient factors when deciding the infection prevention approach, not just in terms of antibiotics but also with regard to patient isolation, PPE (personal protective equipment) usage, patient counseling, and enhancing surgical awareness about infection risks. The goal is a better management pathway and fewer post-surgical infections.

PURPOSE OF MICROBIOLOGICAL ASSESSMENT

1. To identify patients at risks of surgical site infection,
2. To devise an appropriate infection control and preventive strategy based on patient and infrastructural factors,
3. To have a management plan in place in case of development of infection.

PATIENT FACTORS IN MICROBIOLOGICAL ASSESSMENT OF A SURGICAL PATIENT

There are many patient-related factors that may affect the infection-related outcomes of surgical patients. Unless these factors are assessed and addressed adequately, infection risk assessment of the patient will be incomplete, antibiotic prophylaxis may be suboptimal, and decision-making on the part of the patient/relative/doctor/care giver will not be based on all available facts (Table 2.2A).

1. ***Medical history and primary diagnosis:*** This has a significant influence on the way infection is managed. For example, the infection risk during or following surgery (e.g., appendectomy) for a patient with acute myeloid leukemia is substantially different for a patient without any significant medical history. Abdominal surgery in neutropenic patients has a mortality rate of 50 to 70%.[16–18] Inflammation, tissue repair,

TABLE 2.2A Scope of Microbiological Assessment

A. Assessment of patient factors:
1. Previous microbiology
2. Serology for blood borne virus screening
3. Surveillance microbiology
4. Glycaemic control and diabetes
5. Status of immunity of patient
a. Chemotherapy/ radiotherapy/ HIV
6. Obesity and state of nutrition
7. Personal habits of the patient: smoking
8. Personal hygiene
9. Medications: steroids, immuno-suppressive agents
10. Presence of infection remote to the surgical site
11. Socio-economic status of the patient
12. Previous surgery and other interventions
13. Presence of foreign bodies/devices/stents/prosthesis
14. Anatomical site of surgery: clean/clean contaminated/contaminated/dirty
15. Type of surgery:
a. Elective/emergency
b. Clean/clean contaminated/contaminated/dirty
c. Open/laparoscopic/robotic
16. Expected duration of anesthesia and surgery
17. Possibilities for blood transfusion
18. Co-morbidities

coagulation, platelet number and function, metabolic homeostasis and immune defense are involved in surgical wound healing, and conditions that impair these functions have an adverse surgical risk. Laparoscopic appendectomy may be considered in neutropenic patients because it can be related to less postoperative infection, hemorrhagic complications, and a lower mortality rate.[18]

2. *Age:* Both dose of antibiotics, and the method of calculating it (per unit of body weight, or body surface area, or based on creatinine clearance) may depend on patient age. Certain antibiotics are contra-indicated for use in children (e.g., tetracycline) or may be unlicensed for use in children (e.g., quinolones). Surgery in geriatric patients, for example, is associated with greater risk of mortality, and those with lower Braden scale score (which includes sensory perception, moisture on skin, activity, mobility, nutrition, and friction) may have a greater chance of postoperative hospital stay and postoperative complications.[19] It has been seen that after emergency laparotomy, patients over 90 years of age had a less than 10% chance of survival. Other factors in that study associated with similar low survival chances included an American Society of Anesthesiologists (ASA) class V, septic shock, dependent functional status, and abnormal white blood cell count.[20]
3. ***Previous microbiology:*** Previous microbiological data may alert the surgeon/anesthetist to the need for an alternative prophylaxis, e.g., the presence of methicillin resistant *Staphylococcus aureus* (MRSA) or extended spectrum beta lactamase (ESBL) positive organisms in diagnostic or surveillance cultures. Vancomycin/teicoplanin

may have to be used in case of MRSA colonization, and meropenem in case of previous ESBL colonization or infection. For example, Ramirez et al. in a study from Texas, USA noted that MRSA colonization was not an independent risk factor for SSI following major gastro-intestinal (GI) operations, but was strongly predictive of MRSA-associated SSI in these patients. The authors concluded that a preoperative MRSA nasal swab test with decolonization may reduce the incidence of MRSA-associated SSI after major GI surgery.[21] In a septic orthopedic ward in Geneva, Switzerland (between March 2010 and November 2011) ESBL strains were detected in 204 out of 565 anal swab samples from 45 (out of 186) patients, suggesting a significant proportion of carriage.[22] In another hospital in Nigeria, 36 out of 102 Gram-negative isolates from orthopedic wound infections were found to be ESBL producers.[23] The prospect of acquiring more resistant organisms, such as those producing carbapenemases (resistant to meropenem, imipenem, etc.), from high endemic settings including those during medical tourism has also been highlighted.[24] A study in Iowa City, USA showed that surgical programs that implement a bundled intervention including both nasal decolonization and glycopeptide prophylaxis for MRSA carriers decrease rates of surgical site infections caused by *S. aureus* in cardiac and orthopedic surgeries.[25]

4. ***Previous antibiotic and allergy history:*** This may reveal a history of allergy (ranging from rash to anaphylaxis) and probability of antibiotic resistance. A history of allergy (even if minor or non-serious) may significantly affect the prescribing habits of physicians. In a study from Pennsylvania, USA it was reported that the majority of providers were uncomfortable with prescribing penicillins for patients who had had a prior maculopapular rash in reacting to penicillin, although they would use other beta-lactams. In patients with an exfoliative dermatitis caused by penicillin, 46% of responders would not prescribe any beta-lactam. For patients with a positive skin test to penicillin, only 45.1% of non-allergists were comfortable prescribing monobactams versus 62.5% of allergists, and 30.3% of all responders would give a carbapenem.[26] The link between prior antibiotic exposure and subsequent development of antibiotic resistance has been shown in a number of studies. For example, a study from The Netherlands reported that meropenem, ciprofloxacin, and ceftazidime were associated significantly with risk of resistance development in *Pseudomonas aeruginosa* (adjusted hazard ratio, 11.1 for meropenem, 4.1 for ciprofloxacin, 2.5 for ceftazidime).[27] It has been suggested that during antibiotic exposure a small proportion of cells within a given bacterial population (approximately one in a billion) may develop mutations leading to antibiotic resistance, because although the given antibiotic may have exceeded the minimum inhibitory concentration (MIC) it may not have exceeded the mutation prevention concentration (MPC).[28]
5. ***Serology for blood-borne virus screening:*** Patients in many centers are screened pre-operatively for blood borne virus markers (HIV antigen and antibodies, hepatitis B virus [HBV] surface antigen, hepatitis C virus [HCV] antibodies). Although there is debate over whether this screening should be done selectively rather than taking universal precautions at all surgeries, blood-borne virus serology data may help in optimizing the use of extra personal protective equipment and disposable drapes.[29] The results can also help to identify subclinical cases and direct them to appropriate

physicians for further treatment, besides enabling additional infection control interventions (such as screening and vaccination) for at-risk contacts.[29]

6. ***Surveillance microbiology:*** In some centers, especially in those with a significant prevalence of multi-drug resistant organisms (MDROs) such as MRSA or ESBL-producers, screening against these MDROs may be done so that decolonization therapy can be initiated before elective surgery wherever possible (MRSA) and antibiotic prophylaxis optimized (MRSA or ESBL). The screening may involve swab tests from nose/throat/groin/existing wounds (MRSA) or stool/rectal swabs (ESBL and other resistant Gram-negatives).[30] There is good evidence to support preoperative screening for MRSA in centers where prevalence is significant, screening is reliable and affordable, and results are available within a reasonable time frame.[31,32]
7. ***Glycemic control and diabetes:*** The presence of poor serum glucose control in the intra-operative and postoperative period has been recognized as an adverse prognostic factor with regard to surgical site infections. Hanazaki et al. commented that hyperglycemia is associated with abnormalities in function of leukocytes, including granulocyte adherence, impaired phagocytosis, delayed chemotaxis, and depressed bactericidal capacity.[33] These leukocyte deficiencies are the cause of infection and improve with tight glycemic control, which leads to fewer postoperative infections in critically ill surgical patients.[33] Kwon et al. reported from Seattle, USA that patients with perioperative hyperglycemia had significantly increased risks of infection (odds ratio [OR]: 2.0), reoperative interventions (OR: 1.8) or death (OR: 2.71). Increased risk of poor outcomes was observed for patients both with and without diabetes.[34] Ramos et al. reported from Boston, USA that postoperative hyperglycemia increased the risk of postoperative infections by 30% with every 40-point increase from normoglycemia (<110 mg/dL).[35] It was also noted that the increased risk of postoperative infection and length of hospitalization posed by postoperative hyperglycemia is independent of diabetic status.[35] Controlling diabetes before surgery is desirable. This should be done by consultation with a specialist in diabetes management.
8. ***Status of immunity of a patient and immuno-suppressive therapy:*** Patients undergoing surgery (especially those having cancer surgery) may be compromised immunologically due to disease (hematologic malignancy), chemotherapy, radiotherapy or HIV infection. Patients may have a defective immune system for about six months after exposure to immuno-suppressive therapy. This may affect wound healing and ability to fight infections. A South Korean study on head and neck cancers showed that history of prior radiotherapy was an independent risk factor for developing SSI (OR: 2.85, 95% CI: 1.172–6.931).[36] Another study from Italy showed that preoperative radiotherapy influenced perineal healing (after abdomino-perineal resection for rectal cancer) in patients with obesity (body mass index [BMI] more than 30).[37] Neumayer et al. also reported from the Salt Lake City, USA that recent radiotherapy is an independent risk factor for SSI.[38] Gao et al. reported from Beijing, China that Grade III + myelosuppression after adjuvant chemotherapy is an important risk factor for SSI in breast cancer.[39] Warren Peled et al. reported from San Francisco the highest rate of surgical site infections was found in the adjuvant chemotherapy group after mastectomy and immediate breast reconstruction.[40] Drapeau et al. reported from Italy that HIV-HCV co-infection was significantly associated with SSI

occurrence.[41] SSI rates were twice as high in HIV patients than reported for the general population in Italy or Europe.[41] Guild et al. reported from Atlanta, USA that HIV-positive patients with CD4 counts less than 300 per microliter were associated with development of postoperative infection.[42] Good quality evidence is not yet available for antibiotic prophylaxis during various surgeries carried out on such patients, but recognition of this limitation may enable further research.

9. ***Obesity and state of nutrition:*** Malnutrition and obesity are known to be risk factors associated with delayed healing and surgical wound infection. Some studies have reported SSI rates between 45 to 60% among obese patients.[43] The pathophysiology of obesity-related SSI has been suggested to be due to decreased wound oxygen tension, impaired tissue antibiotic penetration, altered immune function, etc.[43] Waisbren et al. reported from Boston, USA that obesity, defined by percentage of body fat, was associated with a five-fold increase in SSI risk. This risk increases as percentage of body fat increases. The percentage of body fat was reported to be a more sensitive and precise measurement of SSI risk than the BMI.[44] In another study from Boston, Giles et al. reported that obesity predicted postoperative SSI. Mortality risk was maximal in the underweight, followed by the morbidly obese and patients with normal weight, while overweight and mild to moderate obesity were associated with the lowest mortality.[45] Another US study by Itani et al. reported that in patients undergoing elective colorectal surgery, the incidence of SSI, specifically superficial incisional SSI, was higher in patients with a BMI greater than or equal to 30 kg/m^2, regardless of the prophylactic antibiotic given.[46] On the other side, Togo et al. reported from Mali that children with anemia or malnutrition were significantly more likely to develop infection postoperatively.[47] Wherever possible, gross under-nutrition should be corrected before surgery. Over nutrition in the form of obesity may have to be dealt with by the use of special surgical techniques during wound closure.
10. ***Personal habits of the patient: Smoking, tobacco chewing, oral hygiene:*** It is important that the dangers of smoking and other habits such as tobacco chewing are explained to the patient. Durand et al. reported from France that smoking was a significant risk factor for organ/space SSI in patients having orthopedic surgery with implant materials.[48] They also observed a significant difference between smokers and non-smokers for surgical wound complications (hematoma, discharge or wound dehiscence).[48] Hawn et al. reported from the USA that current smokers had significantly more postoperative pneumonia, surgical site infection, and deaths.[49] Sato et al. observed in Japan that good oral health care reduced the risk of postoperative surgical wound infections in patients with oral squamous cell carcinoma.[50]
11. ***Personal hygiene:*** Good personal hygiene before surgery is important. Disinfection of the skin before surgery will not have a sufficient effect if the hygiene of the patent remains poor. Patients can be advised to have regular baths with soap and water (enhanced with a medicated bath in case of MRSA colonization) for at least a week before surgery. Nguyen et al. observed a higher incidence of SSI with lack of supplies for wound care and personal hygiene in Vietnam.[51] Patients' hygiene should also include hand hygiene, as was emphasized in a study from New York.[52] Although there has been some debate and uncertainty about the benefits of preoperative bathing or showering with skin antiseptics, there is little doubt that general hygiene of the

patient before and after surgery helps, and all patients should be counseled about the importance of good hygiene.[53]

12. ***Medications such as steroids:*** Steroids may impair surgical wound healing. Ismael reported from Detroit, USA that superficial SSI increased from 2.9% to 5% in those using steroids (OR: 1.724). Deep SSIs increased from 0.8% to 1.8% (OR: 2.353). Organ/space SSIs and dehiscence increased two to three-fold with steroid use (OR: 2.469 and 3.338, respectively), and mortality increased almost four-fold (1.6% to 6.0%; OR: 3.920).[54] A Canadian study reported that the risk of SSI increased with chronic steroid use.[55] Patients may need to taper off long-term steroid use before surgery, and replace them with alternative agents.
13. ***Presence of infection remote from the surgical site:*** All infections remote from the surgical site must be eliminated or controlled before surgery. Edwards suggested in 1976 that a significant reduction in surgical wound infections can be achieved through control of infections at remote sites, particularly those associated with medical devices.[56]
14. ***Socio-economic status of the patient:*** This may give important clues regarding disease risk factors (e.g., tuberculosis) and affordability (in case the patient or relatives are bearing the cost of care). Kaye et al. reported that socio-economic factors (such as private insurance) were associated with lower risk of SSI.[57] Olsen also reported that a lack of private health insurance was marginally associated with increased risk of SSI.[58]
15. ***Previous surgery and other interventions:*** This may influence current surgery due to post-surgical adhesions from previous surgeries. It could also contribute to infection risk in the present by the carriage of infections from the past. A report from Cleveland, USA suggested that previous surgery increased the risk of SSI (OR: 2.4; 95% CI: 1.6 to 3.7).[59] Another report from Dallas, USA found similar results; previous abdominal surgery increased the risk of organ/space SSI (OR: 2.5).[60]
16. ***Presence of foreign bodies/ devices/ stents/ prosthesis:*** Reports from the Sloane-Kettering cancer center in New York observed that preoperative biliary stenting in proximal cholangiocarcinoma increased the incidence of contaminated bile and postoperative infectious complications.[61] An Italian report observed that preoperative biliary drainage was the only independent risk factor for infectious complication in the postoperative course, and concluded that the overall mortality and morbidity after liver resection are not improved by preoperative biliary drainage in patients with jaundice. Pre-hepatectomy biliary drainage increases the incidence of infectious complications.[62] The presence of such materials and devices may indicate the need for alternative prophylaxis while performing surgeries in contaminated areas (gastro-intestinal tract) and the likelihood of materials present in a colonized or contaminated state on which the present surgery needs to be done. As in the case of previous surgery, this may be a risk factor for the presence of specific organism types, including antibiotic-resistant micro-organisms.
17. ***Anatomical site of surgery: clean/ clean contaminated/ contaminated/ dirty:*** The anatomical site of surgery is one of the most important determinants regarding the choice and duration of antibiotic prophylaxis.[8] This is because of the type of micro-organisms present in different environments (mainly *Staphylococcus* sp. on skin surfaces, *Streptococcus* sp. and anaerobes in the mouth, *Enterococcus* sp. along with anaerobes and aerobic Gram-negative bacilli in the gastro-intestinal tract).

18. ***Type of surgery:*** Type of surgery (whether done electively or as an emergency, type of intervention and exposure used [open/laparoscopic/robotic]) may determine the surgical outcomes in terms of infection rates, wound healing and postoperative recovery. There is some debate regarding whether all emergency surgeries carry a worse prognosis than elective surgeries. Ciga et al. observed from Spain that in emergencies involving the descending colon, one-stage surgery may be performed without colonic preparation as safely as elective surgery in selected patients considered suitable for segmental resection of the descending colon and primary anastomosis.[63] However, Lee et al. observed from Taiwan that emergency surgery was independently associated with SSI after cardiac surgery (OR: 9.77).[64] Shigemura et al. reported from Japan that robot-assisted laparoscopic prostatectomy (RALP) tended to have a lower but statistically insignificant rate of postoperative infections than open laparoscopic prostatectomy (ORP).[65] However, Patzkowsky et al. found perioperative outcomes for laparoscopic and robotic hysterectomy for benign indications to be equivalent. Interestingly, in that study the rates of urinary tract infection were higher in the robotic hysterectomy (RH) group compared with laparoscopic hysterectomy (LH 2.7% vs. RH 6.9%).[66] A report from Denmark observed that laparoscopic surgery in obese patients was associated with a surgical site infection rate reduced by 70–80% compared with open surgery across general abdominal surgical procedures.[67]
19. ***Expected duration of anesthesia and surgery:*** This is likely to influence the degree of tissue hypoxia, and also determines the number and frequency of antibiotic doses administered for prophylaxis. Beal et al. in an animal model study on dogs and cats observed that duration of anesthesia is a risk factor for postoperative wound infection, independent of the duration of surgery.[68] Ameh et al. from Nigeria observed that the infection rate was 31% in operations lasting two or more hours and 17.3% in operations lasting less than two hours ($p < 0.05$).[69] As a general rule, if the duration of surgery is more than twice the half-life of the antibiotic, a repeat dose needs to be administered.
20. ***Blood transfusion:*** Blood transfusion is generally essential if the hemoglobin is less than 7 g/dL and is generally unnecessary if the hemoglobin is more than 10 g/dL. Within this range, the need for blood transfusion is debatable but is sometimes given, depending on the clinical situation. In some instances, a blood transfusion may increase infection risk due to immunosuppression. Campbell observed in Ann Arbor, USA that patients with low SSI rates were less likely to have received a blood transfusion.[70] Weber observed from Switzerland that perioperative allogenic blood transfusion and preoperative anemia had an increased risk of SSI.[71] Young et al. reported from the USA that among women undergoing abdominal hysterectomy, blood transfusion was an important risk factor for surgical site infection in patients who experienced blood loss of less than 500 mL.[72]
21. ***Co-morbidities:*** Co-morbidities such as diabetes, hypertension, asthma, chronic obstructive pulmonary diseases (COPD), chronic renal disease, obesity, ASA score (and other scoring systems such as Acute Physiology and Chronic Health Evaluation (APACHE), Sequential Organ Failure Assessment (SOFA), etc.) have a significant influence on intra-operative and postoperative management, morbidity and monitoring. Maizel et al. observed that non-dialysis dependent chronic kidney disease

appeared to be an independent risk factor for death after septic shock.[73] Kaye et al. predicted COPD as an independent predictor of SSI in older people.[57]

Malone et al. reported from Maryland, USA that diabetes (insulin- and non-insulin-dependent), low postoperative hematocrit, weight loss (within six months), and ascites were significantly related with an increased risk of SSI.[74]

ORGANIZATIONAL FACTORS IN THE MICROBIOLOGICAL ASSESSMENT OF A SURGICAL PATIENT (TABLE 2.2B)

1. ***Operating room air quality (ventilation control):*** There is controversy regarding the utility of clean rooms, such as those offering laminar air flow, with regard to reducing SSI in surgical settings. An Australian study reported that the use of laminar air in operating rooms combined with routine antibiotic prophylaxis resulted in an AUD $4.59 million cost increase and 127 quality adjusted life years lost compared with the baseline comparator (antibiotic prophylaxis alone).[75] A study from Berlin unexpectedly reported that operating room ventilation with laminar air flow showed no benefit and was even associated with a significantly higher risk for severe SSI after hip prosthesis.[76] The method of measuring air quality in clean rooms has also been debated. Landrin et al. from France reported that methods of microbiological and particle counting did not correlate with microbiological count (Spearman correlation coefficient = 0.06, $P = 0.6$). Using the receiver operating characteristic (ROC) curve, no particle count value could be predictive of a microbiological count higher than five colony forming units (CFU) per m^3.[77] However, there is little doubt that a clean

TABLE 2.2B Scope of Microbiological Assessment

B	**Assessment of organizational factors:**
1.	Operating room air quality (ventilation control): Class 100/ Class 1000: especially for clean surgeries: e.g., neurosurgery, joint replacement surgery, vascular and cardiothoracic surgery
2.	Quality of CSSD (Central Sterile Supply Department)
3.	Water quality for hand washing, surgical scrub, CSSD and patient bathing/cleaning, CSSD, drinking
4.	Housekeeping standards and quality
5.	Quality of nursing care
6.	Nutritional and dietary support
7.	Competency of surgical team (infection rates, complication rates, experience, mortality rates)
8.	Epidemiology of infection and antimicrobial resistance in the institution
9.	Implementation of the SSI prevention bundle
10.	Multi-disciplinary team approach:
	a. surgical staff
	b. anesthesiologists
	c. nurses
	d. operating room assistants/technologists
	e. infection control personnel
	f. pharmacists
	g. supporting staff (housekeeping)
	h. quality control officers
	i. engineers
	j. medical administration

environment is required for clean surgeries. Gosden et al. stated that periprosthetic infection rates have been shown to correlate with the number of airborne bacteria within 30 cm of the wound.[78] Air quality is critical in clean surgeries, e.g., joint replacement by prosthetic devices, cardiothoracic surgeries involving heart-valve replacement, insertion of vascular prosthetic devices, neuro-surgeries etc. Dharan and Pittet while commenting about the environmental control in operating rooms stated that:

> "Modern operating theaters have conventional plenum ventilation with filtered air where particles $\geq$ 5 micrometers are removed. For orthopedic and other implant surgery, laminar-flow systems are used with high-efficiency particulate air (HEPA) filters where particles $\geq$0.3 micrometers are removed. The use of ultra-clean air has been shown to reduce infection rates significantly in orthopedic implant surgery. Few countries have set bacterial threshold limits for conventionally ventilated operating rooms, although most recommend 20 air changes per hour in order to obtain 50–150 CFU/m^3 of air".[79]

The microbiologist should be knowledgeable about the most recent air microbiology, air particle count figures and how factors such as air velocity, air changes per hour in the operating room and cleanliness of the operating rooms affect air quality. The air quality may be monitored using air particle counters (e.g., Biotest, Merck), anemometers (air velocity measurements) or quantitative microbiological air samplers (e.g., Biotest, Merck using the centrifugal impaction method). There is a lack of consensus about the target level for microbiological air quality, but an important paper from the Hospital Infection Society working party suggests limits of 0.5 CFU/m^3 for ultra-clean operating theaters (clean surgeries) and 180 CFU/m^3 for conventionally ventilated theaters during operations.[80]

2. ***Quality of CSSD:*** The quality of CSSD operations can affect the outcome of surgical wounds. Good planning and design of the CSSD, especially with regard to architecture, water quality, human resources and workflow are important in providing high quality products.[81,82] Poor quality cleaning, disinfection, packaging, and sterilization control has a major impact on surgical site infection risk. A study from Thailand reported a total of 30 quality indicators. These include nine indicators for structure, 12 for process and nine for output of CSSD.[83] Sterilization control can be done by checking physical parameters, chemical indicators and biological indicators.[84]
3. ***Water quality for hand washing, surgical scrub, CSSD and patient bathing/cleaning, drinking:*** This is an area which requires continuous monitoring and assessment. Water quality can be monitored using meters for total dissolved solids (TDS) and chlorine, and microbiological assessment (e.g., using membrane filtration technology). The target levels are 0 CFU of coliforms and pseudomonas per 100 ml and 250 ml of water, respectively, with a total aerobic colony count of less than 2000 CFU/100 ml of water.[85] Outbreaks of atypical mycobacterial infections at laparoscopic port sites have been reported due to poor quality rinse water contaminated with *Mycobacterium chelonae*.[86] Bert et al. reported the outbreak of multi-drug resistant *P. aeruginosa* due to contaminated tap water from a neurosurgery intensive care unit.[87]
4. ***Housekeeping quality and cleaning standards:*** The housekeepers receive some of the lowest pay/remuneration in hospital services, yet their contribution to keeping a hospital aesthetically clean and clinically safe is enormous. Cleaning in a hospital

setting may be assessed using visual inspection, cleaning audits using standards and checklists, microbiological swab tests, ultra-violet illumination, and adenosine triphosphate (ATP) bioluminescence tests.[88–93] The optimal method for cleaning a clinical area, including operating rooms, needs to be based on infrastructure, resources available, and standards used. Dancer, in a paper on cleaning, suggested that cleaning practices should be tailored to clinical risk (cleaning in transplant units, operating rooms and intensive care is likely to be different from that in general wards), given the wide ranges of surfaces, equipment and building design. There may also be confusion between nursing and domestic/housekeeping personnel over the allocation of cleaning responsibilities, and neither may have received sufficient training on specialized cleaning methods.[88] Dancer also observed that fear of infection encouraged the use of powerful disinfectants for the elimination of real or imagined pathogens in hospitals. Disinfectants may offer false assurance against contamination, and their disinfection potential cannot be achieved without prior removal of organic soil. Detergent-based cleaning is cheaper than using disinfectants and much less toxic.[88] In a separate paper, Dancer observed that hand-touch sites are habitually contaminated by hospital pathogens, which are then delivered to patients on hands. The author suggested that prioritizing the cleaning of the hand-touch sites might offer a useful adjunct to the current preoccupation with hand hygiene.[89] Finding appropriate benchmarks for monitoring hospital cleanliness is not easy. Mulvey et al. in a study from Scotland reported that an ATP benchmark value of 100 relative light units offered the closest correlation with microbial growth levels < 2.5 CFU/cm^2 (ROC curve sensitivity: 57%; specificity: 57%).[90] Cooper et al. reported from Wales that visual assessment is not a reliable indicator of surface cleanliness or of cleaning efficacy. Moreover, data obtained using both ATP bioluminescence and aerobic colony counts indicated considerable variability after cleaning.[91] Sherlock et al. reported from Ireland that in a hospital setting with low microbiological counts, the use of chemical tests such as ATP may provide additional information about cleaning efficacy, and ATP trends allow the identification of environmental surfaces that require additional cleaning or cleaning schedule amendments. The accepted standard for ATP test was suggested as 500 relative light units (RLU).[92] However, the ATP based benchmarks may be made more stringent depending on user requirements. Lewis et al. observed that modifying the ATP based standard from 500 RLU to 250 RLU reduced aerobic colony counts from >100 to <2.5 CFU/cm^2, and counts of *S. aureus* reduced from up to 2.5 to <1 CFU/cm^2 (95% of the time).[93]

5. ***Quality of nursing care:*** This is a rate-limiting factor in many clinical settings, including surgical units. Care of central and peripheral venous catheters, prevention of catheter-associated urinary tract infections and pressure sores, prevention of ventilator-associated pneumonia in intensive care settings, maintenance of perineal, oral and dental hygiene are all critically dependent on the quality of nursing care. It was reported in a study from the USA that when Surgical Care Improvement Project (SCIP) guidelines were instituted, a statistically significant reduction in SSI rates was observed. As the compliance with SCIP improved, the rates of superficial SSI decreased.[94] Another study from New Jersey looking at nurse staffing and burnout reported that there was a significant association between patient-to-nurse ratio and

urinary tract and surgical site infections. In a multivariate model controlling for severity of clinical condition, nurse and hospital characteristics, only nurse burnout was significantly associated with urinary tract and surgical site infections. Hospitals in which burnout was reduced by 30% had a total of 6,239 fewer infections, giving an annual cost saving of up to $68 million.[95]

6. ***Nutritional and dietary support:*** A malnourished or an inadequately nourished patient is more vulnerable to infection. The nutritional requirements will vary depending on disease status, degree and nature of disability, specific nutritional needs, etc. The importance given to nutrition in SSI prevention can be judged from three Japanese studies.[96–98] Shinkawa et al. reported the Nutritional Risk Index as an independent predictive factor for the development of surgical site infection after pancreaticoduodenectomy.[96] A Phase III trial is in progress to evaluate the effect of perioperative nutrition enriched with eicosapentaenoic acid on body weight loss after total gastrectomy for T2–T4a gastric cancer.[97] Horie at al reported favorable effects of preoperative enteral immunonutrition (consisting of an enteral diet supplemented with arginine, dietary nucleotides, and omega-3 fatty acids administered for five days [750 ml/day] prior to surgery) on a surgical site infection in patients with colorectal cancer without malnutrition.[98]
7. ***Competency of surgical team (surgical site infection rates, complication rates, experience, mortality rates):*** A surgery with fewer complications such as hemorrhage, tissue damage, anastomotic leak has less chance of getting secondary infection. Although this will vary depending on the nature of the surgery and the anatomical site, the performance of the surgical team is critical in determining the clinical outcome. Many apparently simple and straightforward procedures may develop major infections due to poor surgical techniques; while difficult and complicated procedures may have excellent outcomes due to superior surgical skills. The issue of declining surgical skills with age has been debated by Blasier, who commented:

> "Although there is no federally mandated retirement age for surgeons in the United States, surgeons must realize their skills will decline, a properly planned retirement can be satisfying, and the retired surgeon has much to offer the medical and teaching community".[99]

Datta et al. from Imperial College School of Medicine in London observed that assessment of technical skill using inanimate procedural simulation translates to actual surgical performance within the operating theater.[100] Maan et al. observed that although academic achievement is an effective predictor of successful completion of surgical training programs, intermediate- and high-level visual-spatial perception, as well as psychomotor aptitude, can be used as criteria for assessing candidates for surgical training.[101] Clearly, the assessment of surgical skills needs to be more objective as suggested by Lord Darzi.[102] Neumayer et al. observed that a surgeon's inexperience and older age were significant predictors of recurrence in laparoscopic herniorrhaphy. The odds of recurrence for an inexperienced surgeon aged 45 years or older were 1.72 times that of a younger inexperienced surgeon.[103]

8. ***Epidemiology of infection and antimicrobial resistance in the institution:*** Certain institutions may have higher infection rates than others. Although this may be because

of the nature of the patients (e.g., immunocompromized versus immunocompetent) and procedures (e.g., clean elective cases versus contaminated emergency) there is no doubt that higher infection rates may be reflective of intrinsic differences in infection control practices, and differences in the epidemiology of infections including that caused by antibiotic resistant bacteria. Miller et al. from UCLA Medical Center in Los Angeles noted that glycopeptide prophylaxis minimizes the risk of SSI and cost when MRSA prevalence exceeds 3%. At very low MRSA prevalence (between 3% and 10%), the SSI minimization provided by glycopeptide prophylaxis is small and may be within the error of the model.[104] It was also noted in Connecticut, USA that inadequate antibiotic therapy was associated with longer length of hospital stay and higher costs in complicated SSI. Risk factors for inadequate antibiotic therapy included prior therapy with antibiotics, polymicrobial infection, infection with *P. aeruginosa* or *Enterococcus* spp., and abdominal incisions.[105]

9. ***Implementation of the SSI prevention bundle:*** A bundle is a set of interventions done collectively on all patients in a specific situation (e.g., surgical site infection prevention bundle, ventilator-associated pneumonia (VAP) prevention bundle, catheter-associated urinary tract infection (CAUTI) prevention bundle, central line-associated bloodstream infections (CLABSI) prevention bundle). Implementation of a bundle depends on administrative controls and the quality standards of an institution. The important features of the SSI prevention bundle include hand hygiene, adequate preoperative skin preparation and disinfection, no shaving for hair removal (clippers may be used), appropriate antibiotic prophylaxis.[106] However, observations about the utility of SSI prevention bundles are conflicting. Anthony et al. reported from Texas, USA that an evidence-based intervention bundle did not reduce SSI.[14] However, Lutfiyya et al. reported from the USA that instituting a care bundle composed of core and adjunct strategies significantly decreased the rate of colorectal SSIs.[11] Crolla et al. also reported from The Netherlands that the implementation of the bundle was associated with improved compliance over time and a 36% reduction of the SSI rate after adjustment for confounders.[12]
10. ***Multi-disciplinary team approach:*** Infection prevention and control requires multi-disciplinary effort. An institution which has the mechanism and the system in place to ensure this multi-disciplinary approach has a better chance of taking on the challenge posed by infectious diseases.[107] Ballard et al. reported from Colorado, USA based on a number needed to treat analysis, that one infection in every 16 patients within the high risk group, and one in 30 overall, were prevented up to one year postoperatively using a multi-disciplinary approach in pediatric spine surgery. The relative risk reduction was 43.0% for all patients, 49.3% for high risk patients and 44.6% for low risk patients.[108]

CONCLUSIONS AND FUTURE PROSPECTS

Nothing that has been said so far in this chapter is fundamentally unique or novel regarding the factors that directly or indirectly contribute to surgical wound infections. The principle aim of this chapter is to emphasize the need for comprehensive microbiological assessment before surgery is undertaken on a patient (Table 2.3). It is true that in cases

TABLE 2.3 Examples of Preoperative Microbiological Assessment

	Patient 1	Patient 2	Patient 3
Setting	Tertiary Care Center	General Hospital	Primary Health Center
Patient assessment:			
Age and gender	50 years; male	75 years; female	25 years; female
BMI	31	25	19
Disease/ medical condition	Coronary artery disease	Colon cancer	Pregnancy
Type of surgery	CABG (clean)	Hemicolectomy (contaminated)	Cesarean section (clean contaminated)
Emergency/elective surgery	Elective	Elective	Emergency
Diabetes	Present	Absent	Present
Smoking	Present	Absent	Absent
Steroid	Absent	Present	Absent
On-going infection/ medical condition	Diabetes, high BP	Exacerbation of COPD	Chronic cough
Allergies (medicines, disinfectants)	Sulfa	Penicillin, iodine	None
Previous microbiology	None	ESBL in urine	None
MDRO in surveillance microbiology	Negative	MRSA from nose swab	None
HIV/HBV/HCV screen	HCV reactive	Non-reactive	Not done
Chemo/radiotherapy	None	Recent chemotherapy	None
Personal hygiene	Good	Good	Poor
Nutritional status	Obese	Good	Poor
Previous surgery	None	Hysterectomy 10 yrs ago	Previous C-Section
Patient resources	High income	Middle income	Low income
Organizational factors:			
Housekeeping standards	Good	Satisfactory	Poor
SSI rate	Low	High	Not known
Antibiotic resistance	10% MRSA	70% ESBL	Not known
Water quality	Good	Variable	Uncertain
Hand hygiene compliance	75%	50%	Poor: 20%
Operating room air quality	Class 100	No HEPA filter	Poor: AC (air conditioning) not working

(Continued)

TABLE 2.3 (Continued)

	Patient 1	**Patient 2**	**Patient 3**
CSSD standards	Good	Unsatisfactory	No quality control
Nursing standards	Good	Good	Basic
Surgical experience	Good	Good	Limited
Infection risk	Low to moderate	Moderate to high	Moderate
Suggested actions:			
Antibiotic prophylaxis	Teicoplanin 400 mg IV 2 doses	Meropenem 1 gm 3 doses + Teicoplanin 400 mg IV 2 doses	Cefuroxime 1.5 gm IV after cord clamping
Disinfectant for skin preparation	2% Chlorhexidine + 70% isopropyl alcohol	2% Chlorhexidine + 70% isopropyl alcohol	2% Chlorhexidine + 70% isopropyl alcohol
Preoperative bath and shampoo	Maintain good hygiene	Maintain good hygiene	For at least 3 days
Treatment of infections	Not applicable	Treat COPD exacerbation before surgery	Not applicable
Co-morbidity assessment/ treatment& referral	Stop smoking, control blood glucose and BP – Gen Medicine	To see chest physician	General Medicine referral
Further infection investigation(s)	LFT and HCV viral load followed by HCV genotype, liver imaging	MRSA screening post decolonization therapy	CXR after delivery (suspected tuberculosis)
Decolonisation treatment for MDRO	Not applicable	4% Chlorhexidine body wash with nasal mupirocin	Not applicable
Infection control (for staff):			
Isolation/ barrier precautions	Standard precautions	Required	Standard precautions
Use of enhanced PPE	Standard precautions + enhanced contact precautions during exposure prone interventions	Standard precautions	Standard precautions + enhanced respiratory precautions for suspected tuberculosis

of emergency surgery where time is short, a detailed and relatively time consuming infection risk assessment along the lines outlined in this chapter may not be feasible, however it may be argued that since the organizational factors contributing to surgical wound infections may not change very rapidly (and are hence constant over a limited timeframe), one can still undertake a microbiological assessment of the patient using a standardized checklist (as outlined in Tables 2.2 to 2.4), just as in the case of elective surgical procedures. In the opinion of the authors, this assessment would give a better understanding of the

TABLE 2.4 Auditable Parameters

a.	Proportion of patients with previous positive microbiology getting appropriate antibiotic prophylaxis during surgery
b.	Patients with a positive serology (HIV/HBV/HCV) having appropriate follow up serology and/or referral
c.	Proportion of patients undergoing elective surgery after achieving glycemic control
d.	Proportion of HIV positive patients undergoing elective surgery where anti-retroviral therapy was appropriately managed during perioperative period
e.	Proportion of patients who were previously smokers were offered smoking cessation counseling and actually stopped smoking before surgery
f.	Proportion of patients on steroids before surgery whose steroid dosing was adequately managed in perioperative period
g.	Proportion of patients with infection remote to the surgical site before surgery who were adequately treated with appropriate anti-infective agents before surgery
h.	Proportion of patients with devices/stents/ prosthesis before surgery and getting postoperative infections and multi-drug resistant organisms
i.	Proportion of patients with clean/clean contaminated/contaminated surgery getting postoperative infections despite adequate prophylaxis
j.	Proportion of patients with on-going infection who were adequately treated in the perioperative and postoperative stage
k.	Air particle counts of class 100/1000/10000 operating rooms
l.	Performance of CSSD in terms of failure of physical/chemical/biological indicators of sterilization processes (autoclave/ ethylene oxide/ plasma)
m.	Incidences of unacceptable microbiological quality of hospital water (total aerobic count/pseudomonas/ coliform count) with at least one assessment done every month per area
n.	Toilet and ward cleaning audits
o.	CLABSI/VAP/CAUTI/SSI rates
p.	Infection rates per surgical type and specialty
q.	MRSA/ESBL/carbapenemase related infection rates

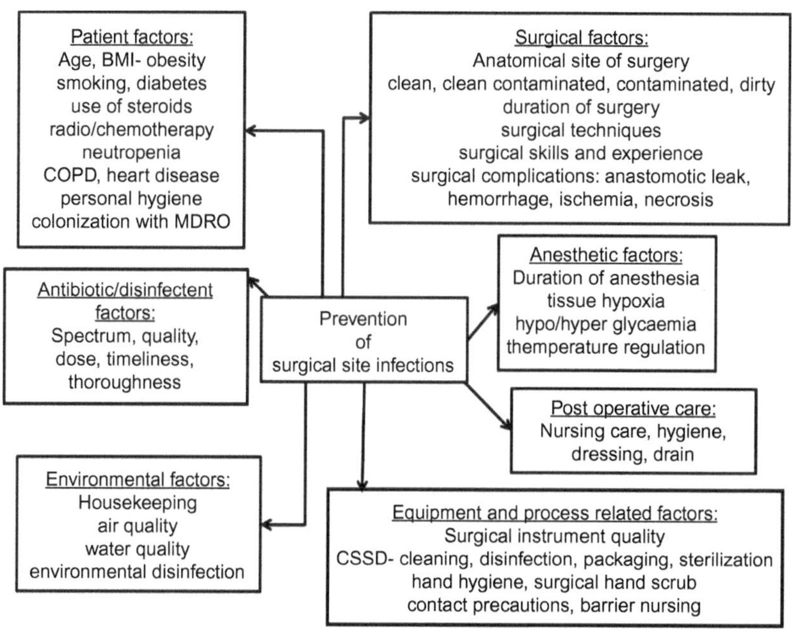

FIGURE 2.1 Factors responsible for surgical site infection. BMI – body mass index, COPD – chronic obstructive pulmonary disease, MDRO – multidrug-resistant organisms, CSSD – central sterile supply department.

infection risks involved and enable optimization of patient counseling, outcome prediction based on prospective studies of patients in various risk groups, health-economic outcome measures, and selection of more evidence-based prophylactic and preventive strategies (see table on auditable parameters: Table 2.4). Currently, the entire approach to infection management of the surgical patient seems to be generalized, and hinges on antibiotic prophylaxis based on anatomical localization. The current chapter is an attempt to emphasize the fact that infection prevention and management of the surgical patient needs to be individualized, and a more nuanced approach is needed taking into account the numerous patient and organizational factors that determine outcome (Figure 2.1).

References

1. Bruce J, Russell EM, Mollison J, Krukowski ZH. The measurement and monitoring of surgical adverse events. *Health Technol Assess* 2001;**5**:1–194.
2. Kao LS, Ghaferi AA, Ko CY, Dimick JB. Reliability of superficial surgical site infections as a hospital quality measure. *J Am Coll Surg* 2011;**213**:231–5.
3. Rosenthal VD, Richtmann R, Singh S, Apisarnthanarak A, Kübler A, Viet-Hung N, et al. Surgical site infections, international nosocomial infection control consortium (INICC) report, data summary of 30 countries, 2005–2010. *Infect Control Hosp Epidemiol* 2013;**34**:597–604.
4. Tanner J, Kiernan M, Leaper D, Baggott R. Reliable surgical site infection surveillance with robust validation is required. *J Hosp Infect* 2013;**84**:270.
5. Mayor S. English hospitals under-report surgical site infections, survey shows. *BMJ* 2013;**346**:f345.
6. van Dishoeck AM, Koek MB, Steyerberg EW, van Benthem BH, Vos MC, Lingsma HF. Use of surgical-site infection rates to rank hospital performance across several types of surgery. *Br J Surg* 2013;**100**:628–36.
7. Anderson DJ, Chen LF, Sexton DJ, Kaye KS. Complex surgical site infections and the devilish details of risk adjustment: important implications for public reporting. *Infect Control Hosp Epidemiol* 2008;**29**:941–6.
8. Ortega G, Rhee DS, Papandria DJ, Yang J, Ibrahim AM, Shore AD, et al. An evaluation of surgical site infections by wound classification system using the ACS-NSQIP. *J Surg Res* 2012;**174**:33–8.
9. Leaper DJ, van Goor H, Reilly J, Petrosillo N, Geiss HK, Torres AJ, et al. Surgical site infection – a European perspective of incidence and economic burden. *Int Wound J* 2004;**1**:247–73.
10. Tillman M, Wehbe-Janek H, Hodges B, Smythe WR, Papaconstantinou HT, et al. Surgical care improvement project and surgical site infections: can integration in the surgical safety checklist improve quality performance and clinical outcomes? *J Surg Res* 2013;**184**:150–6.
11. Lutfiyya W, Parsons D, Breen J. A colorectal "care bundle" to reduce surgical site infections in colorectal surgeries: a single-center experience. *Perm J* 2012;**16**:10–6.
12. Crolla RM, van der Laan L, Veen EJ, Hendriks Y, van Schendel J, Kluytmans J. Reduction of surgical site infections after implementation of a bundle of care. *PLoS One* 2012;**7**:e44599.
13. Johnson B, Starks I, Bancroft G, Roberts PJ. The effect of care bundle development on surgical site infection after hemiarthroplasty: an 8-year review. *J Trauma Acute Care Surg* 2012;**72**:1375–9.
14. Anthony T, Murray BW, Sum-Ping JT, Lenkovsky F, Vornik VD, Parker BJ. Evaluating an evidence-based bundle for preventing surgical site infection: a randomized trial. *Arch Surg* 2011;**146**:263–9.
15. Acklin YP, Widmer AF, Renner RM, Frei R, Gross T. Unexpected increased rate of surgical site infections following implant surgery for hip fractures: problem solution with the bundle approach. *Injury* 2011;**42**:209–16.
16. Brodin P, Gilg S, Lundell L, Mattsson J. Major surgery in a neutropenic patient undergoing allogeneic stem cell transplantation for high risk myelofibrosis. *Int J Hematol* 2012;**96**:798–800.
17. Hayden JT, Wood KM, Pedler S, Lawson A, Skinner R. Invasive aspergillosis of the small bowel in an infant with acute myeloid leukemia and intestinal obstruction. *Pediatr Hematol Oncol* 2009;**26**:84–91.
18. Ustun C. Laparoscopic appendectomy in a patient with acute myelogenous leukemia with neutropenia. *J Laparoendosc Adv Surg Tech A* 2007;**17**:213–5.

19. Cohen RR, Lagoo-Deenadayalan SA, Heflin MT, Sloane R, Eisen I, Thacker JM, et al. Exploring predictors of complication in older surgical patients: a deficit accumulation index and the Braden Scale. *J Am Geriatr Soc* 2012;**60**:1609–15.
20. Al-Temimi MH, Griffee M, Enniss TM, Preston R, Vargo D, Overton S, et al. When is death inevitable after emergency laparotomy? Analysis of the American College of Surgeons National Surgical Quality Improvement Program database. *J Am Coll Surg* 2012;**215**:503–11.
21. Ramirez MC, Marchessault M, Govednik-Horny C, Jupiter D, Papaconstantinou HT. The impact of MRSA colonization on surgical site infection following major gastrointestinal surgery. *J Gastrointest Surg* 2013;**17**:144–52.
22. Agostinho A, Renzi G, Haustein T, Jourdan G, Bonfillon C, Rougemont M, et al. Epidemiology and acquisition of extended-spectrum beta-lactamase-producing Enterobacteriaceae in a septic orthopaedic ward. *Springerplus* 2013;**2**:91.
23. Idowu OJ, Onipede AO, Orimolade AE, Akinyoola LA, Babalola GO. Extended-spectrum beta-lactamase orthopedic wound infections in Nigeria. *J Glob Infect Dis* 2011;**3**:211–5.
24. Kaul TK, Chhina DK. Medical tourism and New Delhi metallo β-lactamase 1 – a concern and threat. *J Anaesthesiol Clin Pharmacol* 2010;**26**:437–8.
25. Schweizer M, Perencevich E, McDanel J, Carson J, Formanek M, Hafner J, et al. Effectiveness of a bundled intervention of decolonization and prophylaxis to decrease Gram positive surgical site infections after cardiac or orthopedic surgery: systematic review and meta-analysis. *BMJ* 2013;**346**:f2743.
26. Prematta T, Shah S, Ishmael FT. Physician approaches to beta-lactam use in patients with penicillin hypersensitivity. *Allergy Asthma Proc* 2012;**33**:145–51.
27. Ong DS, Jongerden IP, Buiting AG, Leverstein-van Hall MA, Speelberg B, Kesecioglu J, et al. Antibiotic exposure and resistance development in *Pseudomonas aeruginosa* and *Enterobacter* species in intensive care units. *Crit Care Med* 2011;**39**:2458–63.
28. Cantón R, Morosini MI. Emergence and spread of antibiotic resistance following exposure to antibiotics. *FEMS Microbiol Rev* 2011;**35**:977–91.
29. Ahmed R, Bhattacharya S. Universal screening versus universal precautions in the context of preoperative screening for HIV, HBV, HCV in India. *Indian J Med Microbiol* 2013;**31**:219–25.
30. Bhattacharya S. Is screening patients for antibiotic-resistant bacteria justified in the Indian context? *Indian J Med Microbiol* 2011;**29**:213–7.
31. Kavanagh K, Abusalem S. Mounting evidence supports universal surveillance for MRSA in preoperative patients. *J Am Coll Surg* 2011;**213**:335–6.
32. Moldovan A, Dobrin L, Cotoara I, Angheloiu B, Proca G, Moldovan B. P194: Preoperative screening of patients – a small step for the hospital, a big step for epidemiology. *Antimicrob Resist Infect Control* 2013;**2**(Suppl 1):P194.
33. Hanazaki K, Maeda H, Okabayashi T. Relationship between perioperative glycemic control and postoperative infections. *World J Gastroenterol* 2009;**15**:4122–5.
34. Kwon S, Thompson R, Dellinger P, Yanez D, Farrohki E, Flum D. Importance of perioperative glycemic control in general surgery: a report from the Surgical Care and Outcomes Assessment Program. *Ann Surg* 2013;**257**:8–14.
35. Ramos M, Khalpey Z, Lipsitz S, Steinberg J, Panizales MT, Zinner M, et al. Relationship of perioperative hyperglycemia and postoperative infections in patients who undergo general and vascular surgery. *Ann Surg* 2008;**248**:585–91.
36. Lee DH, Kim SY, Nam SY, Choi SH, Choi JW, Roh JL. Risk factors of surgical site infection in patients undergoing major oncological surgery for head and neck cancer. *Oral Oncol* 2011;**47**:528–31.
37. Zorcolo L, Restivo A, Capra F, Fantola G, Marongiu L, Casula G. Does long-course radiotherapy influence postoperative perineal morbidity after abdominoperineal resection of the rectum for cancer? *Colorectal Dis* 2011;**13**:1407–12.
38. Neumayer L, Hosokawa P, Itani K, El-Tamer M, Henderson WG, Khuri SF. Multivariable predictors of postoperative surgical site infection after general and vascular surgery: results from the patient safety in surgery study. *J Am Coll Surg* 2007;**204**:1178–87.
39. Gao YX, Xu L, Ye JM, Wang DM, Zhao JX, Zhang LB, et al. Analysis of risk factors of surgical site infections in breast cancer. *Chin Med J (Engl)* 2010;**123**:559–62.
40. Warren Peled A, Itakura K, Foster RD, Hamolsky D, Tanaka J, Ewing C, et al. Impact of chemotherapy on postoperative complications after mastectomy and immediate breast reconstruction. *Arch Surg* 2010;**145**:880–5.

41. Drapeau CM, Pan A, Bellacosa C, Cassola G, Crisalli MP, De Gennaro M, et al. Surgical site infections in HIV-infected patients: results from an Italian prospective multicenter observational study. *Infection* 2009;**37**:455–60.
42. Guild GN, Moore TJ, Barnes W, Hermann C. CD4 count is associated with postoperative infection in patients with orthopaedic trauma who are HIV positive. *Clin Orthop Relat Res* 2012;**470**:1507–12.
43. Hourigan JS. Impact of obesity on surgical site infection in colon and rectal surgery. *Clin Colon Rectal Surg* 2011;**24**:283–90.
44. Waisbren E, Rosen H, Bader AM, Lipsitz SR, Rogers SO Jr, Eriksson E. Percent body fat and prediction of surgical site infection. *J Am Coll Surg* 2010;**210**:381–9.
45. Giles KA, Hamdan AD, Pomposelli FB, Wyers MC, Siracuse JJ, Schermerhorn ML. Body mass index: surgical site infections and mortality after lower extremity bypass from the National Surgical Quality Improvement Program 2005–2007. *Ann Vasc Surg* 2010;**24**:48–56.
46. Itani KM, Jensen EH, Finn TS, Tomassini JE, Abramson MA. Effect of body mass index and ertapenem versus cefotetan prophylaxis on surgical site infection in elective colorectal surgery. *Surg Infect (Larchmt)* 2008;**9**:131–7.
47. Togo A, Coulibaly Y, Dembélé BT, Togo B, Keita M, Kanté L, et al. Risk factors for surgical site infection in children at the teaching hospital Gabriel Touré, Bamako. *J Hosp Infect* 2011;**79**:371–2.
48. Durand F, Berthelot P, Cazorla C, Farizon F, Lucht F. Smoking is a risk factor of organ/space surgical site infection in orthopaedic surgery with implant materials. *Int Orthop* 2013;**37**:723–7.
49. Hawn MT, Houston TK, Campagna EJ, Graham LA, Singh J, Bishop M. The attributable risk of smoking on surgical complications. *Ann Surg* 2011;**254**:914–20.
50. Sato J, Goto J, Harahashi A, Murata T, Hata H, Yamazaki Y, et al. Oral health care reduces the risk of postoperative surgical site infection in inpatients with oral squamous cell carcinoma. *Support Care Cancer* 2011;**19**:409–16.
51. Nguyen D, MacLeod WB, Phung DC, Cong QT, Nguyen VH, Nguyen VH, et al. Incidence and predictors of surgical-site infections in Vietnam. *Infect Control Hosp Epidemiol* 2001;**22**:485–92.
52. Ardizzone LL, Smolowitz J, Kline N, Thom B, Larsen EL. Patient hand hygiene practices in surgical patients. *Am J Infect Control* 2013;**41**:487–91.
53. Webster J, Osborne S. Preoperative bathing or showering with skin antiseptics to prevent surgical site infection. *Cochrane Database Syst Rev* 2012;**9**:CD004985.
54. Ismael H, Horst M, Farooq M, Jordon J, Patton JH, Rubinfeld IS. Adverse effects of preoperative steroid use on surgical outcomes. *Am J Surg* 2011;**201**:305–8 discussion 308–9
55. van Walraven C, Musselman R. The Surgical Site Infection Risk Score (SSIRS): a model to predict the risk of surgical site infections. *PLoS One* 2013;**8**:e67167.
56. Edwards LD. The epidemiology of 2056 remote site infections and 1966 surgical wound infections occurring in 1865 patients: a four year study of 40,923 operations at Rush-Presbyterian-St. Luke's Hospital, Chicago. *Ann Surg* 1976;**184**:758–66.
57. Kaye KS, Sloane R, Sexton DJ, Schmader KA. Risk factors for surgical site infections in older people. *J Am Geriatr Soc* 2006;**54**:391–6.
58. Olsen MA, Higham-Kessler J, Yokoe DS, Butler AM, Vostok J, Stevenson KB, et al. Prevention epicenter program, centers for disease control and prevention. Developing a risk stratification model for surgical site infection after abdominal hysterectomy. *Infect Control Hosp Epidemiol* 2009;**30**:1077–83.
59. Haridas M, Malangoni MA. Predictive factors for surgical site infection in general surgery. *Surgery* 2008;**144**:496–501.
60. Blumetti J, Luu M, Sarosi G, Hartless K, McFarlin J, Parker B, et al. Surgical site infections after colorectal surgery: do risk factors vary depending on the type of infection considered? *Surgery* 2007;**142**:704–11.
61. Hochwald SN, Burke EC, Jarnagin WR, Fong Y, Blumgart LH. Association of preoperative biliary stenting with increased postoperative infectious complications in proximal cholangiocarcinoma. *Arch Surg* 1999;**134**:261–6.
62. Ferrero A, Lo Tesoriere R, Viganò L, Caggiano L, Sgotto E, Capussotti L. Preoperative biliary drainage increases infectious complications after hepatectomy for proximal bile duct tumor obstruction. *World J Surg* 2009;**33**:318–25.
63. Ciga MA, Oteiza F, Fernández L, de Miguel M, Ortiz H. Comparative study of one-stage colectomy of the descending colon in emergency and elective surgery without mechanical preparation. *Dis Colon Rectum* 2010;**53**:1524–9.
64. Lee YP, Feng MC, Wu LC, Chen SH, Chen YH, Chiu CC, et al. Outcome and risk factors associated with surgical site infections after cardiac surgery in a Taiwan medical center. *J Microbiol Immunol Infect* 2010;**43**:378–85.

65. Shigemura K, Tanaka K, Yamamichi F, Muramaki M, Arakawa S, Miyake H, et al. Comparison of postoperative infection between robotic-assisted laparoscopic prostatectomy and open radical prostatectomy. *Urol Int* 2013 Jun 11. Available from: http://dx.doi.org/10.1159/000350334 [Epub ahead of print].
66. Patzkowsky KE, As-Sanie S, Smorgick N, Song AH, Advincula AP. Perioperative outcomes of robotic versus laparoscopic hysterectomy for benign disease. *J SLS* 2013;**17**:100–6.
67. Shabanzadeh DM, Sørensen LT. Laparoscopic surgery compared with open surgery decreases surgical site infection in obese patients: a systematic review and meta-analysis. *Ann Surg* 2012;**256**:934–45.
68. Beal MW, Brown DC, Shofer FS. The effects of perioperative hypothermia and the duration of anesthesia on postoperative wound infection rate in clean wounds: a retrospective study. *Vet Surg* 2000;**29**:123–7.
69. Ameh EA, Mshelbwala PM, Nasir AA, Lukong CS, Jabo PA, Anumah MA, et al. Surgical site infection in children: prospective analysis of the burden and risk factors in a sub-Saharan African setting. Surg Infect (Larchmt) 2009;**10**:105–9.
70. Campbell Jr DA, Henderson WG, Englesbe MJ, Hall BL, O'Reilly M, Bratzler D, et al. Surgical site infection prevention: the importance of operative duration and blood transfusion – results of the first American College of Surgeons – National Surgical Quality Improvement Program Best Practices Initiative. *J Am Coll Surg* 2008;**207**:810–20.
71. Weber WP, Zwahlen M, Reck S, Misteli H, Rosenthal R, Buser AS, et al. The association of preoperative anemia and perioperative allogeneic blood transfusion with the risk of surgical site infection. *Transfusion* 2009;**49**:1964–70.
72. Young H, Berumen C, Knepper B, Miller A, Silverman M, Gilmartin H, et al. Statewide collaboration to evaluate the effects of blood loss and transfusion on surgical site infection after hysterectomy. *Infect Control Hosp Epidemiol* 2012;**33**:90–3.
73. Maizel J, Deransy R, Dehedin B, Secq E, Zogheib E, Lewandowski E, et al. Impact of non-dialysis chronic kidney disease on survival in patients with septic shock. *BMC Nephrol* 2013;**14**:77.
74. Malone DL, Genuit T, Tracy JK, Gannon C, Napolitano LM. Surgical site infections: reanalysis of risk factors. *J Surg Res* 2002;**103**:89–95.
75. Merollini KM, Crawford RW, Whitehouse SL, Graves N. Surgical site infection prevention following total hip arthroplasty in Australia: A cost-effectiveness analysis. *Am J Infect Control* 2013;**41**:803–9.
76. Brandt C, Hott U, Sohr D, Daschner F, Gastmeier P, Henning R. Operating room ventilation with laminar airflow shows no protective effect on the surgical site infection rate in orthopedic and abdominal surgery. *Ann Surg* 2008;**248**:695–700.
77. Landrin A, Bissery A, Kac G. Monitoring air sampling in operating theatres: can particle counting replace microbiological sampling? *J Hosp Infect* 2005;**61**:27–9.
78. Gosden PE, MacGowan AP, Bannister GC. Importance of air quality and related factors in the prevention of infection in orthopedic implant surgery. *J Hosp Infect* 1998;**39**:173–80.
79. Dharan S, Pittet D. Environmental controls in operating theatres. *J Hosp Infect* 2002;**51**:79–84.
80. Hoffman PN, Williams J, Stacey A, Bennett AM, Ridgway GL, Dobson C, et al. Microbiological commissioning and monitoring of operating theatre suites. *J Hosp Infect* 2002;**52**:1–28.
81. Hughes TA. Planning and equipping a CSSD for a new hospital. *Nurs Times* 1965;**61**:399–400.
82. Niléhn B. Role of the environment in hospital infections – design of CSSD. *J Sterile Serv Manage* 1983; **1**:20–2.
83. Sangthong K, Soparat P, Moongtui W, Danchaivijitr S. Development of quality indicators for sterilization practices of the central sterile supply department. *J Med Assoc Thai* 2005;**88**(Suppl 10):S128–32.
84. Jabbari H, Alikhah H, Sahebkaram Alamdari N, Naghavi Behzad M, Mehrabi E, Borzui L, et al. Developing the use of quality indicators in sterilization practices. *Iran J Public Health* 2012;**41**:64–9.
85. Health Protection Agency. The microbiological examination of water samples. Issue no: 2 Issue date: 06.12.05 issued by: Standards unit, Evaluations & standards laboratory. Reference no: QSOP 57i2. Health Protection Agency, UK.
86. Vijayaraghavan R, Chandrashekhar R, Sujatha Y, Belagavi CS. Hospital outbreak of atypical mycobacterial infection of port sites after laparoscopic surgery. *J Hosp Infect* 2006;**64**:344–7.
87. Bert F, Maubec E, Bruneau B, Berry P, Lambert-Zechovsky N. Multi-resistant *Pseudomonas aeruginosa* outbreak associated with contaminated tap water in a neurosurgery intensive care unit. *J Hosp Infect* 1998;**39**:53–62.
88. Dancer SJ. Hospital cleaning in the 21st century. *Eur J Clin Microbiol Infect Dis* 2011;**30**:1473–81.

89. Dancer SJ. The role of environmental cleaning in the control of hospital-acquired infection. *J Hosp Infect* 2009;**73**:378–85.
90. Mulvey D, Redding P, Robertson C, Woodall C, Kingsmore P, Bedwell D, et al. Finding a benchmark for monitoring hospital cleanliness. *J Hosp Infect* 2011;**77**:25–30.
91. Cooper RA, Griffith CJ, Malik RE, Obee P, Looker N. Monitoring the effectiveness of cleaning in four British hospitals. *Am J Infect Control* 2007;**35**:338–41.
92. Sherlock O, O'Connell N, Creamer E, Humphreys H. Is it really clean? An evaluation of the efficacy of four methods for determining hospital cleanliness. *J Hosp Infect* 2009;**72**:140–6.
93. Lewis T, Griffith C, Gallo M, Weinbren M. A modified ATP benchmark for evaluating the cleaning of some hospital environmental surfaces. *J Hosp Infect* 2008;**69**:156–63.
94. Berenguer CM, Ochsner Jr MG, Lord SA, Senkowski CK. Improving surgical site infections: using National Surgical Quality Improvement Program data to institute Surgical Care Improvement Project protocols in improving surgical outcomes. *J Am Coll Surg* 2010;**210**:737–41.
95. Cimiotti JP, Aiken LH, Sloane DM, Wu ES. Nurse staffing, burnout, and health care-associated infection. *Am J Infect Control* 2012;**40**:486–90.
96. Shinkawa H, Takemura S, Uenishi T, Sakai M, Ohata K, Urata Y, et al. Nutritional risk index as an independent predictive factor for the development of surgical site infection after pancreaticoduodenectomy. *Surg Today* 2013;**43**:276–83.
97. Yoshikawa T, Hiki N, Taguri M, Sano T, Nunobe S, Taniguchi H, et al. A Phase III trial to evaluate the effect of perioperative nutrition enriched with eicosapentaenoic acid on body weight loss after total gastrectomy for T2-T4a gastric cancer. *Jpn J Clin Oncol* 2012;**42**:459–62.
98. Horie H, Okada M, Kojima M, Nagai H. Favorable effects of preoperative enteral immunonutrition on a surgical site infection in patients with colorectal cancer without malnutrition. *Surg Today* 2006;**36**:1063–8.
99. Blasier RB. The problem of the aging surgeon: when surgeon age becomes a surgical risk factor. *Clin Orthop Relat Res* 2009;**467**:402–11.
100. Datta V, Bann S, Beard J, Mandalia M, Darzi A. Comparison of bench test evaluations of surgical skill with live operating performance assessments. *J Am Coll Surg* 2004;**199**:603–6.
101. Maan ZN, Maan IN, Darzi AW, Aggarwal R. Systematic review of predictors of surgical performance. *Br J Surg* 2012;**99**:1610–21.
102. Darzi A, Smith S, Taffinder N. Assessing operative skill needs to become more objective. *BMJ* 1999;**318**:887–8.
103. Neumayer LA, Gawande AA, Wang J, Giobbie-Hurder A, Itani KM, Fitzgibbons RJ Jr, et al. CSP #456 Investigators. Proficiency of surgeons in inguinal hernia repair: effect of experience and age. *Ann Surg* 2005;**242**:344–8.
104. Miller LG, McKinnell JA, Vollmer ME, Spellberg B. Impact of methicillin-resistant *Staphylococcus aureus* prevalence among *S. aureus* isolates on surgical site infection risk after coronary artery bypass surgery. *Infect Control Hosp Epidemiol* 2011;**32**:342–50.
105. Eagye KJ, Kim A, Laohavaleeson S, Kuti JL, Nicolau DP. Surgical site infections: does inadequate antibiotic therapy affect patient outcomes? *Surg Infect (Larchmt)* 2009;**10**:323–31.
106. Liu C. The bundled approach to MRSA surgical site infection prevention: is the whole greater than the sum of its parts? *Arch Intern Med* 2011;**171**:73–4.
107. Johnson R, Jameson SS, Sanders RD, Sargant NJ, Muller SD, Meek RMD, et al. Reducing surgical site infection in arthroplasty of the lower limb: A multi-disciplinary approach. *Bone Joint Res* 2013;**2**:58–65.
108. Ballard MR, Miller NH, Nyquist AC, Elise B, Baulesh DM, Erickson MA. A multidisciplinary approach improves infection rates in pediatric spine surgery. *J Pediatr Orthop* 2012;**32**:266–70.

CHAPTER

3

Bacterial Adherence to Biomaterials Used in Surgical Procedures

Jaime Esteban, Ramón Pérez-Tanoira*, Concepción Pérez-Jorge-Peremarch* and Enrique Gómez-Barrena†*

*Dept. of Clinical Microbiology, IIS-Fundación Jiménez Díaz, Madrid, Spain

†Department of Orthopedics, IdiPAZ-Hospital Universitario La Paz, Madrid, Spain

INTRODUCTION

Surgical site infections are one of the most frequent infections in hospital patients, ranking second among all nosocomial infections.[1] They range from the mild and superficial to severe, deep infections that could compromise the life of the patient. Many of them have the presence of a biomaterial as a common pathogenic factor, which can be essential for the development of the disease. Biomaterials are commonly used in surgery (sutures are the most frequently used surgical biomaterial), and in some cases, the use of different types of implants has lead to a revolutionary improvement in patients' quality of life. However, the use of these materials introduces different risks, and infections are among the most significant in terms of morbidity, mortality and economic cost.[2] Minimizing the impact of such infections is crucial, because it is expected that the use of implants will increase in the coming years, and the number of infected patients is expected to increase concurrently. Infected patients need long antibiotic therapies, several surgical procedures, and the infection can even be the actual cause of the patient's death. Hence the implementation of methods to avoid these infections is an extremely important research topic.

Knowledge of the infection process is essential for this research. Regarding this issue, the "race for the surface" theory of Gristina et al.[3] is one of the keys for understanding the pathogenic process. According to this theory, when an implant is located inside the patient, a race starts between bacteria and host cells to reach and colonize the implant's surface. If the

Microbiology for Surgical Infections
DOI: http://dx.doi.org/10.1016/B978-0-12-411629-0.00003-9

cells "win the race", the implant integrates in the body, but if the bacteria "win", they start to multiply and develop the special structure that is called biofilm.[4] These structures protect bacteria from host immunity and other attacks (like antibiotics), and avoid implant integration (because cells cannot cover the implant), causing infection to appear subsequently.

If biofilm development in the biomaterial surface is the key pathogenic phenomenon in implant-related infections, bacterial adherence is the first step in this process. Bacterial adherence is a complex phenomenon that classically has been divided in two phases.[5,6] In the first phase, bacteria reach the biomaterial surface using the long-range interactions. These are different physical forces that act over distances greater than 100 nm, including Brownian motion, hydrophobicity, van der Waals' forces and others. When bacteria reach the surface, short-range interaction forces start to become important. These include ionic attraction, covalent links and hydrophobic forces. When the adherence of bacterial cells is strongly established, the organisms start to produce extracellular polymeric matrix (EPM) that binds the organisms even more tightly, and then the bacterial biofilm is formed and infection appears (Figure 3.1).

However, recent work by Henk et al.[7] has changed the view of the adherence process. These authors describe three adhesion force regimes that explain the bacterial adhesion

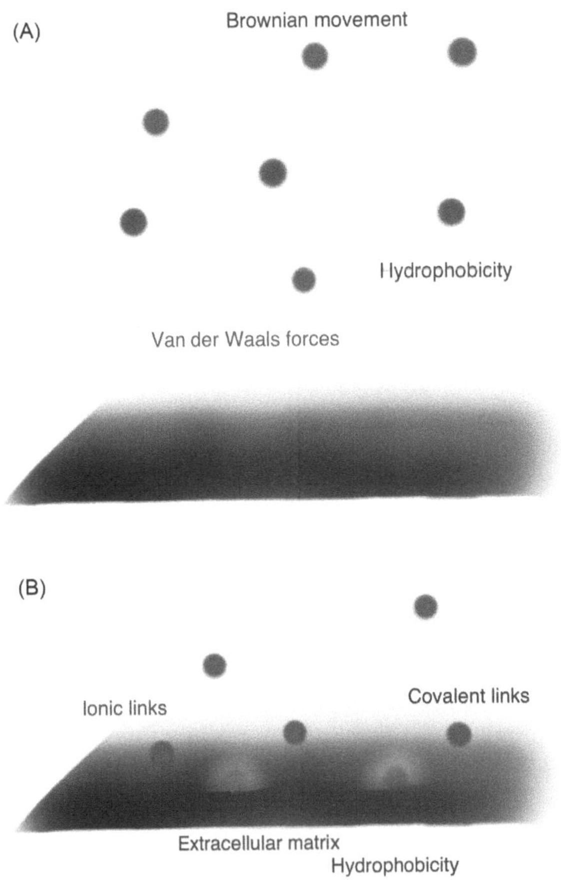

FIGURE 3.1 **Scheme of adherence process.** A: Long-range interactions; B: Short-range interactions.

phenomenon in a more detailed way: the planktonic regime, the interaction regime and the lethal regime. The "planktonic regime" occurs first, characterized by weak adhesion forces, in which bacteria do not change their planktonic phenotype. This regime is followed by the "interaction regime", in which bacteria increase their union with the surface by forming a biofilm. In this regime, phenotypic changes take place, increasing the adhesion forces. The authors explain that the adhesion forces tend to increase during the first minutes after contact, when they change from reversible to irreversible forces. Finally, the "lethal regime" is established, which is characterized by strong adhesion. The forces implied in this phase are apparently related to the bacterial and surface charges. Bacteria usually have a negative surface charge, which interacts with positively charged surfaces. These positively charged surfaces are able to kill bacteria upon contact, with different lethal effects depending on the bacterial species. All these phases are influenced by the nature of the biomaterial, and also by the bacterial structures that we will try to describe in the following paragraphs.

Because we cannot change bacterial properties, recent prevention research has concentrated on the development of new biomaterials that are less prone to bacterial adherence as a strategy for decreasing infection rates. In this chapter, we review the different aspects of bacterial adherence and some of the strategies that have been developed in order to increase the anti-adherent properties of the biomaterials used in surgical procedures.

THE ROLE OF BACTERIAL PROPERTIES IN ADHERENCE

The main mechanisms of bacterial adhesion have been studied for many years. This research has clarified several bacterial properties that are implicated in the adherence process, and these will be described in the following section.

Bacterial Structures

Fimbriae, Flagella and Pili

Flagella are long mobile appendages that facilitate the attachment of the bacterial cell to biomaterial surfaces, and are considered a virulence factor as well as being a biofilm related factor.[8] Fimbriae appear to be related with non-specific adhesion to abiotic surfaces. These bacterial structures stabilize the contact between the bacteria and the surface, interacting with that surface after the first step.[9] The presence of pili (which are like flagella) in bacteria allows them to disseminate and colonize other surfaces.[10] Pili also confer motility, which is closely related with biofilm development, as has been discovered in the biofilm development model of *Pseudomonas aeruginosa*.[11] *P. aeruginosa* attachment to surfaces increases with swimming motility, and decreases with swarming motility,[12] and one of the regulated factors is flagellar synthesis.

Alginate

Production of the exopolysaccharide alginate by *P. aeruginosa* is related to its ability to adhere to solid surfaces. Pseudomona's alginate production is regulated by different factors, one of these being the regulation of the expression of a specific gene (*algC*). It is known that when bacteria are attached to a surface, this specific gene is up-regulated.[13]

Lipopolysaccharides and Cell Wall Components

Lipopolysaccharides (LPS) are one of the major components of the outer membranes of gram-negative bacteria, providing integrity, and in many cases being the molecule that interacts with surfaces. The role of LPS in biofilm formation on biotic and abiotic surfaces is related to an increase in hydrophobicity. Lipopolysaccharides in gram-negative bacteria are composed of different parts; lipid A (the hydrophobic portion, and main virulent factor), O-antigen and core oligosaccharide (core OS). Results obtained by Nakao et al. using a mutant *Escherichia coli*[14] indicate that another component of the LPS, LPS-core-OS, could affect biofilm formation and is closely related to eDNA. The presence of eDNA in the extracellular matrix (ECM) of biofilms could also play a role in bacterial attachment and aggregation, as well as in biofilm development.[14]

The presence of peptidoglycan, lipopeptides and lipoteichoic acids on *Staphylococcus epidermidis* and *Staphylococcus aureus* surfaces are some of the protective mechanisms of this bacteria against host defense mechanisms.[15] For *P. aeruginosa,* the presence of lipopolysaccharides on the surface stimulates the antigenic response within the human body.[16]

Adhesins

The adhesins represent an important group of virulence factors mediating surface adhesive interactions between microbial cells, host cells and ECM.

In the first phase of biofilm development, adhesins are the main factor that allows stable anchorage to surfaces.[17] These structures bind to host proteins adsorbed on the implant surface.[18] The first step is mediated by microbial surface components recognizing adhesive matrix molecules (MSCRAMMs), which interact with host cells and ECM proteins. In the second step, bacterial accumulation and aggregation become the relevant phenomena.[15] Intercellular adhesion is due to the production of polysaccharide intercellular adhesion (PIA).[19]

In *S. aureus* and *S. epidermidis,* the adhesins are part of a protein family that mediates adhesion to several host cell types or matrix protein-covered device surfaces. Fibronectin is a protein usually related to an increase in bacterial attachment due to the fact that there are some bacteria with specific adhesins for fibronectin. For instance, *S. aureus* has two fibronectin adhesins: fibronectin-binding protein A (FnBPA) and fibronectin-binding protein B (FnBPB). These adhesins have different ligands, such as fibronectin, elastin and fibrinogen for FnBPA, and fibronectin and elastin for FnBPB.[17] The adhesins reported to bind to collagen are the collagen-binding proteins (CAN), and also adhesins that promote blood clumping, platelet activation, and aggregation such as clumping factor A (ClfA) and clumping factor B (ClfB). The ligands of ClfA and ClfB differ: Clf A has fibrinogen and complement factor I as its main ligands, while ClfB has fibrinogen, Cytokeratin 10 and Cytokeratin 8.[17] In *S. aureus,* the autolysin that participates in the first phase of adhesion is AtLA, and this is highly homologous with the autolysin of *S. epidermidis,* named AtLE.

For *S. epidermidis,* the main autolysin/adhesin is AtLE. Different studies remark that an AtLE mutant, defective in forming biofilms, is less hydrophobic and forms larger cell clusters when compared with the wild type.[17,18] The extracellular matrix binding protein (Embp) is also involved in the adhesion and accumulation steps of biofilm formation.[15] In this species, the MSCRAMM that binds with fibrinogen is SdrG/Fbe, and the one for fibronectin is Embp. However, other MSCRAMMs (AtlE and AaE) do not bind covalently to vitronectin.[15]

For *S. aureus* and *S. epidermidis* there are other surface proteins, including the biofilm-associated protein (Bap) and the accumulation-associated protein (Aap), which are important in biofilm formation. For *S. epidermidis,* Aap is an extracellular protein essential in the accumulation process on surfaces. It is important to note that Aap needs proteolytic processing in order to obtain the biofilm-active form. Bhp, which is a homologue protein of Bap, seems to be non-essential in biofilm formation.[15]

Another molecule, polysaccharide intercellular adhesion (PIA) or poly N-acetyl glucosamine (PNAG) is indispensable for biofilm formation.[15] Although recent studies have suggested that it is not essential for biofilm formation, it is related to strong biofilms.[20] The expression of PIA is regulated by the *icaADBC* operon. Both *S. epidermidis* and *S. aureus* present this locus, although not in all strains.[21] PIA is composed of linear β-1,6-linked glucosaminyglycan units. The expression of the *ica* operon is dependent on environmental conditions.[17]

The bacterial density in the biofilm formation is regulated by a different system. For instance, among staphylococcal species the regulating system is the *agr* locus. This is activated during the transition between the exponential growth and the stationary growth phases using an auto-regulatory mechanism that involves a modified pheromone peptide that signals the cell density.[17]

Bacterial Hydrophobicity

Different bacterial structures on the surface can influence the bacterial hydrophobicity. It is known that bacteria with hydrophobic properties tend to adhere to hydrophobic surfaces, while bacteria with hydrophilic properties tend to adhere to hydrophilic surfaces. The polysaccharides and proteins that form the bacterial capsule play an important role in this property.[5] While the hydrophilic capsules reduce adhesion, it has been shown that material surface hydrophobicity has even more influence than the bacterial surface hydrophobicity.[6] The adhesins and adhesive subunits associated with the fimbriae could be also involved in hydrophobicity. The important role of the hydrophobicity of bacteria can be modified by the presence of different types of fimbriae on the bacterial surface.[6]

Bacterial Surface Charges

The importance of the bacterial surface charge is another interesting topic that is not yet clearly understood. It is known that bacteria in aqueous suspension are ionized like other particles, and are mostly negatively charged.[5,19] However, bacterial adhesion to surfaces appears not to be significantly affected by the surface charge of the bacteria.

THE ROLE OF BIOMATERIAL PROPERTIES

It is well known that the presence of an implanted device results in increased patient susceptibility to infection.[22,23] It also reduces the number of bacteria required to produce infection.[24] For instance, a study of rats with occluded biliary tracts demonstrated ultrastructural evidence of bacterial colonization and biofilm formation on the biliary mucosal

TABLE 3.1 Types and Frequency of Implant-related Infections

Medical device	Rate of infection
Orthopedic	2–3%
Abdominal	1–18%
Cranial	3.8–7.9%
Cardiac pacemarkers	1–7%
Vascular grafts	1–5%
Interocular lenses	0.13%
External ventricular drainage	6–15%

surfaces, but only in animals that had indwelling biliary implants.[25] In many cases the only solution for an infected implanted device is its surgical removal, due to biofilm development.[5] Because biomaterial-related infections are among the commonest hospital-acquired infections, this problem is an extremely important one, and probably it will increase due to the use of the different medical devices (Table 3.1).

Careful analysis of the data on bacterial adherence and surface modification of the device yields the following five major principles:

1. Different bacteria may adhere differently to the same device material.
2. The same bacteria may adhere differently to different device materials.
3. The same bacteria may adhere differently to the same device material placed under different circumstances, including the medium in which the device is placed (hydrophobic vs. hydrophilic medium), type of flow (dynamic vs. stationary), and temperature.
4. *In vitro* inhibition of bacterial colonization of the device does not ensure anti-infective efficacy *in vivo*.
5. The clinical benefit of a particular surface-modifying approach may vary from one application to another.

Taking into account these five major principles, there are several device-related factors that can affect bacterial adherence, including the source of the material, the surface of the device, and its shape.[25] Moreover, factors influencing bacteria adherence to a biomaterial's surface include the chemical compositions of the material, its surface charge, hydrophobicity and simply the surface roughness or physical configuration. In addition, the surface energy, empty binding sites, and hydrophobic and hydrophilic characteristics can alter rapidly through the adsorption or binding of serum proteins and formation of biofilm.[6,23]

Surface Chemical Composition

Metal Alloys and Clean Metallic Surfaces

These materials represent planar cuts through crystalline structures. They generally exhibit moderate to high surface energies and are therefore reactive and potentially capable of catalyzing chemical reactions with glycoproteinaceous molecules of the conditioning film and possibly directly with the surfaces of bacterial cells. Titanium and chrome-cobalt alloys appear to allow closer and stronger tissue cell binding than polymers or bioglass.[26] Titanium and its alloys (Figure 3.2) are used for orthopedic implants, ossicular constructions and cranioplasties,[27–29] chrome-cobalt alloys and stainless steel are also widely used for orthopedic implants. In these materials, metal ions (such as Fe^{3+}, Mg^{2+}, Cr^{2+}, and Co^{3+}) are available in trace or higher amounts, especially after corrosion, and may serve as cofactors for enzymes engaged in protein and sugar metabolism and DNA replication.[26]

Polymer Substrata

Most medical polymers are amorphous materials. Their structures are not as complex as the higher energy surfaces of metals or ceramics, but are of biological relevance. The polymers in biomaterials are utilized for their physico-chemical properties in an even larger range of applications: polypropylene for sutures, abdominal wall prostheses and intraocular lenses

(A)

(B)

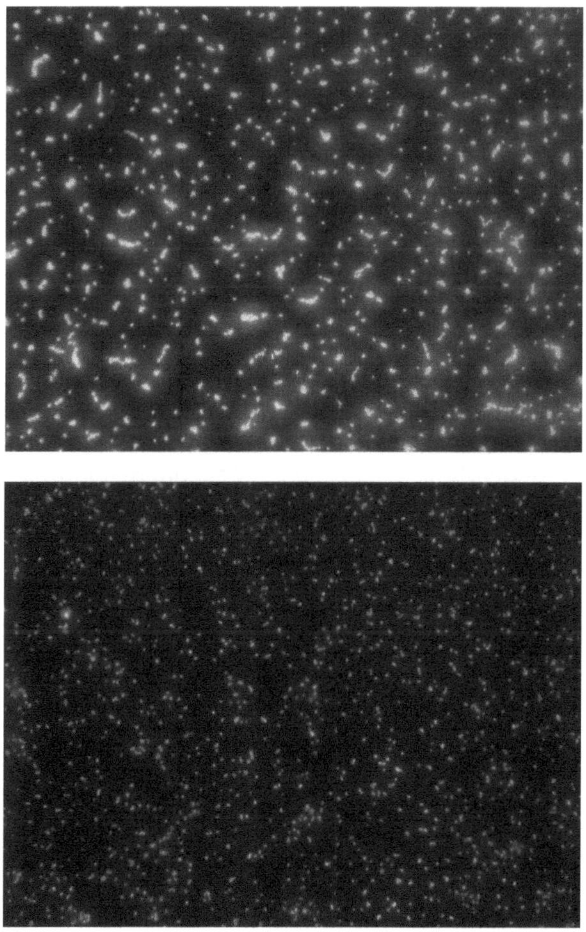

FIGURE 3.2 **Adherence studies on titanium alloy (Magnification x40).**
A: Acridine orange stain, *S. aureus*.
B: Backlight© Live-Dead stain. *S. aureus*.

(IOL), polymethyl-methacrylate (PMMA) for implant fixation in orthopedic surgery [especially for joint prosthesis], or customized linear high density polyethylene are used in cranioplasty[30,31]; silicone in bile draining devices and (rarely) in orthopedic surgery[23]; Plasti-Pore (high density polyethylene) and Proplast (polytetrafluoroethylene-carbon fiber composite) are porous materials that have been employed for the fabrication of ossicular replacement prostheses[32]; and ultra-high molecular weight polyethylene (UHMWPE) in orthopedic surgery.[23] Polyglactin 910(PG910) or polyglicolic acid (PGA) are the polymers used to fabricate absorbable surgical meshes.[33] The need to reduce the amount of mesh material used while retaining the required strength has resulted in the production of composite meshes, in which absorbable and non-absorbable materials are combined.[33]

The polymeric substrate may show an acid or basic character (according to the Lewis definition) owing to the presence of groups of elements which behave as electron acceptors or donors. The adhesion of *E. coli* was investigated on polymers with different acid/basic characteristics: Polyvinyl chloride (PVC) is acid, polymethylmethacrylate (PMMA) is basic, low-density polyethylene (LDPE) is dispersive. The basic character of the *E. coli* cell wall (negative charge and a hydrophilic character) will interact preferably with an acidic counterpart, represented in this case by the PVC surface, followed by LDPE and finally PMMA.[34]

Bioceramics

Bioceramics are often implanted in bone defect areas as bone substitutes or scaffolds due to their osteoinductive behavior, and because they promote bone healing, regeneration or both by promoting cell growth and proliferation into the porous scaffold. In recent years, patients have benefited from the development of better and more aesthetic materials, including all-ceramic dental restorative materials.[35] Hydroxyapatite (HA) and biphasic calcium phosphate (BCP) ceramic materials are also widely employed as bone substitutes due to their porous structure.[23,26]

Zirconia (ZrO_2) ceramics have shown considerable improvement over existing materials in several key areas related to component longevity, including resistance to roughening, wear reduction, frictional behavior and biocompatibility.[36] They are used in dental ceramics for aesthetic restorations.[37] ZrO_2 is an amphoteric metal oxide which exhibits both anion- and cation-exchange properties depending on the pH and composition of the buffer.[38] HA is also amphoteric,[39] but it has a net negative charge because there are more phosphate groups than calcium atoms exposed. Even although Ti is a metal, it is coated by a layer of surface oxide, mainly titanium dioxide, which has physical/chemical characteristics more closely related to ceramics than to metals.[40] These observations may partly explain the similar protein binding properties of Ti, ZrO_2, and HA surfaces: The surfaces of Ti and ZrO_2 showed similar numbers of adherent bacterial cells, likely due to the comparable physical/chemical attributes of ZrO_2 and titanium dioxide surfaces; and the number of *Streptococcus mutans* cells adhering to uncoated Ti and ZrO_2 was significantly higher than to an uncoated HA surface.[40]

However, despite the *in vitro* differences detected in many studies, no difference was detected in infection rates between implant materials (autograft, allograft, cryopreserved bone grafts, polymethylmethacrylate (PMMA), vitallium, intraoperatively fashioned resin and prefabricated resin).[41–43] Implant materials have no effect on the rate of cranioplasty infections,[30] but in other studies with titanium cranioplasty, differences were reported hypothetically due to the influence of the material.[27,28] A recent study using implant sonication and bacterial quantification showed no significant differences between materials used in prosthetic joint surgery. The variability in implants was thought to be due to variation in microorganism aggressiveness and the susceptibility of the infected patient.[44]

Surface Roughness and Morphology

The irregularities of polymeric surfaces, the diameter of the threads and the multifilament meshes promote bacterial adhesion and biofilm deposition.[33,45] For example, when standard micro-rough implant metals such as titanium and titanium alloys, or polymers such as UHMWPE, are compared with smoother variants, there is often a significant decrease in adhesion to the smoother surfaces when the surface features are of a smaller scale than the contaminating bacteria.[23] This may happen because rough surfaces have a larger surface area, and depressions in the roughened surfaces provide more favorable sites for colonization.[5] However, there is no linear dependence of bacterial adhesion with the degree of surface roughness.[46] Once bacteria have adhered to the surface of the biomaterial, it has been shown that expression of the *ica* locus is not affected by the precise substrate upon which biofilm is formed, as was observed for polymers and metals, including stainless steels of varying topographies, and titanium.[23]

Cordero et al. investigated the differences between cobalt-chromium-molybdenum (Co-Cr-Mo) and titanium-aluminum-vanadium (Ti-Al-V) alloys with either a smooth polished surface or a microporous coating. For both materials, the porous surfaces required significantly fewer bacteria to cause a bone infection in a rabbit model.[47] The infection rate with multifilament sutures (or porous materials) is greater than with monofilament sutures (or solid materials) and this fact could be attributed to the porous surfaces, which provide a site for contaminating bacteria; allowing them to multiply while shielding them from host immune defenses.[5,47,48]

Hydrophobicity and Hydrophilicity

Depending on the hydrophobicity of both bacteria and material surfaces, bacteria adhere differently to materials with different hydrophobicities. In general, hydrophilic materials are more resistant to adherent bacteria than hydrophobic materials.[49,50]

Water contact angle (WCA) is a good measure to evaluate the hydrophobic or hydrophilic nature of a surface. A high WCA represents hydrophobicity and a low WCA represents hydrophilicity.[6] Hydrophilic, negatively-charged substrates (such as glass and metal surfaces) and hydrophobic polymers (like Teflon, polyethylene or polystyrene) are less electrostatically charged.[33] Coating substrata surfaces with proteins, such as bovine serum albumin (BSA), bovine glycoprotein or fatty-acid free BSA decreases their hydrophobicity, leading to inhibited bacterial adhesion.[33]

A quantitative value for hydrophobicity, expressed in International System (SI) units, can be obtained through the van Oss approach. According to this theory, hydrophobicity is defined as the free energy of interaction between two entities (i) when immersed in water (w) – $\Delta Giwi$. If $\Delta Giwi < 0$, there is a preferential interaction between entities

(i) rather than between an entity (i) and water, and the substance (i) is considered hydrophobic. By the same reasoning, if $\Delta Giwi > 0$ the substance (i) is hydrophilic.[49]

Environmental Properties and Conditioning Film

One important issue in bacterial adherence is the significant role of some human proteins [like albumin, fibronectin, laminin, and others] in the adhesion process, either by binding to biomaterial surface, or binding to bacteria, or both. In fact, most of the bacteria-protein unions are specific ligand-receptor ones.[5]

Fibronectin (Fn) has been described as a key promoter in the adherence of *S. aureus*.[51] Albumin adsorbed on material surfaces has shown obvious inhibitory effects on bacterial adhesion to polymer, ceramic and metal surfaces.[51] Most studies showed that adsorbed fibrinogen promotes the adherence of bacteria, especially staphylococci, to biomaterials.[51] Thrombin significantly increases bacterial adhesion, since it polymerizes fibrinogen in platelet-poor plasma to fibrin.[52] The adhesion of various coagulase-negative staphylococci onto plasma coated materials is much lower than onto the untreated control surfaces.[51] However, Poor Platelet Plasma-Serum (PPP) with thrombin increases bacterial adhesion. Baumgartner et al. also showed that platelets increased *S. aureus* adhesion in comparison to human albumin (HAS), especially in combination with PPP and thrombin.[52]

One of the predominant microbial species isolated from biliary drain devices is *Enterococcus faecalis*.[53] The growth in bile increases the virulence of *E. faecalis*.[53] Adsorption of fibronectin to the rubber-representing material used for biliary drainage and subsequent adherence of *E. coli* were also studied.[54] In the presence of whole plasma or bile, *in vitro* fibronectin adsorption to the rubber slices was severely inhibited. Bile contains amphiphilic substances, which may induce micelle formation and block hydrophobic binding sites. However, despite these data, *in vivo* studies showed that bile does not prevent bacterial adherence and subsequent development of infection[54]

PREVENTION OF BIOMATERIAL INFECTION

Device factors are the most amenable to modification with the objective of preventing infection. Traditional surface-modifying preventive approaches have largely focused on antimicrobial coating of devices, and have resulted in variable clinical success in preventing device-associated infections[25,55] (Table 3.2). Current antimicrobial-containing materials may be subcategorized on the basis of their means of achieving antimicrobial action:

i) Those that bind antimicrobial agents directly to the implant surface and that act only on local bacteria that contact the surface, e.g., vancomycin covalently linked to titanium[56].
ii) Those that release an antimicrobial agent from the implanted bulk biomaterial, creating a local zone of bacterial killing around the implant, e.g., gentamicin-loaded PMMA[57].
iii) Those that release antibiotic agents from a film or coating over the implant, e.g., silver-coated endotracheal tubes,[58] gentamicin-coated intramedullary nails,[59] and rifampin-minocycline-coated central venous catheters.[60]

TABLE 3.2 Studies performed with different coatings

Type of Coating	Material used as Substrata	References
Lactobacilli	Polymers	a
Antibiotics	Sutures silk and Dacron polytetrafluoroethylene (PTFE) grafts	b
Triclosan	Sutures(polydioxanone PDS), (polyglactin 910) and (poliglecaprone 25) Sutures polyglacitin	c
Anti-fibronectin antibodies synthetic peptides based on the fibronectin and collagen	Stainless steel, pure titanium, or titanium-aluminum-niobium alloy, fibronectin and collagen coated surfaces	d
Diamond-like carbon coatings	Silicon, UHMWPE	e; Del Prado (55)
Silver and HA	Ti-Al-V titanium nanotube surface TiAlZr	Perez-Tanoira et al. (66); f
Antibiotics	Gold and iron oxide (Au) composite particles combined with near-infrared radiation (NIR)	g
Anodic fluoride-TiO(2)nostructured layers	Ti-Al-V	Perez-Jorge C, et al. (67); h
Titanium carboxonitride	PP, PTFE meshes	i
Gold, gold-palladium	PP meshes	j
Silver	Poly(ethylene)terephthalate fabric	k
Collagen	PP mesh	l
Glucosaminoglycan-keratin luminal	PTFE	m
(TiO(2)/Ag) and zinc oxide (ZnO) films	Ti-based implant	n

[a]*Hawthorn LA, Reid G. Exclusion of uropathogen adhesion to polymer surfaces by Lactobacillus acidophilus.* J Biomed Mat Res *1990;***24***(1):39–46.*

[b]*Rodeheaver GT, Kurtz LD, Bellamy WT, Smith SL, Farris H, Edlich RF. Biocidal braided sutures.* Arch Surg *1983;***118***(3):322–7. Matl FD, Obermeier A, Repmann S, Friess W, Stemberger A, Kuehn KD. New anti-infective coatings of medical implants.* Antimicrob Agents Chemother *2008;***52***(6):1957–63. Harvey RA, Greco RS. The noncovalent bonding of antibiotics to a polytetrafluoroethylene-benzalkonium graft.* Ann Surg *1981;***194***(5):642–7. Ney AL, Kelly PH, Tsukayama DT, Bubrick MP. Fibrin glue-antibiotic suspension in the prevention of prosthetic graft infection.* J Trauma *1990;***30***(8):1000–5; discussion 5–6. Okahara K, Kambayashi J, Shibuya T, et al. An infection-resistant PTFE vascular graft; spiral coiling of the graft with ofloxacin-bonded PTFE thread.* Eur J Vasc Endovasc Surg *1995;***9***(4):408–14.*

[c]*Laas E, Poilroux C, Bezu C, et al. Antibacterial-coated suture in reducing surgical site infection in breast surgery: a prospective study.* Int J Breast Cancer *2012;***2012***:819578. Justinger C, Moussavian MR, Schlueter C, Kopp B, Kollmar O, Schilling MK. Antibacterial [corrected] coating of abdominal closure sutures and wound infection.* Surgery *2009;***145***(3):330–4. Suarez Grau JM, De Toro Crespo M, Docobo Durantez F, Rubio Chaves C, Martin Cartes JA, Docobo Perez F. Prevention of surgical infection using reabsorbable antibacterial suture (Vicryl Plus) versus reabsorbable conventional suture in hernioplasty. An experimental study in animals.* Cir Esp *2007;***81***(6):324–9. Ming X, Nichols M, Rothenburger S.* In vivo *antibacterial efficacy of MONOCRYL plus antibacterial suture (Poliglecaprone 25 with triclosan).* Surg Infect (Larchmt) *2007;8(2):209–14. Ming X, Rothenburger S, Nichols MM.* In vivo *and* in vitro *antibacterial efficacy of PDS plus (polidioxanone with triclosan) suture.* Surg Infect (Larchmt) *2008;***9***(4):451–7. Seim BE, Tonnessen T, Woldbaek PR. Triclosan-coated sutures do not reduce leg wound infections after coronary artery bypass grafting.* Interact Cardiovasc Thorac Surg *2012;***15** *(3):411–5. Stadler S, Fleck T. Triclosan-coated sutures for the reduction of sternal wound infections? A retrospective observational analysis.* Interact Cardiovasc Thorac Surg *2011;***13***(3):296–9.*

(Continued)

TABLE 3.2 (Continued)

[d]*Delmi M, Vaudaux P, Lew DP, Vasey H. Role of fibronectin in staphylococcal adhesion to metallic surfaces used as models of orthopedic devices.* J Orthop Res *1994;***12***(3):432–8. Raja RH, Raucci G, Hook M. Peptide analogs to a fibronectin receptor inhibit attachment of Staphylococcus aureus to fibronectin-containing substrates.* Infect immun *1990;***58***(8):2593–8. Ross JM, McIntire LV, Moake JL, Rand JH. Platelet adhesion and aggregation on human type VI collagen surfaces under physiological flow conditions.* Blood *1995;***85***(7):1826–35.*
[e]*Kinnari TJ, Soininen A, Esteban J, et al. Adhesion of staphylococcal and Caco-2 cells on diamond-like carbon polymer hybrid coating.* Journal Biomed Mat Res A *2008;***86***(3):760–8.*
[f]*Das K, Bose S, Bandyopadhyay A, Karandikar B, Gibbins BL. Surface coatings for improvement of bone cell materials and antimicrobial activities of Ti implants.* Journal Biomed Mat Res B *2008;***87***(2):455–60. Ionita D, Grecu M, Ungureanu C, Demetrescu I. Antimicrobial activity of the surface coatings on TiAlZr implant biomaterial.* J Biosci Bioeng *2011;***112***(6):630–4.*
[g]*Huang WC, Tsai PJ, Chen YC. Functional gold nanoparticles as photothermal agents for selective-killing of pathogenic bacteria.* Nanomedicine (Lond) *2007;***2***(6):777–87. Tom RT, Suryanarayanan V, Reddy PG, Baskaran S, Pradeep T. Ciprofloxacin-protected gold nanoparticles.* Langmuir *2004;***20***(5):1909–14. Lima E, Guerra R, Lara V, Guzman A. Gold nanoparticles as efficient antimicrobial agents for Escherichia coli and Salmonella typhi.* Chem Cent J *2013;***7***(1):11. Tran N, Tran PA. Nanomaterial-based treatments for medical device-associated infections.* Chemphyschem *2012;***13***(10):2481–94.*
[h]*Arenas MA, Perez-Jorge C, Conde A, et al. Doped TiO(2) anodic layers of enhanced antibacterial properties.* Colloids and surfaces B, *2013;***105C***:106–12.*
[i]*Lehle K, Lohn S, Reinerth GG, Schubert T, Preuner JG, Birnbaum DE. Cytological evaluation of the tissue-implant reaction associated with subcutaneous implantation of polymers coated with titaniumcarboxonitride* in vivo. Biomaterials. *2004;25(24):5457–66. Scheidbach H, Tannapfel A, Schmidt U, Lippert H, Kockerling F. Influence of titanium coating on the biocompatibility of a heavyweight polypropylene mesh. An animal experimental model.* Eur Surg Res. *2004;36(5):313–7.*
[j]*Saygun O, Agalar C, Aydinuraz K, et al. Gold and gold-palladium coated polypropylene grafts in a S. epidermidis wound infection model.* Journal Surg Res *2006;***131***(1):73–9.*
[k]*Klueh U, Wagner V, Kelly S, Johnson A, Bryers JD. Efficacy of silver-coated fabric to prevent bacterial colonization and subsequent device-based biofilm formation.* Journal Biomed Mat Res *2000;***53***(6):621–31.*
[l]*van't Riet M, de Vos van Steenwijk PJ, Bonthuis F, et al. Prevention of adhesion to prosthetic mesh: comparison of different barriers using an incisional hernia model.* Ann Surg *2003;***237***(1):123–8.*
[m]*Sobinsky KR, Flanigan DP. Antibiotic binding to polytetrafluoroethylene via glucosaminoglycan-keratin luminal coating.* Surgery *1986;***100***(4):629–34.*
[n]*Chang YY, Lai CH, Hsu JT, Tang CH, Liao WC, Huang HL. Antibacterial properties and human gingival fibroblast cell compatibility of TiO2/Ag compound coatings and ZnO films on titanium-based material.* Clin Oral Invest *2012;***16***(1):95–100.*

It is known that there is a significantly higher risk of developing a wound infection after orthopedic surgery when the wound is closed with staples rather than sutures.[61] This fact could be explained if surgical sutures are considered as implants. Coating implants and sutures with a wide-spectrum antibacterial agent, such as triclosan has been used as a method to avoid infection.[62] Potential antiseptics for coating surfaces also include chlorhexidine, polyhexamethylene biguanide (PHMB) and octenidine.[62]

The use of adhesin fragments and antiadhesin antibodies to prevent adhesion has been suggested. Recombinant adhesin fragments cause inhibition by blocking binding sites on the substrate, and antibodies block the binding sites on the adhesin present on the bacterial surface.[63]

On three different novel polyurethanes, Dickinson et al.[51] studied the adherence of *S. aureus* with different protein coatings, and Baumgartner et al.[52] analyzed the role of protein and cellular components of thrombi (Figure 3.2). The tested polyurethanes included a relatively hydrophobic-base polyether urethane and hydrophilic anionomer and cationomer analogs of the base material. These authors demonstrated that the primary role of surface properties can be masked by the presence of the adsorbed protein layer.[51]

One of the strategies gaining renewed attention for combating the threat of bacterial infection is the use of antibacterial nanomaterials (nanophase materials with intrinsic bactericidal properties). Nanomaterials with sizes less than 100 nm possess unique properties (such as high surface-to-volume ratio, high reactivity, and so forth) compared to their bulk micron-structured counterparts, due mainly to size effects and surface phenomena at the nanoscale. Such properties lead to distinctive antibacterial mechanisms including:

1. Production of reactive oxygen species (ROS) or heavy metal ions which are known to damage proteins and DNA in bacteria.
2. Penetration or disruption of cell membranes by nanoparticles.
3. Interruption of electron transduction.

In the next paragraphs, we will discuss some common antibacterial nanomaterials and their bactericidal mechanisms.

a) Metal and metal oxide nanoparticles (from silver-coated endotracheal tubes to silver-coated orthopedic implants). Silver, fluorine, zinc oxide, titanium oxide, iron, iron oxide, copper, Alumina (Al_2O_3), copper oxide (CuO) and magnesium oxide nanoparticles (MgO) also exhibit antibacterial properties.[64–67]

b) Recently, Kang et al. reported strong antimicrobial activity of SWCNTs (single-walled carbon nanotube) against both gram-positive and gram-negative bacteria, and the proposed mechanisms were the development of membrane and oxidative stress that impact morphology, membrane integrity and metabolic activity of bacteria.[68]

c) Fullerenes (C_{60})[69,70] and their derivatives, containing certain peptides, possess antibacterial properties toward *S. aureus* and *E. coli*.

d) The antibacterial activities of chitosan, the N-deacetylated derivative of chitin, have also been studied.[64] Chitosan is more active against fungi and algae but also inhibits a wide range of gram-positive and gram-negative bacteria.[64,71]

e) The antibacterial effects of bioactive glass ceramics have been studied for a wide array of clinically important bacteria for mastoid obliteration.[72–74]

CONCLUSIONS AND FUTURE PROSPECTS

A knowledge of the mechanisms of bacterial adherence is essential regarding the pathogenesis of biomaterial-related infections. This knowledge can lead to the design of new strategies oriented against bacterial structures implicated in the adherence phenomenon. On the other hand, modifications of the biomaterial's surface can be a more practical strategy, because bacterial strains show different adherence properties,[75] even within those that belong to the same species. Improving these strategies could lead to the development of new biomaterials with improved antibacterial properties that can minimize the burden of surgical infections in clinical practice.

Future prospects for these strategies will be the development of biomaterials with modified surfaces that minimize bacterial adhesion and the subsequent development of infection. In the future, perhaps, we can imagine "intelligent" materials that are able to detect bacterial adherence and biofilm development, and have a specific action against these organisms, such as the release of antibiotics or anti-biofilm compounds,[4,76] or even by using electricity to release biofilm,[77] making the organisms susceptible to systemic antibiotics.

Future research must also take into account the intrinsic bacterial heterogeneity, which makes each strain of these organisms different from others of the same species.[75] This phenomenon has extremely important implications, because studies based on one or two strains cannot be representative of all the species, and many clinical strains need to be tested before a conclusion can be reached. Moreover, the study of the complex phenomenon of bacterial adherence will give us many explanations (like for the tendency of some species to adhere to some materials) that will lead to some new strategies of material modifications, and so on.

The human fight against infection is an endless task that started many centuries ago. Nevertheless, many advances have been made, and others are still to come. The development of intelligent antibacterial materials can be seen as science fiction for us in the present, but it is not much different to the idea of a hip prosthesis for a person living in, for example, the Roman Empire.

References

1. Mangram AJ, Horan TC, Pearson ML, Silver LC, Jarvis WR. The-Hospital-Infection-Control-Practices-Advisory-Committee. Guideline for prevention of surgical site infection, 1999. Centers for Disease Control and Prevention (CDC) Hospital Infection Control Practices Advisory Committee. *Am J Infect Control* 1999;**27**:97–132.
2. Darouiche RO. Treatment of infections associated with surgical implants. *New Eng J Med* 2004;**350**:1422–9.
3. Gristina AG, Costerton JW. Bacterial adherence and the glycocalyx and their role in musculoskeletal infection. *Orthop Clin North Am* 1984;**15**:517–35.
4. Costerton JW, Montanaro L, Arciola CR. Biofilm in implant infections: its production and regulation. *Int J Artif Organs* 2005;**28**:1062–8.
5. Katsikogianni M, Missirlis YF. Concise review of mechanisms of bacterial adhesion to biomaterials and of techniques used in estimating bacteria-material interactions. *Eur Cell Mater* 2004;**8**:37–57.
6. An YH, Friedman RJ. Concise review of mechanisms of bacterial adhesion to biomaterial surfaces. *J Biomed Mater Res* 1998;**43**:338–48.
7. Busscher HJ, van der Mei HC. How do bacteria know they are on a surface and regulate their response to an adhering state? *PLoS Pathog* 2012;**8**:e1002440.
8. Morisaki H, Nagai S, Ohshima H, Ikemoto E, Kogure K. The effect of motility and cell-surface polymers on bacterial attachment. *Microbiology* 1999;**145**:2797–802.
9. Otto K, Norbeck J, Larsson T, Karlsson KA, Hermansson M. Adhesion of type 1-fimbriated Escherichia coli to abiotic surfaces leads to altered composition of outer membrane proteins. *J Bacteriol* 2001;**183**:2445–53.
10. Varga JJ, Nguyen V, O'Brien DK, Rodgers K, Walker RA, Melville SB. Type IV pili-dependent gliding motility in the Gram-positive pathogen *Clostridium perfringens* and other Clostridia. Mol Microbiol 2006;**62**:680–94.
11. Klausen M, Heydorn A, Ragas P, Lambertsen L, Aaes-Jørgensen A, Molin S, et al. Biofilm formation by *Pseudomonas aeruginosa* wild type, flagella and type IV pili mutants. *Mol Microbiol* 2003;**48**:1511–24.
12. Ueda A, Wood TK. Connecting quorum sensing, c-di-GMP, pel polysaccharide, and biofilm formation in *Pseudomonas aeruginosa* through tyrosine phosphatase TpbA (PA3885). *PLoS Pathog* 2009;**5**:e1000483.
13. Davies DG, Geesey GG. Regulation of the alginate biosynthesis gene algC in *Pseudomonas aeruginosa* during biofilm development in continuous culture. *App Env Microbiol* 1995;**61**:860–7.
14. Nakao R, Ramstedt M, Wai SN, Uhlin BE. Enhanced biofilm formation by Escherichia coli LPS mutants defective in Hep biosynthesis. *PloS one* 2012;**7**:e51241.
15. Otto M. Molecular basis of *Staphylococcus epidermidis* infections. *Semin Immunopathol* 2011;**34**:201–14.
16. Kipnis E, Sawa T, Wiener-Kronish J. Targeting mechanisms of *Pseudomonas aeruginosa* pathogenesis. *Med Mal Infect* 2006;**36**:78–91.
17. Montanaro L, Speziale P, Campoccia D, Ravaioli S, Cangini I, Pietrocola G, et al. Scenery of *Staphylococcus* implant infections in orthopedics. *Fut Microbiol* 2011;**6**:1329–49.

18. Hall-Stoodley L, Costerton JW, Stoodley P. Bacterial biofilms: from the natural environment to infectious diseases. *Nat Rev Microbiol* 2004;**2**:95–108.
19. Kawamura H, Nishi J, Imuta N, Tokuda K, Miyanohara H, Hashiguchi T, et al. Quantitative analysis of biofilm formation of methicillin-resistant *Staphylococcus aureus* (MRSA) strains from patients with orthopedic device-related infections. *FEMS Immunol Med Microbiol* 2011;**63**:10–5.
20. Rohde H, Burandt EC, Siemssen N, Frommelt L, Burdelski C, Wurster S, et al. Polysaccharide intercellular adhesin or protein factors in biofilm accumulation of *Staphylococcus epidermidis* and *Staphylococcus aureus* isolated from prosthetic hip and knee joint infections. *Biomaterials* 2007;**28**:1711–20.
21. Esteban J, Molina-Manso D, Spiliopoulou I, Cordero-Ampuero J, Fernández-Roblas R, Foka A, et al. Biofilm development by clinical isolates of *Staphylococcus* spp. from retrieved orthopedic prostheses. *Acta Orthopedica* 2010;**81**:674–9.
22. Andriole VT, Nagel DA, Southwick WO. A paradigm for human chronic osteomyelitis. *J Bone Joint Surg Am* 1973;**55**:1511–5.
23. Rochford ET, Richards RG, Moriarty TF. Influence of material on the development of device-associated infections. *Clin Microbiol Inf* 2012;**18**:1162–7.
24. Guggenbichler JP, Assadian O, Boeswald M, Kramer A. Incidence and clinical implication of nosocomial infections associated with implantable biomaterials – catheters, ventilator-associated pneumonia, urinary tract infections. *GMS Krankenhhyg Interdiszip* 2011;**6** Doc18.
25. Darouiche RO. Device-associated infections: a macroproblem that starts with microadherence. *Clin Inf Dis* 2001;**33**:1567–72.
26. Gristina A. Biomaterial-centered infection: microbial adhesion versus tissue integration. 1987. *Clin Orthop Relat Res* 2004;**427**:4–12.
27. Kshettry VR, Hardy S, Weil RJ, Angelov L, Barnett GH. Immediate titanium cranioplasty after debridement and craniectomy for postcraniotomy surgical site infection. *Neurosurgery* 2012;**70**:8–14 discussion –5.
28. Wiggins A, Austerberry R, Morrison D, Ho KM, Honeybul S. Cranioplasty with custom-made titanium plates –14 years experience. *Neurosurgery* 2013;**72**:248–56.
29. Lee JH, Kim SW, Yoon BI, Ha US, Sohn DW, Cho YH. Factors that affect nosocomial catheter-associated urinary tract infection in intensive care units: two year experience at a single center. *Korean J Urol* 2013;**54**:59–65.
30. Lee CH, Chung YS, Lee SH, Yang HJ, Son YJ. Analysis of the factors influencing bone graft infection after cranioplasty. *J Trauma Acute Care Surg* 2012;**73**:255–60.
31. Benzel EC, Thammavaram K, Kesterson L. The diagnosis of infections associated with acrylic cranioplasties. *Neuroradiology* 1990;**32**:151–3.
32. Teichgraeber JF, Spector M, Per-Lee JH, Jackson RT. Tissue response to Plasti-Pore and Proplast otologic implants in the middle ears of cats. *Am J Otol* 1983;**5**:127–36.
33. Engelsman AF, van der Mei HC, Ploeg RJ, Busscher HJ. The phenomenon of infection with abdominal wall reconstruction. *Biomaterials* 2007;**28**:2314–27.
34. Speranza G, Gottardi G, Pederzolli C, Lunelli L, Canteri R, Pasquardini L, et al. Role of chemical interactions in bacterial adhesion to polymer surfaces. *Biomaterials* 2004;**25**:2029–37.
35. Meier R, Hauser-Gerspach I, Luthy H, Meyer J. Adhesion of oral streptococci to all-ceramics dental restorative materials *in vitro*. *J Mater Sci Mater Med* 2008;**19**:3249–53.
36. Heyse TJ, Haas SB, Efe T. The use of oxidized zirconium alloy in knee arthroplasty. *Expert Rev Med Devices* 2012;**9**:409–21.
37. Giordano 2nd R. Zirconia: a proven, durable ceramic for esthetic restorations. *Compend Contin Educ Dent* 2012;**33**:46–9.
38. Nawrocki J, Rigney MP, McCormick A, Carr PW. Chemistry of zirconia and its use in chromatography. *J Chromatogr A* 1993;**657**:229–82.
39. Gibbons RJ, Qureshi JV. Inhibition of adsorption of *Streptococcus mutans* strains to saliva-treated hydroxyapatite by galactose and certain amines. *Infect Immun* 1979;**26**:1214–7.
40. Ellingsen JE. A study on the mechanism of protein adsorption to TiO_2. *Biomaterials* 1991;**12**:593–6.
41. Yadla S, Campbell PG, Chitale R, Maltenfort MG, Jabbour P, Sharan AD. Effect of early surgery, material, and method of flap preservation on cranioplasty infections: a systematic review. *Neurosurgery* 2011;**68**:1124–9 discussion 30.

42. Cheng YK, Weng HH, Yang JT, Lee MH, Wang TC, Chang CN. Factors affecting graft infection after cranioplasty. *J Clin Neurosci* 2008;**15**:1115–9.
43. Tokoro K, Chiba Y, Tsubone K. Late infection after cranioplasty – review of 14 cases. *Neurol Med Chir (Tokyo)* 1989;**29**:196–201.
44. Gomez-Barrena E, Esteban J, Medel F, Molina-Manso D, Ortiz-Pérez A, Cordero-Ampuero J, et al. Bacterial adherence to separated modular components in joint prosthesis: a clinical study. *J Orthop Res* 2012;**30**:1634–9.
45. Kantorski KZ, Scotti R, Valandro LF, Bottino MA, Koga-Ito CY, Jorge AO. Adherence of *Streptococcus mutans* to uncoated and saliva-coated glass-ceramics and composites. *Gen Dent* 2008;**56**:740–7 quiz 8–9, 68.
46. Boyd RD VJ, Jones MV, Bhakoo M. Use of atomic force microscope to determine the effect of substratum surface topography on bacterial adhesion. *Langmuir* 2002;**18**:2343–6.
47. Cordero J, Munuera L, Folgueira MD. Influence of metal implants on infection. An experimental study in rabbits. *J Bone Joint Surg Br* 1994;**76**:717–20.
48. Kinnari TJ, Esteban J, Gomez-Barrena E, Zamora N, Fernandez-Roblas R, Nieto A, et al. Bacterial adherence to SiO_2-based multifunctional bioceramics. *J Biomed Mater Res Part A* 2009;**89**:215–23.
49. Cerca N, Pier GB, Vilanova M, Oliveira R, Azeredo J. Quantitative analysis of adhesion and biofilm formation on hydrophilic and hydrophobic surfaces of clinical isolates of *Staphylococcus epidermidis*. *Res Microbiol* 2005;**156**:506–14.
50. Baillif S, Ecochard R, Hartmann D, Freney J, Kodjikian L. Intraocular lens and cataract surgery: comparison between bacterial adhesion and risk of postoperative endophthalmitis according to intraocular lens biomaterial. *J Fr Ophtalmol* 2009;**32**:515–28.
51. Dickinson RB, Nagel JA, Proctor RA, Cooper SL. Quantitative comparison of shear-dependent *Staphylococcus aureus* adhesion to three polyurethane ionomer analogs with distinct surface properties. *J Biomed Mater Res* 1997;**36**:152–62.
52. Baumgartner JN, Cooper SL. Influence of thrombus components in mediating *Staphylococcus aureus* adhesion to polyurethane surfaces. *J Biomed Mater Res* 1998;**40**:660–70.
53. Waar K, van der Mei HC, Harmsen HJ, Degener JE, Busscher HJ. Adhesion to bile drain materials and physicochemical surface properties of *Enterococcus faecalis* strains grown in the presence of bile. *App Env Microbiol* 2002;**68**:3855–8.
54. Yu JL, Ljungh A, Andersson R, Jakab E, Bengmark S, Wadström T. Promotion of *Escherichia coli* adherence to rubber slices by adsorbed fibronectin. *J Med Microbiol* 1994;**41**:133–8.
55. Del Prado G, Terriza A, Ortiz-Pérez A, Molina-Manso D, Mahillo I, Yubero F, et al. Coatings for UHMWPE: relationship between bacterial adherence and surface properties. *J Biomed Mater Res Part A* 2012;**100**: 2813–20.
56. Antoci Jr V, King SB, Jose B, Parvizi J, Zeiger AR, Wickstrom E, et al. Vancomycin covalently bonded to titanium alloy prevents bacterial colonization. *J Orthop Res* 2007;**25**:858–66.
57. Henry SL, Ostermann PA, Seligson D. The prophylactic use of antibiotic impregnated beads in open fractures. *J Trauma* 1990;**30**:1231–8.
58. Kollef MH, Afessa B, Anzueto A, Veremakis C, Kerr KM, Margolis BD, et al. Silver-coated endotracheal tubes and incidence of ventilator-associated pneumonia: the NASCENT randomized trial. *JAMA* 2008;**300**: 805–13.
59. Schmidmaier G, Lucke M, Wildemann B, Haas NP, Raschke M. Prophylaxis and treatment of implant-related infections by antibiotic-coated implants: a review. *Injury* 2006;**37**:S105–12.
60. Raad I, Darouiche R, Dupuis J, Abi-Said D, Gabrielli A, Hachem R, et al. Central venous catheters coated with minocycline and rifampin for the prevention of catheter-related colonization and bloodstream infections. A randomized, double-blind trial. The Texas Medical Center Catheter Study Group. *Ann Intern Med* 1997;**127**: 267–74.
61. Smith TO, Sexton D, Mann C, Donell S. Sutures versus staples for skin closure in orthopedic surgery: meta-analysis. *BMJ* 2010;**340**:c1199.
62. Leaper D, McBain AJ, Kramer A, Assadian O, Sanchez JL, Lumio J, et al. Healthcare associated infection: novel strategies and antimicrobial implants to prevent surgical site infection. *Ann R Coll Surg Engl* 2010;**92**: 453–8.
63. Mohamed N, Teeters MA, Patti JM, Hook M, Ross JM. Inhibition of Staphylococcus aureus adherence to collagen under dynamic conditions. *Infect Immun* 1999;**67**:589–94.

64. Tran N, Tran PA. Nanomaterial-based treatments for medical device-associated infections. *Chemphyschem* 2012;**13**:2481–94.
65. Fouda MM, El-Aassar MR, Al-Deyab SS. Antimicrobial activity of carboxymethyl chitosan/polyethylene oxide nanofibers embedded silver nanoparticles. *Carbohydr Polym* 2013;**92**:1012–7.
66. Pérez-Tanoira R, Pérez-Jorge C, Endrino JL, Gómez-Barrena E, Horwat D, Pierson JF, et al. Bacterial adhesion on biomedical surfaces covered by micrometric silver Islands. *J Biomed Mater Res Part A* 2012;**100**:1521–8.
67. Perez-Jorge C, Conde A, Arenas MA, et al. *In vitro* assessment of *Staphylococcus epidermidis* and *Staphylococcus aureus* adhesion on TiO(2) nanotubes on Ti-6Al-4V alloy. *J Biomed Mater Res Part A* 2012;**100**:1696–705.
68. Kang S, Pinault M, Pfefferle LD, Elimelech M. Single-walled carbon nanotubes exhibit strong antimicrobial activity. *Langmuir* 2007;**23**:8670–3.
69. Pellarini F, Pantarotto D, Da Ros T, Giangaspero A, Tossi A, Prato M. A novel [60]fullerene amino acid for use in solid-phase peptide synthesis. *Org Lett* 2001;**3**:1845–8.
70. Patel MB, Harikrishnan U, Valand NN, Modi NR, Menon SK. Novel cationic quinazolin-4(3H)-one conjugated fullerene nanoparticles as antimycobacterial and antimicrobial agents. *Arch Pharm (Weinheim)* 2013;**346**: 210–20.
71. Cakmak A, Cirpanli Y, Bilensoy E, Yorganci K, Caliş S, Saribaş Z, et al. Antibacterial activity of triclosan chitosan coated graft on hernia graft infection model. *International journal of pharmaceutics* 2009;**381**:214–9.
72. Sarin J, Grenman R, Aitasalo K, Pulkkinen J. Bioactive glass S53P4 in mastoid obliteration surgery for chronic otitis media and cerebrospinal fluid leakage. *Ann Otol Rhinol Laryngol* 2012;**121**:563–9.
73. Munukka E, Leppäranta O, Korkeamäki M, Vaahtio M, Peltola T, Zhang D, et al. Bactericidal effects of bioactive glasses on clinically important aerobic bacteria. *J Mater Sci Mater Med* 2008;**19**:27–32.
74. Leppäranta O, Vaahtio M, Peltola T, Zhang D, Hupa L, Hupa M, et al. Antibacterial effect of bioactive glasses on clinically important anaerobic bacteria *in vitro*. *J Mater Sci Mater Med* 2008;**19**:547–51.
75. Esteban J, Molina-Manso D, Gomez-Barrena E. Bacterial adherence to vitamin E UHMWPE. Considerations about *in vitro* studies. *J Orthop Res* 2012;**30**:1181 author reply -2.
76. Costerton JW. Biofilm theory can guide the treatment of device-related orthopedic infections. *Clin Orthop Relat Res* 2005;**437**:7–11.
77. Del Pozo JL, Rouse MS, Euba G, Kang CI, Mandrekar JN, Steckelberg JM, et al. The electricidal effect is active in an experimental model of Staphylococcus epidermidis chronic foreign body osteomyelitis. *Antimicrob Agents Chemother* 2009;**53**:4064–8.

SECTION II

INTRA-ABDOMINAL INFECTIONS

CHAPTER

4

Multidrug-Resistant Bacteria in Pancreatic Surgery

Torsten Herzog and Waldemar Uhl

Department of Surgery, St. Josef Hospital, Ruhr-University Bochum, Germany

INTRODUCTION

Developments in pancreatic surgery have reduced mortality but remarkable morbidity levels still remain. Surgical resection has always been and still is the only chance of cure for patients with pancreatic cancer, but until the end of the last century surgical resections of the pancreas, especially for pancreatic head disorders, were associated with a mortality of 20% or even more.[1,2] During the last two decades, mortality after pancreatic surgery was dramatically reduced to less than 5% in several large series of experienced pancreatic centers, while morbidity remains high; approaching 50% even in specialized units.[2–5]

The anatomical situation of the pancreatic head requires a resection of the pancreatic head, the distal bile duct and the duodenum with at least three intestinal anastomoses for the reconstruction. Therefore, the most threatening postoperative surgical complications are related to leaks from intestinal anastomoses which are often complicated by local infection, resulting in intra-abdominal abscess formation, surgical site infections, cholangitis, delayed gastric emptying, or hemorrhage from visceral arteries.[3,4,6] Therefore, the majority of patients with postoperative complications after pancreatic head surgery require antibiotic therapy in combination with interventional drainage or even a re-laparotomy.

After distal pancreatectomy, the most common postoperative surgical complication is a pancreatic fistula, which occurs in approximately 30% of cases, independently of the operative technique that is used for the resection.[7] At least one third of postoperative pancreatic fistulae are complicated by bacterial infections, which require antibiotic treatment, computer tomography (CT) guided interventional drainage, or even a re-operation.[8,9] The outcome for patients with postoperative complications after pancreatectomy depends on the availability of efficient antibiotics.

Microbiology for Surgical Infections
DOI: http://dx.doi.org/10.1016/B978-0-12-411629-0.00004-0

The postoperative management of surgical complications after pancreatectomy is mainly based on three principles; local source control, supportive intensive care management, and adequate antibiotic therapy.[10] Treatment options for source control depend mostly on the local expertise of an experienced team, including specialized pancreatic surgeons, interventional radiologists and gastroenterologists, which form the basis of a pancreatic center. The frequency of surgical revision which is usually associated with high postoperative morbidity and mortality can be reduced, especially if interventional management improves substantially.[11,12] Successful treatment of postoperative complications after pancreatectomy is often accomplished by interventional radiology techniques including CT drainage, percutaneous transhepatic biliary drainage (PTBD) and endovascular treatment with coils or covered stents for postoperative hemorrhage.[13–15] Nonetheless, some patients still require surgical treatment for postoperative complications, especially for hemorrhage.[16] There have also been improvements in intensive care management through the appropriate monitoring of signs and symptoms of acute inflammation. Earlier intervention in intensive care management is able to reduce organ dysfunction by support of physiological organ systems.[17]

Antibiotic therapy is the third principle in the management of postoperative surgical complications after pancreatic surgery, and, despite improvements in the other two principles, there is an increasing concern about postoperative infectious complications, since they are more frequently caused by microorganisms that are resistant to the antibiotic agents in routine use.[10]

A common problem in the treatment of postoperative infectious complications after pancreatic surgery is that it is difficult to anticipate the microorganisms of the infectious focus, because the index operation is usually performed under sterile conditions. Microbiological culture results are only available after a delay, once the management for source control has been initiated and potentially infectious material has been collected and sent for culture. Usually, the results from the microbiology are available within 48 hours, but immediate initiation of an adequate antibiotic therapy is mandatory for the control of sepsis, and any delay of a suitable antibiotic treatment is associated with an increasing risk of postoperative mortality.[18–20] Therefore, the initial antibiotic therapy has to be directed against the most common microorganisms encountered after pancreatectomy. This is usually accomplished by the use of broad spectrum antibiotics, but this can fail if infectious complications are caused by multidrug-resistant (MDR) microorganisms. Furthermore, broad spectrum antibiotics are associated with the further evolution of even more resistant microorganisms and side effects, especially in terms of long lasting antibiotic application.[21] Thus, a differentiated reflection has to be performed, regarding which antibiotic therapy should be used in patients with postoperative complications after pancreatic surgery.

THE IMPACT OF PREOPERATIVE BILIARY DRAINAGE (PBD) ON BILE BACTERIOLOGY AND OUTCOME

Obstructive jaundice is often the first symptom of pancreatic head cancer, but operations on patients with jaundice are associated with a high postoperative mortality, because cholestasis leads to liver dysfunction.[22–24] Therefore, PBD was initially introduced to reduce the rate of postoperative complications and mortality after pancreatic head resection. Several techniques for PBD have been investigated and internal biliary drainage via endoscopically placed bile duct stents, which are usually introduced via endoscopic retrograde cholangiopancreaticography (ERCP), have been found to be better than PTBD. After endoscopic bile duct stenting, the entero-hepatic circulation of bile acids is ensured and the risk of post-interventional complications after ERCP is low, while PTBD is often associated with bile leaks, biliary sepsis and hemorrhage.[25,26] Unfortunately, the expected effect of a reduction in the rate of postoperative complications after placement of an endoscopic bile duct stent could not be verified by the majority of studies.[27] Indeed, PBD seems to lead to a higher rate of bacterial bile duct infections.[28]

A recently published, randomized, double blinded, multicenter trial tried to answer the question of whether PBD should be performed or not, but the results remain controversial.[29] The authors concluded that PBD should not be performed, because patients who underwent PBD had a significantly higher rate of overall complications, but they did not analyze post-surgical morbidity, but rather any complication that occurred during their trial. The rate of complications associated with PBD was 46% which is extremely high. A detailed analysis of postoperative complications showed that patients with PBD had only a slightly higher post-surgical morbidity, but the difference was not significant.[29,30]

It is certainly desirable that PBD should not be performed routinely, especially if it is associated with complications in almost every second patient, but data from health service research in high volume centers show that currenty almost 50% of patients are referred from external institutions after PBD has already been performed.[30] Thus, there is an important discrepancy between the theoretical knowledge about avoiding PBD and the reality in high volume centers.[30] Furthermore, even if PBD is not routinely recommended, it can be indicated if surgical resection can only be performed after a delay, or if patients have septic cholangitis.[31]

Therefore, physicians who have to deal with patients who have already undergone PBD should be aware that this is associated with bacteribilia.[32] The etiology of bacteribilia is usually explained by bacterial dislocation from the upper gastrointestinal tract into the bile ducts, which are usually sterile. The rate of bile duct infections is approximately 25% in patients with obstructive jaundice and rises up to 70% after ERCP, while it is almost 100% in patients with an endoscopically placed bile duct stent.[32–34]

Initially bile bacteribilia was considered to be a concomitant effect without further consequences on post-surgical outcome, but there is now evidence that PBD is associated with a higher rate of postoperative infectious complications which are the result of biliary spillage from infected bile ducts.[35–39] Only a few studies have focused on bile microbiology, but recent investigations have shown that the risk of a postoperative infectious complication depends on the microbiology of the microorganisms in intraoperative bile duct cultures.[30]

While a routine perioperative antibiotic prophylaxis is usually effective against the majority of microorganisms, PDB, especially in combination with an antibiotic therapy, leads toward a selection of MDR microorganisms in bile ducts which are not covered by the routine antibiotic prophylaxis.[30] Therefore, intraoperative bile duct cultures should be taken in all patients undergoing pancreatic surgery, because analysis of intraoperative bile duct cultures enables the identification of hospital specific microbiological surveillance data. The identity of the microorganisms that cause postoperative infectious complications can be anticipated from this data, so bile duct cultures should be routinely collected during the operation, immediately after bile duct dissection. These microbiological results can be effectively used to guide the antibiotic prophylaxis and postoperative antibiotic therapy for patients with postoperative complications.

THE ETIOLOGY OF MDR INFECTIONS

Intra-abdominal infections requiring antibiotic treatment are usually classified into community acquired and health care associated or hospital acquired infections. The treatment for both types of infection is based on the same principles, which include adequate source control, intensive care management and antibiotic therapy. While most community acquired infections will respond to a standardized antibiotic therapy, treatment of patients with postoperative complications is more complex.

By definition, all patients with postoperative infectious complications after pancreatectomy have hospital acquired infections. Although infections among these patients include the same microorganisms as community acquired infections, there is a higher frequency of opportunistic nosocomial pathogens, including methicillin resistant *Staphylococcus aureus* (MRSA), vancomycin resistant *Enterococcus* (VRE), *Pseudomonas aeruginosa*, resistant *Acinetobacter* ssp. and Enterobacteriaceae with extended spectrum β-lactamase (ESBL).[40–42] Overall, patients with infections caused by these resistant microorganisms have a significantly higher morbidity and mortality.[43]

After pancreatic surgery the majority of patients have at least a perioperative antibiotic prophylaxis during the index operation, but patients with more than one postoperative complication, immune suppression, and/or long term intensive care unit stay will receive broad spectrum antibiotics which are often administered during the whole intensive care unit stay. This long term antibiotic treatment is often inappropriate and inadequate, leading to an increasing rate of antibiotic resistance in surgical patients.[44]

Other risk factors associated with the development of infections with MDR microorganisms are preoperative antibiotic treatment, long postoperative antibiotic exposure, and postoperative changes in antibiotic therapy. Furthermore, the number of severe co-morbidities and increased APACHE II score, long preoperative and postoperative hospitalization, and incidence of hospital acquired infections have been identified as being associated with a higher risk of postoperative infections involving MDR microorganisms.[45–48]

The number of patients with risk factors for infectious complications with MDR microorganisms will further increase in the future, because there are more and older patients with more severe co-morbidities who have prior broad spectrum antibiotic exposure. Therefore, before a new antibiotic treatment is initiated, patients with postoperative complications after pancreatic surgery require adequate risk stratification, and physicians have to consider whether an infection with a MDR microorganism could be present in each individual patient.

MDR MICROORGANISMS

Essential challenges for the treatment of infections with MDR microorganisms are to avoid further resistant developments in order to preserve effective antibiotics, and to develop new antimicrobial substrates. At the beginning of the 20th century the rate of MDR microorganisms isolated from patients after pancreatic surgery significantly increased. The most common resistant gram-positive pathogens in pancreatic surgery include MRSA and VRE; mostly *Enterococcus faecium* and *Enterococcus faecalis*. The most common MDR gram-negative aerobe microorganisms include *Pseudomonas aeruginosa*, *Stenotrophomonas maltophilia*, and *Acinetobacter* spp. The largest group of MDR microorganisms is found among facultative anaerobe gram-negative pathogens with ESBL, including *Escherichia coli*, *Klebsiella* spp., *Enterobacter* spp. and *Morganella morganii*.

MRSA

S. aureus is the most common cause of postoperative surgical site infections worldwide.[49] Although most isolates from surgical wounds show infections with susceptible *S. aureus*, the frequency of MRSA has increased from less than 5% to 30% with an even further increase to almost 40% among hospitals with more than 500 beds, and today, there are even series where the majority of *Staphylococcus* isolates are MRSA.[50,51]

MRSA is the most common resistant gram-positive microorganism worldwide that is responsible for hospital acquired infections. Surgical infectious complications and sepsis with MRSA are associated with high morbidity and mortality. In general and orthopedic surgery, MRSA mostly causes surgical site infections, while in abdominal and pancreatic surgery there is a high frequency of intra-abdominal infectious complications and abscesses with MRSA.[49,52,53] These infectious complications have also been related to bacteribilia with associated postoperative surgical site infections with MRSA.[54] After pancreatectomy, MRSA infection leads to an increased rate of surgical site infections and chest infections resulting in a longer hospital stay.[55] Prior antibiotic therapy can reduce the rate of postoperative wound infections with MRSA, but preoperative antibiotic therapy leads to more intra-abdominal surgical site infections with *Enterococcus* spp.[53]

Intra-abdominal infectious complications with MRSA have a higher risk of postoperative hemorrhage.[55,56] The etiology of delayed visceral hemorrhage is usually explained by local arterial vessel wall erosion as a result of a pancreatic fistula, but secretion of bacterial enzymes may also be responsible for hemorrhage, especially if resistant bacteria are present or if the initiated antibiotic treatment fails.[14,56] Thus, even with successful endovascular treatment, postoperative hemorrhage remains the most threatening postoperative complication after pancreatectomy, and is associated with a high mortality. Thus, anticipation and risk stratification for postoperative infectious complications with MRSA should be considered and treated efficiently.

Enterococcus spp.

The most common two species include *E. faecalis* and *E. faecium*, while other *Enterococcus* spp. are only rarely found in pancreatic surgery. *Enterococcus* spp. have a high intrinsic activity against many penicillins and all cephalosporins with exceptions among new developments actually undergoing clinical evaluation.[57] There are different reports of postoperative morbidity caused by the two strains, but usually *E. faecium* is less susceptible, and infections with *E. faecium* have been associated with a grave prognosis, especially in the presence of vancomycin resistance.[58]

For patients with community acquired intra-abdominal infections there is no treatment recommendation for *Enterococcus* spp., while antibiotic treatment is mandatory for hospital acquired infections and especially for postoperative surgical complications.[59] Postoperative infections with *Enterococcus* spp. occur more frequently after PBD and prior antibiotic therapy, while infections with VRE are related to prior vancomycin application, which is common in the United States but unusual in Europe.[30,60]

After pancreatic surgery, most infectious complications with *Enterococcus* spp. are related to bile bacteribilia with *Enterococcus* spp. and in the literature there are very variable hospital surveillance data for *Enterococcus* bacterbilia, ranging between almost 0% and more than 60% in other series.[30,61,62]

Bile bacteribilia with *Enterococcus* spp. depends on the presence of PBD, preoperative antibiotic exposure and the duration of PBD.[30,61] Although a preoperative drainage period of less than 30 days is also associated with a high rate of bacteribilia, bile duct infections with MDR microorganisms, especially *Enterococcus* spp., are significantly more frequent after a drainage period of more than 30 days.[30] After preoperative antibiotic treatment, the rate of bile duct infections with *Enterococcus* spp. increases from 16% to more than 60%.[61] Furthermore *Enterococcus* spp. have been found to be responsible for the majority of postoperative septic complications in patients with malignant biliary obstruction.[61]

Taken together, preoperative antibiotic therapy and PBD predispose for bacteribilia with *Enterococcus* spp. Postoperative infectious complications with *Enterococcus* spp. are frequently found in patients after pancreatic surgery and despite former suggestions they require adequate antibiotic therapy.

Gram-Negative Aerobe MDR Microorganisms

The most common gram-negative aerobe MDR microorganisms include *P. aeruginosa*, *S. maltophilia* and *Acinetobacter* spp. In complicated intra-abdominal infections *P. aeruginosa* occurs in 8% of patients, while the rate can be much higher in patients with appendicitis.[63–65] After pancreatic surgery postoperative infectious complications with *P. aeruginosa* are less common than infections with other MDR microorganisms, but *P. aeruginosa* is frequently found in blood cultures of patients with cholangitis, postoperative surgical site infections and postoperative peritonitis.[66] Furthermore, *P. aeruginosa* is one of the most leading microorganisms that are responsible for hospital acquired infections, especially hospital acquired and ventilator associated pneumonia in the intensive care unit.[67] Coverage of *Pseudomonas* spp. is required in patients with postoperative infectious complications after pancreatic surgery, if these patients develop further infectious problems, especially respiratory complications.

Postoperative infectious complications with MDR *Acinetobacter* spp. are one of the most dangerous healthcare acquired infections and the most threatening scenario is an outbreak on intensive care units.[68,69] A complete eradication in intensive care units is extremely difficult, because *Acinetobacter* ssp. are remarkably resistant to various environmental stresses including soaps, disinfectants and dry conditions.[70] In pancreatic surgery postoperative infections with *Acinetobacter* spp. are usually hospital or intensive care unit acquired infections which require antibiotic therapy according to the microbiological resistance analyses and strict isolation to prevent a further spread.[71,72]

Postoperative infectious complications with MDR *S. maltophilia* are frequently found in patients with immune suppression, especially after liver transplantation.[73,74] After pancreatic surgery these species are usually only found after long term intensive care unit stay, although *S. maltophilia* is also often found in cholangitis secondary to biliary obstruction.[75]

EXTENDED SPECTRUM β-LACTAMASE (ESBL) PRODUCING ENTEROBACTERIACEAE

During the last decade a remarkable increase in MDR pathogens with ESBL-producing Enterobacteriaceae, including *E. coli*, *Klebsiella* spp., *Enterobacter* spp. and *Morganella morganii* has occurred, especially in critically ill patients.[76,77] The acquisition of ESBL-producing species has been associated with prior broad spectrum antibiotic therapy, and long preoperative and postoperative hospitalization.[44,78] Therefore, ESBL-producing Enterobacteriaceae have been frequently found in patients with immune suppression who have had prior antibiotic exposure, especially after liver or kidney transplantation.[73,74,79] The rate of postoperative infectious complications with ESBL-producing Enterobacteriaceae, especially *E. coli* and *Klebsiella* spp. is still increasing, even in patients without multiple risk factors.[76,80] Therefore, Enterobacteriaceae with ESBL are of particular interest to pancreatic surgeons, because these species and *Enterococcus* spp. represent the most frequently isolated microorganisms in postoperative infectious complications.[81]

The threat of postoperative infections with ESBL-producing microorganisms arises from the potential for ESBL gene exchange between different microorganisms, resulting in the further development of even more resistant microorganisms.[82] Furthermore, resistance to β-lactams is often associated with other resistance mechanisms, finally resulting in complex panresistant phenotypes.[82] Thus, antibiotic treatment options for postoperative infections are extremely limited and this problem will further increase in the future. Without further investigation from the pharmacological industry these gram-negative bacteria will be responsible for most cases of treatment failure in pancreatic surgery in the future.

TREATMENT OPTIONS

Piperacillin/ Sulbactam

A combination of the β-lactam antibiotic Piperacillin with the β-lactam inhibitor Sulbactam is active against most microorganisms encountered in patients with community acquired intra-abdominal infections, including activity against Enterobacteriaceae with ESBL and *P. aeruginosa*, although resistance has increased recently.[59] There is no activity against MRSA and *Enterococcus* spp., thus, this combination is only indicated for postoperative complications with a low risk for infections with gram-positive resistant microorganisms, otherwise it should be combined with antibiotic agents with activity against gram-positive microorganisms.[83]

New β-Lactams (Ceftobiprole)

Ceftobiprole is a fifth generation, broad spectrum Cephalosporin with activity against most gram-positive resistant microorganisms, including MRSA, but it has no activity against *E. faecium* or VRE. Ceftobiprole is susceptible to some β-lactamases with uncertain activity against several Enterobacteriaceae.[57,84] It shows good tissue penetration into soft tissues with activity against gram-positive microorganisms, but has insufficient activity against the majority of gram-negative microorganisms.[85] Therefore, Ceftobiprole cannot be recommended for patients with postoperative complications after pancreatic surgery.

Quinolones (Moxifloxacin)

Moxifloxacin is a broad spectrum fluoroquinolone with good activity against aerobic and anaerobic gram-negative pathogens isolated from patients with intra-abdominal infections. It has no activity against the majority of resistant gram-positive microorganisms.[86] Moxifloxacin shows good tissue penetration into soft tissues.[87] The most common problem with fluoroquinolone application is the high rate of antibiotic resistance, which exceeds 50% in some regions, depending on the local application rate.[88]

Glycopeptides (Vancomycin)

Vancomycin is a glycopeptide with activity against most gram-positive microorganisms, including MRSA but excepting VRE.[89] Vancomycin resistance has been associated with former exposure to vancomycin, especially for VRE.[60,90] Tissue penetration into soft tissues is poor.[89] Vancomycin was the most common antibiotic agent used against postoperative infections with MRSA for some years, when no other antibiotics were available. A major problem with vancomycin application is nephrotoxicity, especially in combination with other nephrotoxic antibiotics.[91] With the exception of microbiologically proven infections with vancomycin-susceptible gram-positive bacteria, it should never be used as a single agent. As new antibiotic agents with gram-positive activity have been developed, vancomycin has no relevance in the treatment of postoperative infectious complications in patients who have undergone pancreatic surgery.

Lipopeptides (Daptomycin)

Daptomycin is a cyclic lipopeptide antibiotic with a good soft tissue penetration. It has activity against most gram-positive bacteria including MRSA and VRE, but no activity against gram-negative microorganisms.[92] Therefore, with the exception of microbiologically-proven infectious complications with gram-positive microorganisms, Daptomycin should be used as part of a combination therapy for patients with postoperative complications after pancreatic surgery.

Oxazolidinones (Linezolid)

Linezolid is an oxazolidinone antibiotic that has activity against the majority of gram-positive resistant microorganisms, including MRSA and VRE.[93] Linezolid shows good penetration into soft tissues and is indicated for infections with gram-positive resistant microorganisms in cases of complicated skin and soft tissue infections.[94] It should never be administered as a single agent because it has no activity against gram-negative microorganisms. Thus, a combination with antibiotic agents with activity against gram-negative microorganisms is mandatory.

Carbapenemes (Doripenem)

Doripenem is a carbapenem antibiotic with broad spectrum activity against the majority of gram-negative Enterobacteriaceae including Enterobacteriaceae with ESBL and *P. aeruginosa*. It has activity against several gram-positive bacteria, but it has no activity against MRSA, VRE, or *E. faecium*.[95] Doripenem shows good tissue penetration into soft tissues, especially peritoneal fluids.[96] Carbapenems are the most important antibiotic agents for the treatment of postoperative complications after pancreatic surgery, because they have a good penetration into pancreatic tissue.[97,98]

Glycylcyclines (Tigecycline)

Tigecycline is a glycylcycline antibiotic with broad spectrum activity against the majority of Enterobacteriaceae with ESBL, and carbapenem resistant bacteria including MRSA and VRE.[99] It shows good tissue penetration into soft tissues, but has no activity against *P. aeruginosa*.[100] Thus, a combination with an antibiotic with activity against *P. aeruginosa* should be used once an infection with *P. aeruginosa* is suspected. Recently the American Food and Drug Administration (FDA) issued a warning concerning increased mortality with tigecycline application, which was verified by a recent meta-analysis.[101] Therefore, caution should be used when tigecycline is administered for postoperative infectious complications, but the debate on associated adverse events continues.[102] It remains an effective alternative for infections with MDR microorganisms that do not respond to other antibiotic regimes, especially for infections with carbapenem-resistant bacteria and Enterobacteriaceae with ESBL.[59,84,103]

Most of the above mentioned antibiotics can be used effectively in the treatment of postoperative infectious complications after pancreatic surgery, but there is no antibiotic agent which is obviously superior to the rest. Table 4.1 summarizes the activities of the antibiotics that are commonly used for the treatment of postoperative infectious complications with the most frequently encountered MDR pathogens after pancreatic surgery.

TABLE 4.1 Antibiotic agents against MDR microorganisms (modified by[103])

	MRSA	VRE	ESBL	*Acinetobacter*	*Pseudomonas aeruginosa*
Piperacillin/Sulbactam	Ø	Ø	X	(X)	X
β-lactams (Ceftobiprole)	X	Ø	(X)	X	Ø
Quinolones (Moxifloxacin)	Ø	Ø	(X)	(X)	(X)
Glycopeptides (Vancomycin)	(X)	Ø	Ø	Ø	Ø
Lipopeptides (Daptomycin)	X	X	Ø	Ø	Ø
Oxazolidinones (Linezolid)	X	X	Ø	Ø	Ø
Carbapenemes (Doripenem)	(X)	(X)	X	(X)	X
Glycylcycline (Tigecycline)	X	X	X	(X)	Ø

X = effective
Ø = not effective
(X) = partial activity

TREATMENT STRATEGIES – RECOMMENDATIONS FOR MDR MICROORGANISMS – TARAGONA STRATEGY

Several guidelines for the treatment of postoperative infectious complications have been published, which aim to facilitate the selection of the appropriate antibiotic agent according to the severity of illness of the patient, but most of these guidelines represent rather fixed plans, without possibilities for choosing from the large number of differently-acting antibiotic agents available.[59] The best description of the most important criteria that should be considered for the selection of an antibiotic regime in patients with postoperative infectious complications is the "Taragona Strategy", which was published first in 2003 by Sandiumenge et al.[104] For patients with postoperative complications after pancreatic surgery a modified treatment regime, based on the reflections which refer to the Tarragona strategy, highlights five recommendations for the selection of antibiotic therapy[103]:

1. Hit hard and early
   The initial antibiotic therapy should be initiated as soon as possible after the diagnosis of a postoperative infectious complication has been made. The antibiotic spectrum should be broad enough to include the vast majority of possible microorganisms that could be responsible for the infectious complication.
2. Look at your patient
   A differentiated reflection with respect to the patients' risk factors for the development of infections with MDR microorganisms is mandatory when selecting a suitable antibiotic agent. Patients with hospital acquired postoperative complications have a high risk of infections with resistant microorganisms. The highest risk of infections with MDR microorganisms exists for patients who have experienced prior broad spectrum antibiotic therapy, former long term hospitalization, severe co-morbidities, and/or long postoperative hospital stay.

3. Listen to your hospital
   The development of antibiotic resistance is a dynamic process which depends on the local policy for antibiotic prescription and antibiotic consumption. Continuous application of one single agent augments selection pressure, giving an increased risk of the development of MDR microorganisms. Therefore, antibiotic treatment regimes should be based on the full diversity of differently acting antibiotic agents, which should be updated regularly with respect to the local hospital surveillance data.
4. Get to the point
   Antibiotics cannot be active if they do not reach an effective concentration at the site of the infection. Therefore, the chosen antibiotic substrates should have pharmacodynamic and pharmacokinetic characteristics that permit high concentrations at the site of the infection.
5. Focus, focus, focus
   After stabilization, regular re-evaluation of the antibiotic treatment regime is mandatory to avoid long term inappropriate and ineffective broad spectrum antibiotic therapy which will augment selection pressure among pathogens, resulting in the further development of new, even more resistant species. Therefore, de-escalation according to microbiology results is recommended after the patient has been stable for three days.

CONCLUSIONS AND FUTURE PROSPECTS

Morbidity and mortality after pancreatic surgery has reduced in the last two decades, but today there is an increasing fear of postoperative infectious complications with MDR microorganisms. Risk factors for the development of postoperative infectious complications with MDR microorganisms after pancreatic surgery include patients with severe co-morbidities, with a long preoperative or postoperative hospital stay, with prior broad spectrum antibiotic exposure. PBD increases the risk of postoperative infectious complications, because bile duct stenting leads to bacteribilia, which is often caused by MDR microorganisms. Intraoperative bile duct cultures are essential for generating hospital-specific surveillance data. Cultures from postoperative infectious complications are mandatory to identify the pathogens that are responsible for the infection, and this information can be used to adapt the antibiotic treatment with the option of de-escalation. The majority of the actually available antibiotics do not cover all possible MDR microorganisms. Therefore, combinations of different spectrum antibiotics are often required before the results from the microbiology are available. The initial antibiotic treatment should be broad enough to cover the most common microorganisms, be administered as soon as possible based on the individual patient's risk factors and local antibiotic susceptibility patterns, reach effective concentrations at the site of the infectious focus, and be re-evaluated after three days with the option of de-escalation. For the majority of gram-positive pathogens Ceftobiprole, Linezolid and Daptomycin are effective, while there is an increasing rate of gram-negative Enterobacteriaceae with ESBL which develop resistance against Piperacillin/Sulbactam and Carbapenems. Without new antibiotic developments from the pharmacological industry these pathogens will be responsible for the majority of treatment failure among patients with postoperative complications after pancreatic surgery in the future.

References

1. Brunschwig A. One stage pancreatoduodenectomy. *Surg Gynecol Obstet* 1947;**85**:161–4.
2. Winter JM, Cameron JL, Campbell KA, Arnold MA, Chang DC, Coleman J, et al. 1423 pancreaticoduodenectomies for pancreatic cancer: A single-institution experience. *J Gastrointest Surg* 2006;**10**:1199–210; discussion 1210–1.
3. Yeo CJ, Cameron JL, Sohn TA, Lillemoe KD, Pitt HA, Talamini MA, et al. Six hundred fifty consecutive pancreaticoduodenectomies in the 1990s: pathology, complications, and outcomes. *Ann Surg* 1997;**226**:248–57; discussion 257–60.
4. Gouma DJ, van Geenen RC, van Gulik TM, de Haan RJ, de Wit LT, Busch OR, et al. Rates of complications and death after pancreaticoduodenectomy: risk factors and the impact of hospital volume. *Ann Surg* 2000;**232**:786–95.
5. Cameron JL, Riall TS, Coleman J, Belcher KA. One thousand consecutive pancreaticoduodenectomies. *Ann Surg* 2006;**244**:10–5.
6. Trede M, Schwall G. The complications of pancreatectomy. *Ann Surg* 1988;**207**:39–47.
7. Diener MK, Seiler CM, Rossion I, Kleeff J, Glanemann M, Butturini G, et al. Efficacy of stapler versus hand-sewn closure after distal pancreatectomy (DISPACT): a randomised, controlled multicenter trial. *Lancet* 2011;**377**:1514–22.
8. Knaebel HP, Diener MK, Wente MN, Buchler MW, Seiler CM. Systematic review and meta-analysis of technique for closure of the pancreatic remnant after distal pancreatectomy. *Br J Surg* 2005;**92**:539–46.
9. Vin Y, Sima CS, Getrajdman GI, Brown KT, Covey A, Brennan MF, et al. Management and outcomes of post-pancreatectomy fistula, leak, and abscess: results of 908 patients resected at a single institution between 2000 and 2005. *J Am Coll Surg* 2008;**207**:490–8.
10. Lichtenstern C, Schmidt J, Knaebel HP, Martin E, Buchler MW, Weigand MA. Postoperative bacterial/fungal infections: a challenging problem in critically ill patients after abdominal surgery. *Dig Surg* 2007;**24**:1–11.
11. Buchler MW, Wagner M, Schmied BM, Uhl W, Friess H, Z'Graggen K. Changes in morbidity after pancreatic resection: toward the end of completion pancreatectomy. *Arch Surg* 2003;**138**:1310–4; discussion 1315.
12. Standop J, Glowka T, Schmitz V, Schafer N, Overhaus M, Hirner A, et al. Operative re-intervention following pancreatic head resection: indications and outcome. *J Gastrointest Surg* 2009;**13**:1503–9.
13. Sohn TA, Yeo CJ, Cameron JL, Geschwind JF, Mitchell SE, Venbrux AC, et al. Pancreaticoduodenectomy: role of interventional radiologists in managing patients and complications. *J Gastrointest Surg* 2003;**7**:209–19.
14. Herzog T, Suelberg D, Belyaev O, Uhl W, Seemann M, Seelig MH. Treatment of acute delayed visceral hemorrhage after pancreatic surgery from hepatic arteries with covered stents. *J Gastrointest Surg* 2011;**15**:496–502.
15. Link BC, Yekebas EF, Bogoevski D, Kutup A, Adam G, Izbicki JR, et al. Percutaneous transhepatic cholangio-drainage as rescue therapy for symptomatic biliary leakage without biliary tract dilation after major surgery. *J Gastrointest Surg* 2007;**11**:166–70.
16. Limongelli P, Khorsandi SE, Pai M, Jackson JE, Tait P, Tierris J, et al. Management of delayed postoperative hemorrhage after pancreaticoduodenectomy: a meta-analysis. *Arch Surg* 2008;**143**:1001–7; discussion 1007.
17. Vincent JL. Critical care – where have we been and where are we going? *Crit Care* 2013;**17(Suppl 1)**:S2.
18. Garnacho-Montero J, Garcia-Garmendia JL, Barrero-Almodovar A, Jimenez-Jimenez FJ, Perez-Paredes C, Ortiz-Leyba C. Impact of adequate empirical antibiotic therapy on the outcome of patients admitted to the intensive care unit with sepsis. *Crit Care Med* 2003;**31**:2742–51.
19. MacArthur RD, Miller M, Albertson T, Panacek E, Johnson D, Teoh L, et al. Adequacy of early empiric antibiotic treatment and survival in severe sepsis: experience from the MONARCS trial. *Clin Infect Dis* 2004;**38**:284–8.
20. Rivers E, Nguyen B, Havstad S, Ressler J, Muzzin A, Knoblich B, et al. Early goal-directed therapy in the treatment of severe sepsis and septic shock. *N Engl J Med* 2001;**345**:1368–77.
21. Barie PS. Multidrug-resistant organisms and antibiotic management. *Surg Clin North Am* 2012;**92**:345–91; ix–x.
22. Armstrong CP, Dixon JM, Taylor TV, Davies GC. Surgical experience of deeply jaundiced patients with bile duct obstruction. *Br J Surg* 1984;**71**:234–8.
23. Dixon JM, Armstrong CP, Duffy SW, Elton RA, Davies GC. Factors affecting mortality and morbidity after surgery for obstructive jaundice. *Gut* 1984;**25**:104.

24. Dixon JM, Armstrong CP, Duffy SW, Davies GC. Factors affecting morbidity and mortality after surgery for obstructive jaundice: a review of 373 patients. *Gut* 1983;**24**:845–52.
25. Speer AG, Cotton PB, Russell RC, Mason RR, Hatfield AR, Leung JW, et al. Randomised trial of endoscopic versus percutaneous stent insertion in malignant obstructive jaundice. *Lancet* 1987;**2**:57–62.
26. Gouma DJ, Coelho JC, Schlegel JF, Li YF, Moody FG. The effect of preoperative internal and external biliary drainage on mortality of jaundiced rats. *Arch Surg* 1987;**122**:731–4.
27. Sewnath ME, Karsten TM, Prins MH, Rauws EJ, Obertop H, Gouma DJ. A meta-analysis on the efficacy of preoperative biliary drainage for tumors causing obstructive jaundice. *Ann Surg* 2002;**236**:17–27.
28. Grizas S, Stakyte M, Kincius M, Barauskas G, Pundzius J. Etiology of bile infection and its association with postoperative complications following pancreatoduodenectomy. *Medicina (Kaunas)* 2005;**41**:386–91.
29. van der Gaag NA, Rauws EA, van Eijck CH, Bruno MJ, van der Harst E, Kubben FJ, et al. Preoperative biliary drainage for cancer of the head of the pancreas. *N Engl J Med* 2010;**362**:129–37.
30. Herzog T, Belyaev O, Hessam S, Suelberg D, Janot M, Schrader J, et al. Bacteribilia with resistant microorganisms after preoperative biliary drainage – the influence of bacteria on postoperative outcome. *Scand J Gastroenterol* 2012;**47**:827–35.
31. Bonin EA, Baron TH. Preoperative biliary stents in pancreatic cancer. *J Hepatobiliary Pancreat Sci* 2011;**18**:621–9.
32. Herzog T, Belyaev O, Muller CA, Mittelkotter U, Seelig MH, Weyhe D, et al. Bacteribilia after preoperative bile duct stenting: a prospective study. *J Clin Gastroenterol* 2009;**43**:457–62.
33. Neve R, Biswas S, Dhir V, Mohandas KM, Kelkar R, Shukla P, et al. Bile cultures and sensitivity patterns in malignant obstructive jaundice. *Indian J Gastroenterol* 2003;**22**:16–8.
34. Gregg JA, De Girolami P, Carr-Locke DL. Effects of sphincteroplasty and endoscopic sphincterotomy on the bacteriologic characteristics of the common bile duct. *Am J Surg* 1985;**149**:668–71.
35. Howard TJ, Yu J, Greene RB, George V, Wairiuko GM, Moore SA, et al. Influence of bactibilia after preoperative biliary stenting on postoperative infectious complications. *J Gastrointest Surg* 2006;**10**:523–31.
36. dos Santos JS, Junior WS, Modena JL, Brunaldi JE, Ceneviva R. Effect of preoperative endoscopic decompression on malignant biliary obstruction and postoperative infection. *Hepatogastroenterology* 2005;**52**:45–7.
37. Garcea G, Chee W, Ong SL, Maddern GJ. Preoperative biliary drainage for distal obstruction: the case against revisited. *Pancreas* 2010;**39**:119–26.
38. Jethwa P, Breuning E, Bhati C, Buckles J, Mirza D, Bramhall S. The microbiological impact of pre-operative biliary drainage on patients undergoing hepato-biliary-pancreatic (HPB) surgery. *Aliment Pharmacol Ther* 2007;**25**:1175–80.
39. Limongelli P, Pai M, Bansi D, Thiallinagram A, Tait P, Jackson J, et al. Correlation between preoperative biliary drainage, bile duct contamination, and postoperative outcomes for pancreatic surgery. *Surgery* 2007;**142**:313–8.
40. Mazuski JE. Antimicrobial treatment for intra-abdominal infections. *Expert Opin Pharmacother* 2007;**8**:2933–45.
41. Singh AV, Mishra B, Thakur A. Multidrug-resistant Gram-negative bacteria in postoperative infections. *J Indian Med Assoc* 2009;**107**(148–50):163.
42. Nicoletti G, Nicolosi D, Rossolini GM, Stefani S. Intra-abdominal infections: etiology, epidemiology, microbiological diagnosis and antibiotic resistance. *J Chemother* 2009;**21(Suppl 1)**:5–11.
43. Mazuski JE, Solomkin JS. Intra-abdominal infections. *Surg Clin North Am* 2009;**89**:421–37.
44. Seguin P, Fedun Y, Laviolle B, Nesseler N, Donnio PY, Malledant Y. Risk factors for multidrug-resistant bacteria in patients with post-operative peritonitis requiring intensive care. *J Antimicrob Chemother* 2010;**65**:342–6.
45. Seguin P, Laviolle B, Chanavaz C, Donnio PY, Gautier-Lerestif AL, Campion JP, et al. Factors associated with multidrug-resistant bacteria in secondary peritonitis: impact on antibiotic therapy. *Clin Microbiol Infect* 2006;**12**:980–5.
46. Swenson BR, Metzger R, Hedrick TL, McElearney ST, Evans HL, Smith RL, et al. Choosing antibiotics for intra-abdominal infections: what do we mean by "high risk"? *Surg Infect (Larchmt)* 2009;**10**:29–39.
47. Burnett RJ, Haverstock DC, Dellinger EP, Reinhart HH, Bohnen JM, Rotstein OD, et al. Definition of the role of enterococcus in intraabdominal infection: analysis of a prospective randomized trial. *Surgery* 1995;**118**:716–21; discussion 721–3.
48. Montravers P, Gauzit R, Muller C, Marmuse JP, Fichelle A, Desmonts JM. Emergence of antibiotic-resistant bacteria in cases of peritonitis after intraabdominal surgery affects the efficacy of empirical antimicrobial therapy. *Clin Infect Dis* 1996;**23**:486–94.

49. Sisirak M, Zvizdic A, Hukic M. Methicillin-resistant Staphylococcus aureus (MRSA) as a cause of nosocomial wound infections. *Bosn J Basic Med Sci* 2010;**10**:32–7.
50. Kujath P, Kujath C. Complicated skin, skin structure and soft tissue infections - are we threatened by multi-resistant pathogens? *Eur J Med Res* 2010;**15**:544–53.
51. Panlilio AL, Culver DH, Gaynes RP, Banerjee S, Henderson TS, Tolson JS, et al. Methicillin-resistant staphylococcus aureus in US hospitals, 1975–1991. *Infect Control Hosp Epidemiol* 1992;**13**:582–6.
52. Kassavin DS, Pascarella L, Goldfarb MA. Surgical site infections: incidence and trends at a community teaching hospital. *Am J Surg* 2011;**201**:749–53.
53. Uchiyama K, Takifuji K, Tani M, Ueno M, Kawai M, Ozawa S, et al. Prevention of postoperative infections by administration of antimicrobial agents immediately before surgery for patients with gastrointestinal cancers. *Hepatogastroenterology* 2007;**54**:1487–93.
54. Takara D, Sugawara G, Ebata T, Yokoyama Y, Igami T, Nagino M. Preoperative biliary MRSA infection in patients undergoing hepatobiliary resection with cholangiojejunostomy: incidence, antibiotic treatment, and surgical outcome. *World J Surg* 2011;**35**:850–7.
55. Sanjay P, Fawzi A, Kulli C, Polignano FM, Tait IS. Impact of methicillin-resistant Staphylococcus Aureus (MRSA) infection on patient outcome after pancreatoduodenectomy (PD) – a cause for concern? *Pancreas* 2010;**39**:1211–4.
56. Sanjay P, Fawzi A, Fulke JL, Kulli C, Tait IS, Zealley IA, et al. Late post pancreatectomy haemorrhage. Risk factors and modern management. *JOP* 2010;**11**:220–5.
57. Barbour A, Derendorf H. Resistance and the management of complicated skin and skin structure infections: the role of ceftobiprole. *Ther Clin Risk Manag* 2010;**6**:485–95.
58. Chou YY, Lin TY, Lin JC, Wang NC, Peng MY, Chang FY. Vancomycin-resistant enterococcal bacteremia: comparison of clinical features and outcome between Enterococcus faecium and Enterococcus faecalis. *J Microbiol Immunol Infect* 2008;**41**:124–9.
59. Solomkin JS, Mazuski JE, Bradley JS, Rodvold KA, Goldstein EJ, Baron EJ, et al. Diagnosis and management of complicated intra-abdominal infection in adults and children: guidelines by the Surgical Infection Society and the Infectious Diseases Society of America. *Clin Infect Dis* 2010;**50**:133–64.
60. Behrman SW, Bahr MH, Dickson PV, Zarzaur BL. The microbiology of secondary and postoperative pancreatic infections: implications for antimicrobial management. *Arch Surg* 2011;**146**:613–9.
61. Nomura T, Shirai Y, Hatakeyama K. Enterococcal bactibilia in patients with malignant biliary obstruction. *Dig Dis Sci* 2000;**45**:2183–6.
62. Jagannath P, Dhir V, Shrikhande S, Shah RC, Mullerpatan P, Mohandas KM. Effect of preoperative biliary stenting on immediate outcome after pancreaticoduodenectomy. *Br J Surg* 2005;**92**:356–61.
63. Dinubile MJ, Friedland I, Chan CY, Motyl MR, Giezek H, Shivaprakash M, et al. Bowel colonization with resistant gram-negative bacilli after antimicrobial therapy of intra-abdominal infections: observations from two randomized comparative clinical trials of ertapenem therapy. *Eur J Clin Microbiol Infect Dis* 2005;**24**:443–9.
64. Sotto A, Lefrant JY, Fabbro-Peray P, Muller L, Tafuri J, Navarro F, et al. Evaluation of antimicrobial therapy management of 120 consecutive patients with secondary peritonitis. *J Antimicrob Chemother* 2002;**50**:569–76.
65. Lin WJ, Lo WT, Chu CC, Chu ML, Wang CC. Bacteriology and antibiotic susceptibility of community-acquired intra-abdominal infection in children. *J Microbiol Immunol Infect* 2006;**39**:249–54.
66. Christou NV, Turgeon P, Wassef R, Rotstein O, Bohnen J, Potvin M. Management of intra-abdominal infections. The case for intraoperative cultures and comprehensive broad-spectrum antibiotic coverage. The Canadian Intra-abdominal Infection Study Group. *Arch Surg* 1996;**131**:1193–201.
67. Barbier F, Andremont A, Wolff M, Bouadma L. Hospital-acquired pneumonia and ventilator-associated pneumonia: recent advances in epidemiology and management. *Curr Opin Pulm Med* 2013;**19**:216–28.
68. Koeleman JG, Parlevliet GA, Dijkshoorn L, Savelkoul PH, Vandenbroucke-Grauls CM. Nosocomial outbreak of multi-resistant Acinetobacter baumannii on a surgical ward: epidemiology and risk factors for acquisition. *J Hosp Infect* 1997;**37**:113–23.
69. Consales G, Gramigni E, Zamidei L, Bettocchi D, De Gaudio AR. A multidrug-resistant Acinetobacter baumannii outbreak in intensive care unit: antimicrobial and organizational strategies. *J Crit Care* 2011;**26**:453–9.
70. Borgmann S, Wolz C, Grobner S, Autenrieth IB, Heeg P, Goerke C, et al. Metallo-beta-lactamase expressing multi-resistant Acinetobacter baumannii transmitted in the operation area. *J Hosp Infect* 2004;**57**:308–15.

71. Mioljevic V, Simic AP, Radenkovic D, Galun D, Palibrk I, Ostojic S, et al. Acinetobacter spp. colonization and infection risk factors in surgical patients. *Acta Chir Iugosl* 2011;**58**:81–7.
72. Urban C, Segal-Maurer S, Rahal JJ. Considerations in control and treatment of nosocomial infections due to multidrug-resistant Acinetobacter baumannii. *Clin Infect Dis* 2003;**36**:1268–74.
73. Shi SH, Kong HS, Xu J, Zhang WJ, Jia CK, Wang WL, et al. Multidrug resistant gram-negative bacilli as predominant bacteremic pathogens in liver transplant recipients. *Transpl Infect Dis* 2009;**11**:405–12.
74. Zhong L, Men TY, Li H, Peng ZH, Gu Y, Ding X, et al. Multidrug-resistant gram-negative bacterial infections after liver transplantation – spectrum and risk factors. *J Infect* 2012;**64**:299–310.
75. Kaya M, Bestas R, Bacalan F, Bacaksiz F, Arslan EG, Kaplan MA. Microbial profile and antibiotic sensitivity pattern in bile cultures from endoscopic retrograde cholangiography patients. *World J Gastroenterol* 2012;**18**:3585–9.
76. Hawser SP, Bouchillon SK, Hoban DJ, Badal RE. *In vitro* susceptibilities of aerobic and facultative anaerobic Gram-negative bacilli from patients with intra-abdominal infections worldwide from 2005–2007: results from the SMART study. *Int J Antimicrob Agents* 2009;**34**:585–8.
77. Fridkin SK, Gaynes RP. Antimicrobial resistance in intensive care units. *Clin Chest Med* 1999;**20**:303–16.
78. Paterson DL, Bonomo RA. Extended-spectrum beta-lactamases: a clinical update. *Clin Microbiol Rev* 2005;**18**:657–86.
79. Linares L, Cervera C, Cofan F, Ricart MJ, Esforzado N, Torregrosa V, et al. Epidemiology and outcomes of multiple antibiotic-resistant bacterial infection in renal transplantation. *Transplant Proc* 2007;**39**:2222–4.
80. Brink AJ, Botha RF, Poswa X, Senekal M, Badal RE, Grolman DC, et al. Antimicrobial susceptibility of gram-negative pathogens isolated from patients with complicated intra-abdominal infections in South African hospitals (SMART Study 2004–2009): impact of the new carbapenem breakpoints. *Surg Infect (Larchmt)* 2012;**13**:43–9.
81. Chen YH, Hsueh PR. Changing bacteriology of abdominal and surgical sepsis. *Curr Opin Infect Dis* 2012;**25**:590–5.
82. Sartelli M. A focus on intra-abdominal infections. *World J Emerg Surg* 2010;**5**:9.
83. Dupont H. The empiric treatment of nosocomial intra-abdominal infections. *Int J Infect Dis* 2007;**11(Suppl 1)**: S1–6.
84. Hasper D, Schefold JC, Baumgart DC. Management of severe abdominal infections. *Recent Pat Antiinfect Drug Discov* 2009;**4**:57–65.
85. Barbour A, Schmidt S, Rout WR, Ben-David K, Burkhardt O, Derendorf H. Soft tissue penetration of cefuroxime determined by clinical microdialysis in morbidly obese patients undergoing abdominal surgery. *Int J Antimicrob Agents* 2009;**34**:231–5.
86. Malangoni MA, Song J, Herrington J, Choudhri S, Pertel P. Randomized controlled trial of moxifloxacin compared with piperacillin-tazobactam and amoxicillin-clavulanate for the treatment of complicated intra-abdominal infections. *Ann Surg* 2006;**244**:204–11.
87. Wirtz M, Kleeff J, Swoboda S, Halaceli I, Geiss HK, Hoppe-Tichy T, et al. Moxifloxacin penetration into human gastrointestinal tissues. *J Antimicrob Chemother* 2004;**53**:875–7.
88. Dalhoff A. Global fluoroquinolone resistance epidemiology and implictions for clinical use. *Interdiscip Perspect Infect Dis* 2012;**2012**:976273.
89. Daschner FD, Frank U, Kummel A, Schmidt-Eisenlohr E, Schlosser V, Spillner H, et al. Pharmacokinetics of vancomycin in serum and tissue of patients undergoing open-heart surgery. *J Antimicrob Chemother* 1987;**19**:359–62.
90. Gearhart M, Martin J, Rudich S, Thomas M, Wetzel D, Solomkin J, et al. Consequences of vancomycin-resistant Enterococcus in liver transplant recipients: a matched control study. *Clin Transplant* 2005;**19**:711–6.
91. Hermsen ED, Hanson M, Sankaranarayanan J, Stoner JA, Florescu MC, Rupp ME. Clinical outcomes and nephrotoxicity associated with vancomycin trough concentrations during treatment of deep-seated infections. *Expert Opin Drug Saf* 2010;**9**:9–14.
92. Steenbergen JN, Alder J, Thorne GM, Tally FP. Daptomycin: a lipopeptide antibiotic for the treatment of serious Gram-positive infections. *J Antimicrob Chemother* 2005;**55**:283–8.
93. Dodds TJ, Hawke CI. Linezolid versus vancomycin for MRSA skin and soft tissue infections (systematic review and meta-analysis). *ANZ J Surg* 2009;**79**:629–35.

94. Itani KM, Dryden MS, Bhattacharyya H, Kunkel MJ, Baruch AM, Weigelt JA. Efficacy and safety of linezolid versus vancomycin for the treatment of complicated skin and soft-tissue infections proven to be caused by methicillin-resistant *Staphylococcus aureus*. *Am J Surg* 2010;**199**:804–16.
95. Lucasti C, Jasovich A, Umeh O, Jiang J, Kaniga K, Friedland I. Efficacy and tolerability of IV doripenem versus meropenem in adults with complicated intra-abdominal infection: a phase III, prospective, multicenter, randomized, double-blind, noninferiority study. *Clin Ther* 2008;**30**:868–83.
96. Ikawa K, Morikawa N, Urakawa N, Ikeda K, Ohge H, Sueda T. Peritoneal penetration of doripenem after intravenous administration in abdominal-surgery patients. *J Antimicrob Chemother* 2007;**60**:1395–7.
97. Foitzik T, Fernandez-del Castillo C, Ferraro MJ, Mithofer K, Rattner DW, Warshaw AL. Pathogenesis and prevention of early pancreatic infection in experimental acute necrotizing pancreatitis. *Ann Surg* 1995; **222**:179–85.
98. Villatoro E, Mulla M, Larvin M. Antibiotic therapy for prophylaxis against infection of pancreatic necrosis in acute pancreatitis. *Cochrane Database Syst Rev* 2010; CD002941.
99. Kresken M, Leitner E, Brauers J, Geiss HK, Halle E, von Eiff C, et al. Susceptibility of common aerobic pathogens to tigecycline: results of a surveillance study in Germany. *Eur J Clin Microbiol Infect Dis* 2009;**28**:83–90.
100. Nicolau DP. Management of complicated infections in the era of antimicrobial resistance: the role of tigecycline. *Expert Opin Pharmacother* 2009;**10**:1213–22.
101. Yahav D, Lador A, Paul M, Leibovici L. Efficacy and safety of tigecycline: a systematic review and meta-analysis. *J Antimicrob Chemother* 2011;**66**:1963–71.
102. Bodmann KF, Heizmann WR, von Eiff C, Petrik C, Loschmann PA, Eckmann C. Therapy of 1,025 severely ill patients with complicated infections in a German multicenter study: safety profile and efficacy of tigecycline in different treatment modalities. *Chemotherapy* 2012;**58**:282–94.
103. Herzog T, Chromik AM, Uhl W. Treatment of complicated intra-abdominal infections in the era of multi-drug resistant bacteria. *Eur J Med Res* 2010;**15**:525–32.
104. Sandiumenge A, Diaz E, Bodi M, Rello J. Therapy of ventilator-associated pneumonia. A patient-based approach based on the ten rules of "The Tarragona Strategy". *Intensive Care Med* 2003;**29**:876–83.

CHAPTER

5

Anastomotic Leakage after Colonic and Rectal Surgery: Operative and Conservative Management

Elia Poiasina, Michela Giulii Capponi and Luca Ansaloni

General Surgery Department, Papa Giovanni XXIII Hospital, Bergamo, Italy

INTRODUCTION

Leakage from an anastomosis in the gastrointestinal tract is a major complication that is often associated with increased morbidity, mortality and prolonged hospital stay. The frequency and consequences of anastomotic failure vary according to the site within the gastrointestinal tract. Anastomotic leak may be used as an indicator of the quality of surgical care. However, comparisons of leak rates that should be made between and within surgical centers are vitiated by lack of standard definitions and methods of measurement. Moreover, little consensus is gained about the management of this threatening complication.[1,2]

Among several unanswered questions, it is not known whether and when the primary anastomosis can be preserved, or if or when fecal diversion is necessary. In addition, the majority of the published studies do not distinguish between colonic and rectal resection. Rectal surgery shows a higher incidence of leakage but this can often be treated conservatively, especially when a protective ileostomy has primarily been constructed.[3,4] In these last cases, literature shows several experiences of local treatment with the use of transanal or percutaneus drains, flap-reconstruction or transanal vacuum-assisted closure (VAC) for the resolution of presacral abscess cavities.

Even if recommendations about the correct treatment of anastomotic leakage (AL) are of low value due to the fact that randomized clinical trials are extremely difficult to perform in this particular group of patients, nevertheless conservative management seems to be a safe option in selected cases.

Microbiology for Surgical Infections
DOI: http://dx.doi.org/10.1016/B978-0-12-411629-0.00005-2

DEFINITION OF ANASTOMOTIC LEAKAGE

Although anastomotic leakage (AL) is still a major challenge for colorectal surgeons, there is no consensus on its definition.

Some key points can be fixed.

- Firstly, the level of colorectal anastomosis does not influence the definition of a leak.
- Usually there are no chronological elements in the definitions of AL, as it may become apparent even after discharge from hospital.
- The definitions of leak after bowel surgery usually include peritonitis (localized or generalized), fecal or purulent drainage from the wound and/or drain, presence of an abscess and fever.
- The criteria most often used to define AL combine clinical and radiological features.

Interestingly, in a review of the definition and measurement of AL, Bruce et al. identified 29 separate definitions of lower gastrointestinal AL, because of differences in the use of clinical features, radiological investigation or both.[5]

An attempt to formulate a standard definition of AL was proposed by the UK Surgical Infections Study Group (SISG) for the clinical audit to form the basis for meaningful comparisons.[6] Leak was defined as:

> "the leak of luminal contents from a surgical join between two hollow viscera. The luminal contents may emerge either through the wound or at the drain site, or they may collect near the anastomosis, causing fever, abscess, septicemia, metabolic disturbance and/or multiple-organ-failure. The escape of luminal contents from the site of the anastomosis into an adjacent localized area, detected by imaging, in the absence of clinical symptoms and signs, should be recorded as a subclinical leak".

Although this definition was proposed in 1991, it has not been widely adopted, perhaps because it is too general to be applied in different clinical situations.

In a recent paper, Trencheva et al. defined an AL as:

a. Leakage of bowel content and/or gas from the surgical connection between the two bowel ends into the abdomen or pelvis with either spillage and/or fluid collection around the anastomotic site or extravasation through a wound, drain site, or anus;
b. Clinical manifestation causing fever, abscess, septicemia, peritonitis, and/or organ failure; and
c. Confirmation by imaging technique (e.g., radiograph, endoscopy, computerized tomography scan, magnetic resonance imaging, sonography) or by digital rectal examination or anoscopy and/or proctoscopy for low rectal anastomoses.

An AL detected by imaging study only but not clinically manifested is an "asymptomatic" AL.[7]

In our opinion it is reasonable to distinguish leaks as clinically significant or insignificant, as this feature affects the management and clinical outcomes. In this regard we consider Bruce's proposal of a three point definition of AL to be useful, as it includes signs and symptoms, level of severity and components of clinical management.[5]

TABLE 5.1 Components of Definition of Anastomotic Leak in Lower Gi* Leak[5]

Grade	Symptoms and signs	Management
Radiological	Detected only on routine study. No clinical signs.	No change in management
Clinical minor	Presence of luminal contents through the drain or wound site causing local inflammation, e.g., fever (temperature >38°C), leucocytosis (white cell count >10000/L), fecal discharge from wound or drain or abscess. Leak may also be detected on imaging studies.	No change in management or intervention but may have prolonged hospital stay
Clinical major	As clinical minor. Severe disruption to anastomosis. Leak may also be detected on imaging studies.	Change in management and intervention required

**GI = gastrointestinal*

In essence, any leak that requires a change in management, including prescription of antibiotics, with subsequent delay in the reintroduction of oral intake or discharge from hospital, would be considered a major clinical leak (Table 5.1).

EPIDEMIOLOGY

As there is no consensus on AL definition, the prevalence of anastomotic leak varies from 0.5% to 30% after colon and rectal resections[8–15]; but the incidence of clinically significant AL is between 1% and 12% overall, and up to 10% to 14% in lower colorectal resections.[12–15] In fact, postoperative contrast studies conducted regardless of symptoms after colorectal surgery show much higher rates of AL although many of these have no clinical consequences.[16]

Leaks rates are broadly similar for all types of bowel anastomosis; stapled or sutured.[15,17–19] Published leakage rates for different types of colonic resections vary considerably, from 1 to 22% for ileocecal resection and from 2 to 11% for subtotal colectomy.[20,21] In addition, Veyrie et al. registered a statistically significant difference in the prevalence of AL between right and left colectomy.[22] Some factors may explain this difference: first, in the left colon there could be greater concentrations of intraluminal and mucosal germs, causing higher intraluminal sepsis. Second, the absence of peritoneal cover in the distal rectum causes/induces anastomoses to heal less well,[23] especially when the vascularization of the rectal stump is compromised[24] for instance, after preoperative radiation therapy, total mesorectal excision[15] or anastomosis in the narrow male pelvis.[4]

RISK FACTORS

Recognizing the independent risk factors on which the surgeon can eventually act preventively should help decrease the rate of AL, attenuate its severity, and guide its management. In any case, the large spectrum of colorectal diseases, the variety of surgical and medical treatment modalities, different surgical anastomotic technologies, and lack of a standard definition for AL contribute to the complexity of defining predisposing factors for leakage. Several studies have looked at this issue. Some authors investigated AL rates in patients undergoing colectomy for mixed pathologies (cancer, diverticular disease, other inflammatory diseases),[26] or under different circumstances (either emergency surgery, elective surgery, or both)[27]; in variable anastomotic sites (a combination of several ileocolonic, colocolonic, or, most often, colorectal, proximal, and distal); in different settings (single center[2,25] or multicenter[26]), with data culled either prospectively[2,26] or retrospectively.[25]

We can distinguish patient-related risk factors and operative risk factors. The latter vary according to the site of operation performed for the same pathology.

Some of the ***patient-specific risk factors*** are: under-nutrition (serum albumin less than 30 g/dl), obesity, male sex, chronic obstructive pulmonary disease, ischemic heart disease, diabetes mellitus, American Society of Anesthesiologists (ASA) score ≥ 3, systemic hypertension, tobacco and alcohol abuse and prolonged use of high-dose steroids,[15,28,29] preoperative pelvic irradiation, inflammatory bowel diseases. In addition, cancer itself has been reported as a risk factor for AL.[30,31]

Operative risk factors identified are poor colonic preparation, presence of peritonitis, adverse effect during operation, intraoperative blood loss/transfusion, anastomotic ischemia or tension, presacral hematoma or fluid collection with subsequent infection and pelvic drainage.[15,25,29,32–34]

The most important risk factor for leakage is the height of anastomosis from the anal verge: the lower the anastomosis is, particularly below 6 cm, the greater is the risk.[15,32] Nevertheless, AL also occurs in patients with no obvious risk factors.[35,36]

It seems helpful to add some details to this discussion.

Weight loss[19,37] is an independent risk factor for AL, and affects mortality[26] and overall morbidity.[23,38] Weight loss is probably related to hypoalbuminemia.[39] Both weight loss and hypoalbuminemia are potential markers for poor nutrition, which can play a role in anastomotic healing.[38,40] One prospective randomized study[37] showed that preoperative nutritional support for undernourished patients (weight loss $\geq 10\%$ of usual body weight) with cancer who planned to undergo major gastrointestinal surgery could reduce postoperative morbidity, hospital stay and costs, but it did not influence mortality. Malnutrition should be identified as early as possible before surgery, and appropriate measures instituted preoperatively to improve nutrition.[16]

On the contrary, there is no evidence that preoperative **bowel preparation** reduces the rate and consequences of leaks. This trend is in agreement with the most recent meta-analysis involving controlled studies on oral mechanical preparation; five of nine studies included polyethylene glycol, which was found to be not only useless but also potentially dangerous.[41]

There is little agreement on prophylactic use of **drains** in anastomoses in elective colorectal surgery despite many randomized clinical trials.[42,43] The results of these trials are contradictory, and the quality and statistical power of the individual studies have been questioned.

The use of abdominal drains following the creation of anastomoses outside the pelvis has not been shown to be of value in decreasing the rate or severity of anastomotic leaks.[44]

Drains may also give false assurance as well as causing pain and acting as conduits for infection, even resulting in fistula formation. Instead, once AL has occurred, it is generally agreed that drains should be used for therapeutic purposes. Instead a large prospective study suggested that pelvic drain may reduce symptomatic AL after low colorectal and coloanal anastomoses.[45] Pelvic drainage may be important, as blood and serous fluid are likely to collect in the pelvis and may become infected, forming an abscess which discharges through the anastomosis.[16] A correctly sited pelvic drain can prevent this situation.[46]

There is no evidence that **laparoscopically** assisted colorectal surgery results in different AL rates to those achieved by open surgery.[47–49]

In 1998, a meta-analysis of 13 randomized clinical trials found no difference in leak rates between **stapled** and **hand-sewn anastomoses.**[50] The Cochrane review has confirmed that there is insufficient evidence to demonstrate any superiority of stapled over hand-sewn techniques in colorectal anastomoses, regardless of the level of anastomosis.[51]

Consistent with the results of Rose et al., the AL prevalence in **anastomoses less than 10 cm from the anal verge** reaches 16%.[52]

Palliative resection also seems to be an independent risk factor for AL. Disseminated or metastatic cancer, which is found to be an independent risk factor for mortality,[38,39,53,54] may not only influence the immunological defense systems but also may decrease the patient's healing capacity.[55]

Postoperative complication rates reported in **Crohn's disease** tend to be higher than those in surgery for non-inflammatory diseases. However, AL remains a rare event unless severe malnutrition or intra-abdominal infection is present.[56]

Sex also has been identified as a risk factor of AL in rectal surgery: the relative risk in males was 2.3 times higher than for females. One possible explanation for this is the narrower pelvis of males, which makes surgical dissection and the creation of an anastomosis more technically demanding.[16]

Instead, **surgeon's training** is minimized by Trencheva et al., who did not find significant differences among surgeons with different lengths of experience and degrees of board certification, as already discussed by Prystowsky.[7,53]

Finally **intraoperative complications** increase the chances of AL by 4.1 times**.** These are defined as unexpected adverse events occurring during the operative period, such as bowel injuries, bleeding requiring transfusion, stapling device malfunction, hypotension less than 20% of the baseline measurement, oxygen saturation less than 90% for more than 5 minutes, and metabolic acidosis.[7]

A recent prospective study on 616 patients accurately examined all of the above-mentioned predictors for AL, and introduced some interesting issues.[7] In summary, Trencheva et al. found the following variables to be independent risk factors for AL: Charlson Comorbidity Index (CCI) of 3 or more, level of anastomoses less than 10 cm from the anal verge, high inferior mesenteric artery ligation (above left colic artery), intraoperative complications/adverse events, and male sex.

In particular, Trencheva et al. used the CCI to evaluate preoperative comorbidities. This is a validated index for measuring the risk of one-year mortality and considers both the number and the severity level of comorbidities.[7] The study also considers the effect of blood vessel ligation on AL risk: patients with high ligation of the inferior mesenteric artery had a 3.8 times higher chance of leaking than those with low ligation. This finding represents the extent of the surgical procedure in patients with left-sided surgery and with neoplasm where the oncologic radicality demands a more extensive lymphadenectomy.

The effectiveness of a temporary **diverting stoma** in reducing the incidence of AL has been a matter of controversy. A protective ileostomy is not used routinely in colonic surgery.

Some studies have demonstrated a reduction in leakage rates in patients with a covering stoma,[30,57] but others showed no clear benefit from using it.[27,58] Moreover, a protective stoma is also associated with increased hospital stay and cost, as stoma reversal can cause morbidity and even mortality.[25,27,58]

More accurate assessments of the risk of AL would be very helpful in tailoring treatment in colorectal surgery. Even if much has been published in the literature regarding the risk of leakage, evidence-based tools to aid surgeons in deciding whether or not to create a defunctioning stoma after left-sided colorectal surgery are lacking, and a clinical assessment for anastomotic leakage by the operating surgeon has a low predictive value and underestimates leakage risk.[43,59] If it is true that a protective stoma may reduce the risk of clinically relevant leaks[16] and urgent reoperations,[60] probably without reducing leakage rate, nevertheless it does not impact mortality,[59,61] and it can represent an overtreatment for most patients.

To better assess the risks, a **Colon Leakage Score [CLS]** has been developed and tested in left-sided colorectal surgery.[59] Application of this score correctly predicted which patients should undergo primary anastomosis and which should receive a proximal non-functional stoma or a definitive stoma (Table 5.2). Ideally, the CLS would be calculated preoperatively, and only few intraoperative variables (blood loss, duration of operation, distance of anastomosis to anal verge and additional procedures) will be needed to complete it. Its purpose is patient stratification: it could help to define subgroups with high, intermediate and low risk of anastomotic leakage. For the authors, a risk lower than 3% (CLS 11) does not justify a non-functional stoma. However, it can be left to the surgeon and the patient to determine their own cut-off CLS value considering their preferences concerning the risk of anastomotic leakage versus (temporary) stomata. Up to now the CLS is the only scoring system that predicts anastomotic leakage and which could potentially be used to inform intraoperative decision making, but its clinical effectiveness depends upon the validity of the assumption that defunctioning stomas reduce clinically significant anastomotic leaks without increasing other complications or undesirable outcomes.

CLINICAL SIGNS AND SYMPTOMS OF ANASTOMOTIC LEAKAGE

Early sequelae of the leakage of fecal material from the large bowel are:

- Generalized peritonitis
- Localized peritonitis/abscess
- Fecal fistula (feces from the drain, rectovaginal fistula, enterocutaneous fistula)

Postoperative AL may be symptomatic (clinical leaks) or asymptomatic (subclinical leaks). Subclinical leaks are more benign in their natural history than clinical leaks.

TABLE 5.2 Colon Leakage Score[59]

Risk Factors	Score	Risk Factors	Score
Age		Emergency surgery	
< 60	0	No	0
60–69	1	Bleeding	2
70–79	2	Obstruction	3
> 80	4	Perforation	4
Gender		Distance of anastomosis to anal verge (cm)	
Female	0	> 10	0
Male	1	5–10	3
		< 5	6
ASA		Additional procedures	
I	0	No	0
II	1	Yes	1
III	3		
IV	6		
BMI		Blood loss (cc)	
19–24	0	< 500	0
25–30	1	500–1000	1
> 30 / < 19 or weight loss > 5 kg/6 months	3	1001–2000	3
		> 2000	6
Intoxication		Duration of operation (h: min)	
No	0	< 2:00	0
Smoking	1	2:00–2:59	1
Alcohol	1	3:00–3:59	2
Steroids	4	> 4:00	4
Neoadjuvant therapy			
No	0		
Radiotherapy	1		
Chemoradiation	2		

AL typically becomes clinically apparent between the 5th and the 8th postoperative day, but many exceptions exist, and some studies report an even longer interval; up to 45 days postoperatively.[62]

Although it has been described in only a few previous studies, symptomatic anastomotic leakage diagnosed after hospital discharge following low anterior resection of the rectum for cancer is not uncommon: Matthiessen et al. has shown that 7.7% of patients undergoing rectal anterior resection have symptomatic anastomotic leakage diagnosed after hospital discharge.[63]

Clinical signs of systemic inflammatory response syndrome, fever ileus and pain are frequent but have low positive predictive value for symptomatic AL. However, any unexplained sepsis should raise suspicion of an anastomotic leak until proven otherwise. Even cardiac symptoms often masquerade as anastomotic leakage.[64] Rickert et al. found that 16% of patients exhibited cardiac symptoms at the time of diagnosis of AL.[65]

Only a quarter of patients with AL may present with discharge of stool or pus through the drain or the wound, so the first way to recognize an AL is to suspect it. Some clinical signs help us to be alert. Alves et al., in examining a series of non-diverted large bowel resections, stated that, when compared to those without leaks, those patients with AL confirmed by laparotomy are more likely to have: fever (>38°C) on day 2, absence of bowel action till day 4, diarrhea before day 7, >400 ml of fluid in the abdominal drain by day 3, renal failure on day 3, leukocytosis after day 7.[66]

DIAGNOSIS

The prompt diagnosis of AL before its clinical manifestation is of paramount value for the patient, although this may be difficult for the clinician because its signs and symptoms can be subtle.

Correct surgical technique is aimed at creating a well perfused, tension free anastomosis. An anastomotic defect may be detected intraoperatively by a transanal air testing (better using a flexible sigmoidoscope) at the end of the primary left colic or rectal resection. Leaks demonstrated can be repaired and a covering stoma employed. There is no evidence that simply wrapping the anastomosis in omentum decreases the rate or severity of anastomotic leaks.[67]

During the postoperative course, the diagnosis of AL relies on clinical and instrumental examination.

Laboratory Diagnosis

Various biochemical markers have been assessed to suggest the presence of AL.

A prolonged increase in C-reactive protein (CRP) has been found to be an early indicator of anastomotic leakage after low anterior resection in recent studies.[68–70] White blood cell count, which is usually an indicator of inflammation, is increased in only 61% of patients when AL occurs.[65]

Furthermore, intraperitoneal biochemical markers such as the lactate/pyruvate ratio of the citric acid cycle and some cytokines have been reported to be increased in patients before symptomatic leakage becomes clinically apparent, compared with patients without symptomatic leakage.[35] Reduced blood flow induces a switch from aerobic to anaerobic metabolism; the level of lactate rises and pyruvate decreases, resulting in an increased lactate/pyruvate ratio and decreased glucose levels – which may be early signs of symptomatic AL.[35]

Some studies have shown increased levels of the peritoneal cytokines, including TNF-α, IL-1, IL-6, and IL-10, in patients with postoperative complications reflecting the severity of stress caused by abdominal operations.[71] More specifically, peritoneal IL-6, IL-10, and TNF-α levels were significantly higher in patients with AL compared with patients without AL.[71] It has been shown that, during the first postoperative day, the peritoneal concentrations of cytokines reflect the severity of stress caused by abdominal operations. It is suggested that decreasing peritoneal cytokine levels occur in a normal postoperative course, whereas increasing levels indicate an abnormal postoperative course. The overexpression of peritoneal cytokines as a local inflammatory response in response to microbial invasion might be a very early event in the development of AL.[71–75]

The bacterial flora in the human colon is normally a stable ecological environment. After perforation or spillage of the colon, more than 400 different species of bacteria will contaminate the peritoneal cavity,[76,77] but, of the large number of species of bacteria that invade the peritoneum, only a few will survive outside their native intraluminal environment. If an infection results, it will be polymicrobial in nature.[78]

It has long been thought that cleaning the bowel preoperatively reduces the bacterial load. However, while reducing fecal mass, preoperative bowel preparation does not alter the concentration of fecal organisms intraluminally. The first stage of microbial infection is *colonization*, which is defined as the presence of a micro-organism in an internal organ that is normally sterile. Infection is a microbiologically-proven clinical diagnosis of inflammation, local and/or generalized. This includes not only clinical signs, but also the presence of micro-organisms at $\geq 10^5$ CFU/milliliter (CFU = colony forming units) in diagnostic samples obtained from an internal organ, or the isolation of a micro-organism from peritoneal fluid.[57,76,79,80] High bacterial loads of up to 10^8–10^9 aerobic and anaerobic germs per gram of feces can contaminate the pelvic cavity as a consequence of an AL.

Ruiter et al.[76] reported that the composition of the microbial flora present in the abdominal fluid of patients critically ill with abdominal sepsis varied with the location of the perforation. In lower gastrointestinal perforation, the most frequently isolated facultative anaerobic and aerobic organisms were *Escherichia coli*, *Klebsiella* and *Pseudomonas* species. The predominant obligate anaerobes were *Bacteroides*.[76,78] The vast majority of postoperative infectious complications after colorectal surgery are caused by colonic flora. The organisms that predominate in peritonitis are the endotoxin-generating facultative anaerobes, such as *E. coli*, and the obligate anaerobes such as *Bacteroides fragilis*, which are involved in the later phases of the infection, while *E. coli* is responsible for the acute peritonitis phase of infection.[57,76,79,80]

Instrumental Diagnosis

Anastomotic leakage can be diagnosed radiologically (extraluminal contrast/extravasation in computed tomography [CT] scan or contrast enema), endoscopically or surgically during re-laparotomy.

Radiological investigation can be performed by contrast enema or CT scan with rectal contrast, but the value of early postoperative contrast radiology in the assessment of colorectal anastomotic integrity has not yet been proven in asymptomatic patients.[81] Akyol et al. showed that when such a study was conducted on day 7 postoperatively in 233 patients with a left-sided anastomosis, 11 patients who had normal contrast enemas developed a clinically significant AL in the following days.[81] At present, the standard method for diagnosing anastomotic defect is Hydro-CT-scanning using rectal contrast. This will show the leak in up to 90% of patients and can also diagnose abscesses.[66,70] Endoscopy, formerly seen as contraindicated in the early postoperative course, is gradually gaining popularity as it may be more sensitive than the other diagnostic modalities. It must be pointed out that the detection rate of the AL is seriously influenced by the location of the leakage; while distal leaks were identified in more than 90% of cases, proximal leaks were only seen on imaging in about half of the cases and none of the diagnostic modalities has proven to be better than any of the others in proximal anastomosis.[65]

During the course of symptomatic AL, microbiological cultures with antibiogram will guide the specific antibiotic therapy.

TREATMENT AND PROGNOSIS

In 2013 WSES guidelines for the management of intra-abdominal infections, Sartelli et al. stated that:

> "As a general principle, every verified source of infection should be controlled as soon as possible. The level of urgency of treatment is determined by the affected organ(s), the relative speed at which clinical symptoms progress and worsen, and the underlying physiological stability of the patient. The procedure used to treat the infection depends on the anatomical site of infection, the degree of peritoneal inflammation, the generalized septic response, the patient's underlying condition and the available resources of the treatment center."[82]

These concepts are indeed applicable to AL. Currently, the management of AL depends on the experience and the view of the operating surgeon. Most studies differentiate between limited leakage with mild clinical manifestations and large leaks with a severely ill patient. There is some agreement regarding patients with a high risk profile: reoperation can be life-saving,[83–85] but patients should probably not undergo direct repair or a new anastomosis without at least a protective stoma. However, in the literature detailed information on exactly which operation is performed in these cases is scarce. When AL is asymptomatic, a wait and see strategy is usually recommended; in 60% of cases the leakage heals under conservative treatment. However, the surgeon should keep in mind that, in rectal surgery, even an asymptomatic leak leads to poorer function. These patients are likely to have a reduced neorectal capacity, more evacuation problems, and show a trend towards fecal urgency and incontinence when compared to those without a history of AL.[86]

If AL begins with peritonitis and threatening sepsis, emergency re-laparotomy is necessary but the related mortality reaches 12–27%.[15,25,87–89]

In general, AL increases the risk of mortality four to seven-fold[22] and also worsens oncological results.[90] In fact, AL has been associated with a higher local recurrence rate after curative treatment of colorectal malignancies.[2,91–93] Reoperation is associated with surgical injury, which induces a systemic endocrine-metabolic and immune response and may be associated with transient immunocompromised status. This depends on the severity of surgical stress and is mediated by several cytokines. In colon cancer, cytokines may induce the initiation and progression of colitis-associated colonic carcinogenesis and the adhesion of colonic tumor cells to the peritoneum during the development of peritoneal metastasis. On the other hand, in the immediate postoperative period, colonic tumor cells have been noted to be circulating in the blood, bone marrow and lymph nodes. Reoperation may be also associated with other factors such as blood transfusion and delays in the beginning of adjuvant therapy, factors that may influence oncological outcomes.[94] Furthermore, AL has been associated with a 10–100% risk of permanent stoma.[95]

The mean length of hospital stay for patients with AL has been described as being between 36 and 39 days, approximately four times longer than for patients without leaks. When multiple reoperations and stoma creation are necessary to control the leak, health risks and health care costs significantly increase by up to five times those of patients with no leak.[13,25]

Following recovery, restoration of bowel continuity or abdominal wall reconstruction after laparostomy will potentially add morbidity to the patients and certainly costs to the healthcare system.[25,66,96,97]

Finally 1.2–4.2% of colonic anastomoses result in anastomotic strictures.[98] Even if these can usually be treated by simple dilatation, radiological or endoscopic balloon dilatation, surgery sometimes becomes unavoidable. In such cases, resection of the stricture with re-anastomosis is the goal, but the creation of a permanent end colostomy as the only possible solution is a fearsome eventuality.

Maggiori et al. suggest a simple classification of AL into four grades of treatment[99]:

Grade 1: no treatment (anticipating spontaneous healing);
Grade 2: medical treatment (antibiotics);
Grade 3: drainage of abscess (percutaneous or transanal);
Grade 4: abdominal reoperation.

For asymptomatic AL, no treatment seems to be required other than to delay stoma closure until imaging shows that the AL has healed, if stoma is packaged during primary surgery, or to start parenteral nutrition and lengthen hospital stay if there is no stoma. For symptomatic AL with moderate sepsis, antibiotics without radiological or transanal drainage are likely to be sufficient. Transanal drainage is not required if there is no obvious abscess or if an abscess is located in the upper pelvis when percutaneous radiological drainage is feasible. In patients treated by transanal drainage, the drain can be removed on about the seventh day without CT imaging if the patient is well. In patients with severe sepsis, including peritonitis, abdominal reoperation is necessary (with the package of diverting stoma), but the aim should be to preserve anastomosis whenever possible. Hartmann's procedure should be performed only in cases of generalized fecal peritonitis and complete circumferential anastomotic dehiscence or where there is ischemia or necrosis of the bowel. Rickert's study has identified three factors on multivariate analysis which have a significant influence on the surgeon's decision to dismantle the anastomosis: when more than a half of the circumference of the anastomosis has broken down, when generalized peritonitis is present, or if the patient is high risk.[65]

Early classification of patients presenting with peritonitis and intra-abdominal sepsis by means of objective scoring systems is desirable to select those for aggressive surgery and to compare the results of different treatment regimens. However, none of the existing scoring systems has fulfilled all expectations. At present, a combination of the APACHE II

(Acute Physiology And Chronic Health Evaluation) and the MPI (Mannheim Peritonitis Index) provides the best scoring system to fit clinical goals.[100]

Limited Leaks

Limited leaks may be defined as AL with localized peritonitis with or without abscess formation. They have a milder clinical presentation and generally occur later than the larger ones. CT scanning with water-soluble rectal contrast can confirm the size of a leak and the presence of an abscess. When the leak of contrast agent is contained, then management can be conservative with close observation, parenteral nutrition and intravenous antibiotics. If an abscess of reasonable size is diagnosed, this should be drained under radiological guide. Where expertise is unavailable or radiologically-guided drainage fails, a further operation is required. Success rates for CT scan-guided drainage of intra-abdominal abscesses are at least 80%.[101] Transrectal endoscopic drainage may be an alternative option for selected cases of pelvic fluid collection. The technique also allows the visualization of the local cavity, the cleavage of multi-loculated abscesses and saline irrigation if necessary. The use of Transanal Endoscopic Microsurgery (TEM) instrumentation allows safe access to the peritoneal cavity.[102] Ramesh et al. suggest an endoscopic ultrasound-guided transcolonic and transrectal drainage of abdomino-pelvic abscesses, drained by a standard single-step EUS-guided technique with deployment of double-pigtail stents, with or without catheters. The procedure seems to have a high rate of treatment success, defined as resolution of abscess on follow-up computed tomography at two weeks, with symptom improvement.[103]

On the contrary, when the leak becomes evident as a fistula between a rectal anastomosis and the vagina it is unlikely to close spontaneously and a delayed operative solution will be necessary.

Limited leaks enclose also entero-cutaneous fistulas following right colon surgery for Crohn's disease. In the absence of peritonitis (demanding laparotomy and stoma formation), treatment is conservative. The patient is resuscitated, oral intake limited and parenteral nutrition commenced. Stable radiological investigations can be used to determine the course of the fistula and rule out eventual distal obstruction. When the fistula has been fully described, definitive management may be considered. Anti-TNF alpha antibody therapy has a role in treating Crohn's fistulas once infection has been adequately dealt with[104]: it can result in closure of the draining fistula in 36% of patients.[105] Unfortunately this is less likely with distal fistulas. Another consequence of a limited AL is the formation of a presacral abscess. Usually this follows the partial or total mesorectal excision technique, even with a diverting ileostomy, as a large presacral cavity remains that is incompletely filled by the neorectum. Early closure of the anastomotic leak might prevent chronic pelvic sepsis and its adverse sequelae.

Verlaan et al. report on a novel technique for early closure of the anastomotic gap using a combination of endosponge drainage and closure by sutures or a novel endoscopic clip. First, the abscess cavity is examined and rinsed with saline solution (0.9%) using a small-caliber flexible gastroscope. Then a sponge is installed transanally and connected to a low-vacuum suction bottle. The endosponge is changed every 3–4 days. Just before removal, a lidocaine or saline solution is introduced into the sponge to facilitate painless extraction. When the cavity is considered clean and the edema of the neorectum has subsided, the anastomotic defect is re-sutured transanally under general anesthesia. A 5 week endosponge therapy is generally required and even then small fistulas remain. The earlier the endosponge therapy is started, the more likely the cavity will close, because the neorectum is still flexible at that stage.

Large Leaks

When a life-threatening sepsis follows a large AL, the patient needs resuscitation, intravenous antibiotics and emergency re-laparotomy to lavage the contaminated abdominal cavity. In most cases a defunctioning stoma is formed. If a right sided colonic resection has leaked, the bowel can be re-anastomosed with adequate drainage but no stoma, in favorable conditions. In some cases re-anastomosis is not immediately possible and a laparostomy with a second look surgery should be considered. An AL after left colonic resection is traditionally treated with resection of the anastomosis and an end colostomy (Hartmann's procedure), which is often not reversed.[70] The simple local oversewing of the disrupted anastomosis runs the risk of further failure with persistence of the septic focus and high mortality. With increasing numbers of bowel resections being undertaken laparoscopically, the question of whether postoperative complications in such patients could also be tackled laparoscopically arises. While such an approach is certainly feasible, there is still little literature to let us compare it with the open approach. In any case, laparoscopic management of serious anastomotic complications should possibly be undertaken only by surgeons with considerable laparoscopic experience.[16] As there are no comparative studies available which examine the results of all the different therapeutic approaches for anastomotic leakage, the decision on which operation should be performed is up to the individual surgeon. There is, however, some agreement that severely ill septic patients should undergo resection without creation of an anastomosis. The advantages of an end stoma are control of the sepsis, a shorter operation time, and avoidance of the risk of subsequent failure but on the other hand, restoration of intestinal continuity may later again contribute to overall morbidity.

CONCLUSIONS AND FUTURE PROSPECTS

Postoperative cardiovascular and cardiorespiratory complications have replaced AL as the leading cause of postoperative mortality after colectomy in most studies.[5,28] Notwithstanding these findings, AL remains a major challenge for the surgeon because of its early or late consequences; increased intra- and extra-abdominal morbidity,[27] reoperation[25] and longer hospitalization,[4] temporary or definitive stomas, worse functional or oncological outcomes and increased mortality rate.[106]

AL risk factors are malnutrition, sepsis, cardiovascular disease, perioperative blood transfusion, steroid use, smoking and alcohol abuse. Furthermore AL is more common in low anastomoses. The use of covering stomas for low rectal surgery does not reduce leak rates but mitigates the consequences of such leaks. Prompt recognition of this complication

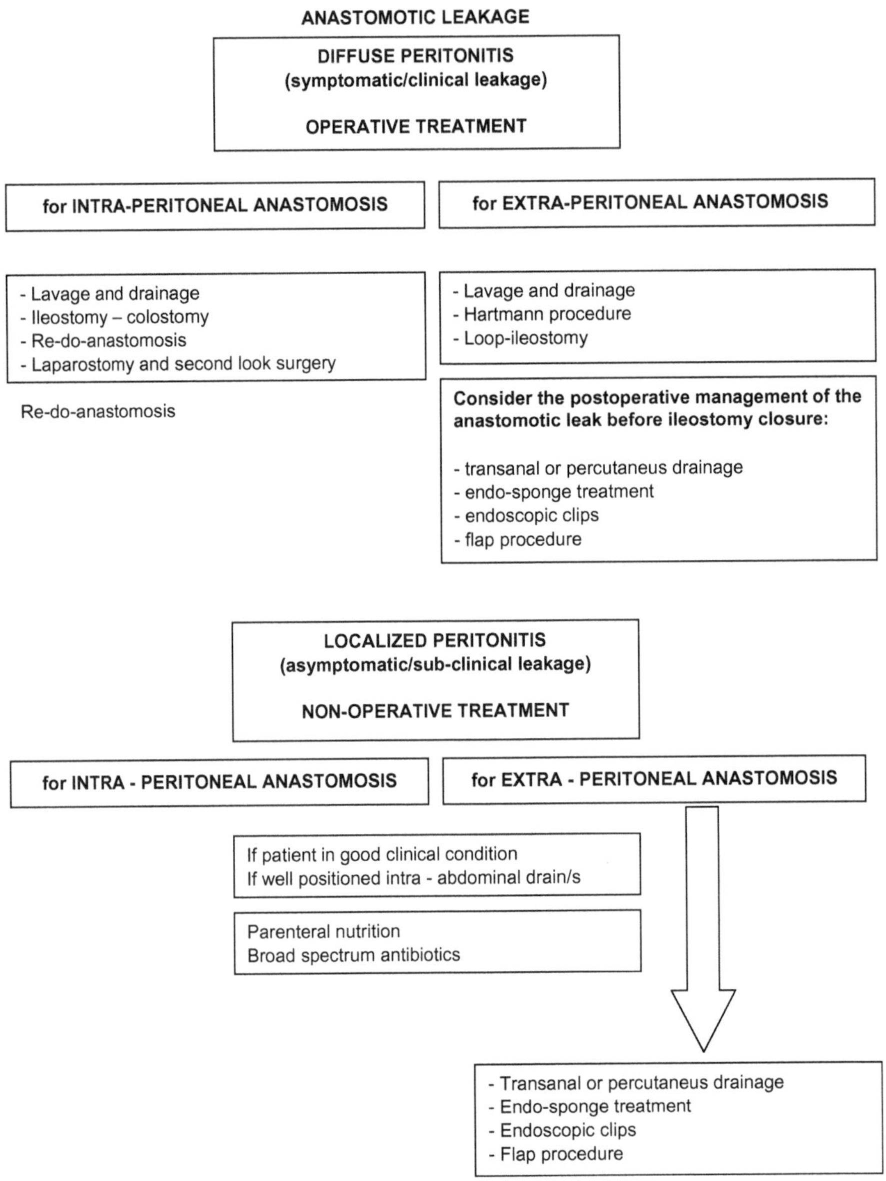

FIGURE 5.1 **Anastomotic leakage treatment algorithm.**

is crucial, but there are no solidly established guidelines for its management. A recent shift towards conservative management has been registered. Leaks of low clinical significance can be reasonably managed conservatively with radiologically guided drainage. However, management of major clinical leaks often requires a laparotomy with peritoneal lavage and stoma formation. Figure 5.1 shows a proposed AL treatment algorithm.

Knowledge of the factors predisposing to AL is the key to its early detection, and improves decision making for surgical treatment time, anticipating postoperative complications, and managing preoperative risk factors. In the future, it might be interesting to study independent risk factors separately for one type of colectomy (e.g., left or right), in benign and malignant disease. Such studies could possibly collect independent risk factors and analyze them by multivariate analysis in order to work out a predictive score for the need to create a protective stoma.

References

1. Akyol AM, McGregor JR, Galloway DJ, George WD. Anastomotic leaks in colorectal cancer surgery: a risk factor for recurrence? *Int J Colorect Dis* 1991;**6**:179–83.
2. Walker KG, Bell SW, Rickard MJ, Mehanna D, Dent OF, Chapuis PH, et al. Anastomotic leakage is predictive of diminished survival after potentially curative resection for colorectal cancer. *Ann Surg* 2004;**240**:255–9.
3. Karanjia ND, Corder AP, Holdsworth PJ, Heald RJ. Risk of peritonitis and fatal septicaemia and the need to defunction the low anastomosis. *Br J Surg* 1991;**78**:196–8.
4. Rullier E, Laurent C, Garrelon JL, Michel P, Saric J, Parneix M. Risk factors for anastomotic leakage after resection of rectal cancer. *Br J Surg* 1998;**85**:355–8.
5. Bruce J, Krukowski ZH, Al-Khairy G, Russell EM, Park KG. Systematic Review of the definition and measurement of anastomotic leak after gastrointestinal surgery. *Br J Surg* 2001;**88**:1157–68.
6. Peel AL, Taylor EW. Proposed definitions for the audit of postoperative infection: a discussion paper. Surgical Infection Study Group. *Ann R Coll Surg Engl* 1991;**73**:385–8.
7. Trencheva K, Morrissey KP, Wells M, Mancuso CA, Lee SW, Sonoda T, et al. Identifying important predictors for anastomotic leak after colon and rectal resection. *Ann Surg* 2013;**257**:108–13.
8. Nasirkhan MU, Abir F, Longo W, Kozol R. Anastomotic disruption after large bowel resection. *World J Gastroenterol* 2006;**12**:2497–504.
9. Boccola MA, Buettner PG, Rozen WM, Siu SK, Stevenson AR, Stitz R, et al. Risk factors and outcomes for anastomotic leakage in colorectal surgery: a single-institution analysis of 1576 patients. *World J Surg* 2011;**35**:186–95.
10. Fouda E, El Nakeeb A, Magdy A, Hammad EA, Othman G, Farid M. Early detection of anastomotic leakage after elective low anterior resection. *J Gastrointest Surg* 2011;**15**:137–44.
11. Milsom JW, de Oliveira Jr O, Trencheva KI, Pandey S, Lee SW, Sonoda T. Long-term outcomes of patients undergoing curative laparoscopic surgery for mid and low rectal cancer. *Dis Colon Rectum* 2009;**52**:1215–22.
12. Vignali A, Fazio VW, Lavery IC, Milsom JW, Church JM, Hull TL, et al. Factors associated with the occurrence of leaks in stapled rectal anastomoses: a review of 1,014 patients. *J Am Coll Surg* 1997;**185**:105–13.
13. Akiyoshi T, Ueno M, Fukunaga Y, Nagayama S, Fujimoto Y, Konishi T, et al. Incidence of and risk factors for anastomotic leakage after laparoscopic anterior resection with intracorporeal rectal transection and double-stapling technique anastomosis for rectal cancer. *Am J Surg* 2011;**202**:259–64.
14. Kockerling F, Rose J, Schneider C, Scheidbach H, Scheuerlein H, Reymond MA, et al. Laparoscopic colorectal anastomosis: risk of postoperative leakage. Results of a multicenter study. Laparoscopic Colorectal Surgery Study Group (LCSSG). *Surg Endosc* 1999;**13**:639–44.
15. Karanjia ND, Corder AP, Bearn P, Heald RJ. Leakage from stapled low anastomosis after total mesorectal excision for carcinoma of the rectum. *Br J Surg* 1994;**81**:1224–6.
16. Chambers WM, Mortensen NJMcC. Postoperative leakage and abscess formation after colorectal surgery. *Best Practice & Research Clinical Gastroenterology* 2004;**18**:865–80.
17. Katz M. *Assumptions of multiple linear regression, multiple logistic regression, and proportional hazard regression in multivariable analysis. A practical guide for clinicians.* New York: Cambridge University Press; 2001:36–59.
18. De Silva Lustosa SA, Matos D, Atallah AN, Castro AA. Stapled versus handsewn methods for colorectal anastomosis surgery: a systematic review of randomized controlled trials. *Sao Paulo Med J* 2002;**120**:132–6.
19. Makela JT, Kiviniemi H, Laitinen S. Risk factors for anastomotic leakage after left-sided colorectal resection with rectal anastomosis. *Dis Colon Rectum* 2003;**46**:653–60.

20. Khan AA, Wheeler JMD, Cunningham C, George B, Kettlewell M, Mortensen NJ. The management and outcome of anastomotic leaks in colorectal surgery. *Colorect Dis* 2007;**10**:587–92.
21. Lipska MA, Bissett IP, Parry BR, Merrie AEH. Anastomotic leakage after low gastrointestinal anastomosis: men are at higher risk. *ANZ J Surg* 2006;**76**:579–85.
22. Veyrie N, Ata T, Muscari F, Msika S, Hay JM, Fingerhut A, et al. French Associations for Surgical Research. Anastomotic leakage after elective right versus left colectomy for cancer: prevalence and independent risk factors. *J Am Coll Surg* 2007;**205**:785–93.
23. Goligher JC, Graham NG, De Dombal FT. Anastomotic dehiscence after anterior resection of rectum and sigmoid. *Br J Surg* 1970;**57**:109–18.
24. Yamamoto T, Allan RN, Keighley MR. Risk factors for intra-abdominal sepsis after surgery in Crohn's disease. *Dis Colon Rectum* 2000;**43**:1141–5.
25. Alves A, Panis Y, Trancart D, Regimbeau JM, Pocard M, Valleur P. Factors associated with clinically significant anastomotic leakage after large bowel resection: multivariate analysis of 707 patients. *World J Surg* 2002;**26**:499–502.
26. Alves A, Panis Y, Mathieu P, Mantion G, Kwiatkowski F, Slim K. Postoperative mortality and morbidity in French patients undergoing colorectal surgery: results of a prospective multicenter study. *Arch Surg* 2005;**140**:278–83 [discussion; 284].
27. Golub R, Golub RW, Cantu Jr R, Stein HD. A multivariate analysis of factors contributing to leakage of intestinal anastomoses. *J Am Coll Surg* 1997;**184**:364–72.
28. Fielding LP, Stewart-Brown S, Blesovsky L, Kearney G. Anastomotic integrity after operations for large-bowel cancer: a multicenter study. *Br Med J* 1980;**281**:411–4.
29. Platell C, Barwood N, Dorfmann G, Makin G. The incidence of anastomotic leaks in patients undergoing colorectal surgery. *Colorectal Dis* 2006;**9**:71–9.
30. Konishi T, Watanabe T, Kishimoto J, Nagawa H. Risk factors for anastomotic leakage after surgery for colorectal cancer: results of prospective surveillance. *J Am Coll Surg* 2006;**202**:439–44.
31. Collee JG, Marr W. Specimen collection, culture containers and media. In: Collee JG, Fraser AG, Marmion BP, Simmons A, editors. *Mackie and MacCartney Practical Medical Microbiology. New York*: Churchill Livingstone; 1996. pp. 95–111.
32. Millan M, García-Granero E, Flor B, García-Botello S, Lledo S. Early prediction of anastomotic leak in colorectal cancer surgery by intramucosal pH. *Dis Colon Rectum* 2006;**49**:595–601.
33. Yeh CY, Changchien CR, Wang JY, Chen JS, Chen HH, Chiang JM, et al. Pelvic drainage and other risk factors for leakage after elective anterior resection in rectal cancer patients: a prospective study of 978 patients. *Ann Surg* 2005;**241**:9–13.
34. Sagar PM, Hartley MN, Macfie J, Mancey-Jones B, Sedman P, May J. Randomized trial of pelvic drainage after rectal resection. *Dis Colon Rectum* 1995;**38**:254–8.
35. Matthiessen P, Strand I, Jansson K, Törnquist C, Andersson M, Rutegård J, et al. Is early detection of anastomotic leakage possible by intraperitoneal microdialysis and intraperitoneal cytokines after anterior resection of the rectum for cancer? *Dis Colon Rectum* 2007;**50**:1918–27.
36. Poon RT, Chu KW, Ho JW, Chan CW, Law WL, Wong J. Prospective evaluation of selective defunctioning stoma for low anterior resection with total mesorectal excision. *World J Surg* 1999;**23**:463–8.
37. Braga M, Gianotti L, Nespoli L, Radaelli G, Di Carlo V. Nutritional approach in malnourished surgical patients: a prospective randomized study. *Arch Surg* 2002;**137**:174–80.
38. Harmon JW, Tang DG, Gordon TA, Bowman HM, Choti MA, Kaufman HS, et al. Hospital volume can serve as a surrogate for surgeon volume for achieving excellent outcomes in colorectal resection. *Ann Surg* 1999;**230**:404–13.
39. Longo WE, Virgo KS, Johnson FE, Oprian CA, Vernava AM, Wade TP, et al. Risk factors for morbidity and mortality after colectomy for colon cancer. *Dis Colon Rectum* 2000;**43**:83–91.
40. Gibbs J, Cull W, Henderson W, Daley J, Hur K, Khuri SF. Preoperative serum albumin level as a predictor of operative mortality and morbidity: results from the National VA Surgical Risk Study. *Arch Surg* 1999;**134**:36–42.
41. Guenaga KF, Matos D, Castro AA, Atallah AN, Wille-Jørgensen P. Mechanical bowel preparation for elective colorectal surgery. *Cochrane Database Syst Rev* 2005: CD001544.
42. Jesus EC, Karliczek A, Matos D, Castro AA, Atallah AN. Prophylactic anastomotic drainage for colorectal surgery. *Cochrane Database Syst Rev* 2004;**18**:CD002100.
43. Karliczek A, Jesus EC, Matos D, Castro AA, Atallah AN, Wiggers T. Drainage or non-drainage in elective colorectal anastomosis: a systematic review and meta-analysis. *Colorectal Dis* 2006;**8**:259–65.

44. Merad F, Yahchouchi E, Hay JM, Fingerhut A, Laborde Y, Langlois-Zantain O. Prophylactic abdominal drainage after elective colonic resection and suprapromontory anastomosis: a multicenter study controlled by randomization. French Associations for Surgical Research. *Arch Surg* 1998;**133**:309–14.
45. Peeters KC, Tollenaar RA, Marijnen CA, Klein Kranenbarg E, Steup WH, Wiggers T. Dutch Colorectal Cancer Group. Risk factors for anastomotic failure after total mesorectal excision of rectal cancer. *Br J Surg* 2005;**92**:211–6.
46. Fazio VW. Sump suction and irrigation of the presacral space. *Dis Colon Rectum* 1978;**21**:401–5.
47. Chapman AE, Levitt MD, Hewett P, Woods R, Sheiner H, Maddern GJ. Laparoscopic-assisted resection of colorectal malignancies. A systematic review. *Ann Surg* 2001;**234**:590–606.
48. Braga M, Vignali A, Gianotti L, Zuliani W, Radaelli G, Guarin P, et al. Laparoscopic versus open colorectal surgery. A randomised trial on short-term outcome. *Ann Surg* 2002;**236**:759–67.
49. Surgical Therapy Study Group. The clinical outcomes of surgical therapy study group. A comparison of laparoscopically assisted and open colectomy for colon cancer. *N Engl J Med* 2004;**350**:2050–9.
50. MacRae HM, McLeod RS. Handsewn vs. stapled anastomoses in colon and rectal surgery: a meta-analysis. *Dis Colon Rectum* 1998;**41**:180–9.
51. Neutzling CB, Lustosa SA, Proenca IM, da Silva EM, Matos D. Stapled versus handsewn methods for colorectal anastomosis surgery. *Cochrane Database Syst Rev* 2012;**15**:CD003144.
52. Rose J, Schneider C, Yildirim C, Geers P, Scheidbach H, Köckerling F. Complications in laparoscopic colorectal surgery: results of a multicentric trial. *Tech Coloproctol* 2004;**8**:S25–8.
53. Prystowsky JB, Bordage G, Feinglass JM. Patient outcomes for segmental colon resection according to surgeon's training, certification, and experience. *Surgery* 2002;**132**:663–72.
54. Hannan EL, Radzyner M, Rubin D, Dougherty J, Brennan MF. The influence of hospital and surgeon volume on in-hospital mortality for colectomy, gastrectomy, and lung lobectomy in patients with cancer. *Surgery* 2002;**131**:6–15.
55. Choi SH, Chung EJ, Whang DY, Lee SS, Jang YS, Kim CW. Alteration of signal-transducing molecules in tumor-infiltrating lymphocytes and peripheral blood T lymphocytes from human colorectal carcinoma patients. *Cancer Immunol Immunother* 1998;**45**:299–305.
56. Pettit S, Irving M. The operative management of fistulating Crohn's disease – experience with100 consecutive cases. *Surg Gynaecol Obstet* 1988;**167**:223–8.
57. Matthiessen P, Hallbo O, Rutegard J, Simert G, Sjödahl R. Defunctioning stoma reduces symptomatic anastomotic leakage after low anterior resection of the rectum for cancer a randomized multicenter trial. *Annals of Surgery* 2007;**246**:207–14.
58. Karanjia ND, Heald RJ. Anterior resection without a defunctioning colostomy: questions of safety. *Br J Surg* 1992;**79**:1109–10.
59. Dekker JW, Liefers GJ, de Mol van Otterloo JC, Putter H, Tollenaar RAEM. Predicting the risk of anastomotic leakage in left-sided colorectal surgery using a colon leakage score. *J Surg Research* 2011;**166**: e27–34.
60. Montedori A, Cirocchi R, Farinella E, Sciannameo F, Abraha I. Covering ileo- or colostomy in anterior resection for rectal carcinoma. *Cochrane Database Syst Rev* 2010: CD006878.
61. Kao LS, Millas SG. Commentary of: Predicting the risk of anastomotic leakage in left-sided colorectal surgery using a Colon Leakage Score. *J Surg Research* 2012;**173**:246–8.
62. Hyman N, Manchester TL, Osler T, et al. Anastomotic leaks after intestinal anastomosis: it's later than you think. *Ann Surg* 2007;**245**:254–8.
63. Matthiessen P, Lindgren R, Hallböök O, Rutegård J, Sjödahl R. Rectal Cancer Trial on Defunctioning Stoma Study Group Symptomatic anastomotic leakage diagnosed after hospital discharge following low anterior resection for rectal cancer. Rectal Cancer Trial on Defunctioning Stoma Study Group. *Colorectal Dis* 2010;**12**: e82–7.
64. Sutton CD, Marshall LJ, Williams N, Williams N, Berry DP, Thomas WM, et al. Colorectal anastomotic leakage often masquerades as a cardiac complication. *Colorect Dis* 2004;**6**:21–2.
65. Rickert A, Willeke F, Kienle P, Post S. Management and outcome of anastomotic leakage after colonic surgery. *Colorectal Dis* 2010;**12**:e216–23.
66. Alves A, Panis Y, Pocard M, Regimbeau JM, Valleur P. Management of anastomotic leakage after nondiverted large bowel resection. *J Am Coll Surg* 1999;**189**:554–9.

67. Merad F, Hay JM, Fingerhut A, Flamant Y, Molkhou JM, Laborde Y. Omentoplasty in the prevention of anastomotic leakage after colonic or rectal resection: a prospective randomised study in 712 patients. French associations for surgical research. *Ann Surg* 1998;**227**:179–86.
68. Platt JJ, Romanathan ML, Crosbie RA, Anderson JH, McKee RF, Horgan PG, Mc Millan DC. C-reactive protein as a predictor of postoperative infective complications after curative resection in patients with colorectal cancer. *Ann Surg Oncol* 2012;**19**:4168–77.
69. Garcia-Granero A, Frasson M, Flor-Lorente B, Blanco F, Puga R, Carratalà A, et al. Procalcitonin and C-reactive protein as early predictors of anastomotic leak in colorectal surgery: a prospective observational study. *Dis Colon Rectum* 2013;**56**:475–83.
70. Matthiessen P, Henriksson M, Hallböök O, Grunditz E, Norén B, Arbman G. Increase of serum C-reactive protein is an early indicator of subsequent symptomatic anastomotic leakage after anterior resection. *Colorectal Dis* 2008;**10**:75–80.
71. Ugrasx B, Girisx M, Erbil Y, et al. Early prediction of anastomotic leakage after colorectal surgery by measuring peritoneal cytokines: Prospective study. *Int J Surg* 2008;**6**:28–35.
72. Herwig R, Glodny B, Kuhle C, Schlüter B, Brinkmann OA, Strasser H, et al. Early identification of peritonitis by peritoneal cytokine measurement. *Dis Colon Rectum* 2002;**45**:514–21.
73. Sherwood ER, Toliver-Kinsky T. Mechanisms of the inflammatory response. *Best Pract Res Clin Anaesthesiol* 2004;**18**:385–405.
74. Scheingraber S, Bauerfeind F, Bohme J, Dralle H. Limits of peritoneal cytokine measurements during abdominal lavage treatment for intra-abdominal sepsis. *Am J Surg* 2001;**181**:301–8.
75. Jansson K, Redler B, Truedsson L, Magnuson A, Matthiessen P, Andersson M, et al. Intraperitoneal cytokine response after major surgery: higher postoperative intraperitoneal versus systemic cytokine levels suggest the gastrointestinal tract as the major source of the postoperative inflammatory reaction. *Am J Surg* 2004;**187**:372–7.
76. Ruiter JDe Weel J, Manusama E, Kingma WP, van der Voort PH. The epidemiology of intra-abdominal flora in critically ill patients with secondary and tertiary abdominal sepsis. *Infection* 2009;**37**:522–7.
77. Brook I. Microbiology and management of abdominal infections. *Dig Dis Sci* 2008;**53**:2585–91.
78. Brook I, Edith HF. Aerobic and anaerobic microbiology in intra-abdominal infections associated with diverticulitis. *J Med Microbiol* 2000;**49**:827–30.
79. Prescott LM, Klein DA, Harley JP. Normal microbiota and non-specific host resistance. In: Prescott LM, Klein DA, Harley JP, editors. *Microbiology*. McGraw Hill; 2005. pp. 673–704.
80. Sarginson RE, Taylor N, van Saene HKF. Glossary of terms and definitions. *Curr Anaesth Crit Care* 2001;**12**:2–5.
81. Akyol AM, McGregor JR, Galloway DJ, George WD. Early postoperative contrast radiology in the assessment of colorectal anastomotic integrity. *Int J Colorectal Dis* 1992;**7**:141–3.
82. Sartelli M, Viale P, Catena F, Ansaloni L, Moore E, Malangoni M, et al. WSES guidelines for management of intra-abdominal infections. *World J Emerg Surg* 2013;**8**:3.
83. Eckmann C, Kujath P, Schiedeck TH, Shekarriz H, Bruch HP. Anastomotic leakage following low anterior resection: results of a standardized diagnostic and therapeutic approach. *Int J Colorectal Dis* 2004;**19**:128–33.
84. Luna-Perez P, Rodriguez-Martinez HA, Chavez-Mercado L. Treatment of anastomotic leakage following low anterior resection for rectal adenocarcinoma. *Rev Invest Clin* 1999;**51**:23.
85. Soeters PB, de Zoete JPJGM, Dejong CHC, Williams NS, Baeten CGMI. Colorectal surgery and anastomotic leakage. *Dig Surg* 2002;**19**:150–5.
86. Nesbakken A, Nygaard K, Lunde OC. Outcome and late functional results after anastomotic leakage following mesorectal excision for rectal cancer. *Br J Surg* 2001;**88**:400–4.
87. Buchs NC, Gervaz P, Secic M, Bucher P, Mugnier-Konrad B, Morel P. Incidence, consequences, and risk factors for anastomotic dehiscence after colorectal surgery: a prospective monocentric study. *Int J Colorectal Dis* 2008;**23**:265–70.
88. Volk A, Kersting S, Held HC, Saeger HD. Risk factors for morbidity and mortality after single-layer continuous suture for ileocolonic anastomosis. *Int J Colorectal Dis* 2011;**26**:321–7.
89. Thornton M, Joshi H, Vimalachandran C, Heath R, Carter P, Gur U, et al. Management and outcome of colorectal anastomotic leaks. *Int J Colorectal Dis* 2011;**26**:313–20.
90. Rondelli F, Reboldi P, Rulli A, Barberini F, Guerrisi A, et al. Loop ileostomy versus loop colostomy for fecal diversion after colorectal or coloanal anastomosis: a meta-analysis. *Int J Colorectal Dis* 2009;**24**:479–88.

91. Law WL, Choi HK, Lee YM, Ho JWC, Leung CS. Anastomotic leakage is associated with poor long-term outcome in patients after curative colorectal resection for malignancy. *J Gastrointest Surg* 2007;**11**:8–15.
92. Bell SW, Walker KG, Rickard MJ. Anastomotic leakage after curative anterior resection results in a higher prevalence of local recurrence. *Br J Surg* 2003;**90**:1261–6.
93. Petersen S, Freitag M, Hellmich G, Ludwig K. Anastomotic leakage: impact on local recurrence and survival in surgery of colorectal cancer. *Int J Colorectal Dis* 1998;**13**:160–3.
94. Khoury W, Lavery IC, Kiran RP. Impact of early reoperation after resection for colorectal cancer on long-term oncological outcomes. *Colorectal Dis* 2011;**14**:e117–23.
95. Brisinda G, Vanella S, Cadeddu F, Civello IM, Brandara F, Nigro C, et al. End-to-end versus end-to-side stapled anastomoses after anterior resection for rectal cancer. *J Surg Oncol* 2009;**99**:75–9.
96. Isbister WH. Unplanned return to the operating room. *Aust NZ J Surg* 1998;**68**:143–6.
97. VanGeldere D, Fa-Si-Oen P, Noach LA, PJGM Rietra, Peterse JL, Boom RPA. Complications after colorectal surgery without mechanical bowel preparation. *J Am Coll Surg* 2002;**194**:40–7.
98. Dinneen MD, Motson RW. Treatment of colonic anastomotic strictures with "through the scope" balloon dilators. *J R Soc Med* 1991;**84**:264–6.
99. Maggiori L, Bretagnol F, Lefevre JH, Ferron M, Vicaut E, Panis Y. Conservative management is associated with a decreased risk of definitive stoma after anastomotic leakage complicating sphincter-saving resection for rectal cancer. *Colorectal Dis* 2011;**13**:632–7.
100. Bosscha K, Reijnders K, Hulstaert PF, Algra A, van der Werken C. Prognostic scoring systems to predict outcome in peritonitis and intra-abdominal sepsis. *Br J Surg* 1997;**84**:1532–4.
101. Schechter S, Eisenstat T, Oliver G, Rubin RJ, Salvati EP. Computerized tomographic scan guided drainage of intra-abdominal abscesses. *Dis Colon Rectum* 1994;**37**:984–8.
102. Martins BC, Marques CF, Nahas CS, Hondo FY, Pollara W, Nahas SC, et al. A novel approach for the treatment of pelvic abscess: transrectal endoscopic drainage facilitated by transanal endoscopic microsurgery access. *Surg Endosc* 2012;**26**:2667–70.
103. Ramesh J, Bang JY, Trevino J, Varadarajulu S. Comparison of outcomes between endoscopic ultrasound-guided transcolonic and transrectal drainage of abdominopelvic abscesses. *J Gastroenterol Hepatol* 2013;**28**:620–5.
104. Present DH, Rutgeerts P, Targan S, Hanauer SB, Mayer L, van Hogezand RA, et al. Infliximab for the treatment of fistulas in patients with Crohn's disease. *N Engl J Med* 1999;**340**:1398–405.
105. Sands BE, Anderson FH, Bernstein CN, Chey WY, Feagan BG, Fedorak RN, et al. Infliximab maintenance therapy for fistulizing Crohn's disease. *N Engl J Med* 2004;**350**:876–85.
106. Isbister WH. Anastomotic leak in colorectal surgery: a single surgeon's experience. *ANZ J Surg* 2001;**71**:508–10.

CHAPTER

6

Acute Appendicitis: An Open Issue. Current Trends in Diagnostic and Therapeutic Options

Michele Pisano, Michela Giulii Capponi and Luca Ansaloni

General Surgery Dept. Papa Giovanni XXIII Hospital, Bergamo, Italy

INTRODUCTION

Acute appendicitis (AA) is a very common disease and represents the most common cause of abdominal emergency. It is usually seen in the second and third decades of life, but can occur at any age.[1] A male preponderance exists (ratio 1.4:1); the overall lifetime risk of developing AA is 8.6% for males and 6.7% for females.[2]

Although the etiology of AA is still poorly understood, the most common hypothesis refers to appendix obstruction followed by impairment of wall appendix barrier and subsequent wall perforation and/or abscess formation.[1] However, some studies suggest that uncomplicated and complicated appendicitis are different entities that require a different treatment.[3,4] This might explain why up to 20% of AA resolves spontaneously, as was observed 125 years ago by Fitz.[5]

Worldwide, the standard of care for AA is appendectomy, which is to date generally considered a routine and safe operation. Appendicitis can be notoriously difficult to diagnose, as its clinical presentation can be subtle, and the rate of unnecessary appendectomies is approximately 15%[6,7]; the routine use of ultrasonography (US) reduces this to 10%, and preoperative computer tomography (CT) further decreases it to 5–10%, but not to zero.[8,9] Furthermore, most European surgeons do not routinely perform imaging studies as they trust the clinical diagnosis.[10]

Despite the fact that surgery represents the gold standard treatment of AA, several attempts have been made to investigate the role of antibiotic treatment. The conservative management of AA was considered a long time ago. In 1959, Coldrey reported on a

Microbiology for Surgical Infections
DOI: http://dx.doi.org/10.1016/B978-0-12-411629-0.00006-4

five-year experience with 471 unselected patients who were non-operatively treated for AA, with low morbidity and mortality rates.[11] In 1977, 425 unselected patients were treated by traditional Chinese medicine and antibiotic treatment and recurrent appendicitis was described in 7/100 patients.[12]

Since then the question has remained open.

CLINICAL SIGNS AND SYMPTOMS

Appendicitis is exceedingly common and its manifestations are diverse, although most cases are relatively easy to recognize. Abdominal pain is the primary complaint. Typically, its onset is gradual in the periumbelical region, then it migrates to the right lower quadrant and becomes constant. Anorexia, nausea, vomiting and fever can be associated. Diarrhea may be present, as may urinary symptoms.

The initial pain represents a referred pain resulting from the visceral innervation of the mid-gut and the localized pain is caused by involvement of the parietal peritoneum after progression of the inflammatory process. It generally localizes at McBurney's point. Focal tenderness is elicited by palpation in the region. Rovsing's sign, psoas muscle irritation or the obturator sign can be occasionally assessed.

This classical presentation can be influenced by the age of the patient and anatomical position of the appendix; for this reason the typical presentation occurs only in one half to two thirds of all patients.

In about 20% of cases, AA is complicated, leading to local or diffuse peritonitis.[13] Perforation should be suspected whenever the patient's temperature exceeds 38°C.

DIAGNOSIS

Laboratory and Clinical Scores

Traditionally AA has been considered a clinical diagnosis; it can mainly be reached through a good history and proper clinical examination.[14,15] However, it still represents a challenge, even for experienced surgeons, as it can present with variable and atypical presentations.

No specific laboratory diagnostic test exists: **leukocytosis** and the elevation of inflammatory markers are supportive, but not diagnostic for AA. However, a meta-analysis showed that when both elevated white blood cell (WBC) count and elevated **C-reactive protein (CRP) level** are present, the positive likelihood ratio for AA increases five-fold.[16]

The lack of a noninvasive, sensitive and specific means of making or confirming diagnosis of AA and the considerable expense and definably morbidity associated with a negative appendectomy has stimulated the development of various scoring systems and algorithms to improve early diagnosis, discriminating AA from nonspecific abdominal pain. Many of them are difficult to apply in an emergency clinical setting, especially in low resources countries.[17,18]

The **Alvarado Score** is the most well known and most performed in validated studies, as it is simple, effective and easy to use.[19] It provides an accurate and consistent tool for ruling out appendicitis and identifying patients at higher risk. The score arose from a retrospective review of patients operated on due to suspicion of AA: Alvarado identified three symptoms (migration of pain in right iliac fossa, anorexia and nausea/vomiting), three physical signs (tenderness, rebound pain and elevation of temperature) and two laboratory findings (leukocytosis and shift of the formula to the left) to be useful in AA diagnosis, and assigned them a number according to their diagnostic weight (Table 6.1). Counting the numbers of each indicator present for the patient we obtain a score: Alvarado found that:

> "A score of 5 or 6 is compatible with the diagnosis of AA, a score of 7 or 8 indicates a probable appendicitis, a score of 9 or 10 indicates a very probable appendicitis."[19]

He also suggested that the score could be used as a guide to decide whether to observe or to operate on the patient with suspected AA; a patient with a score of 7 or more requires surgery.

Despite its relatively accuracy in the extremes of the score range, once again this score does not have 100% diagnostic power/certainty; it has a sensitivity of 71% and a specificity of 68%.[19] Furthermore some limits have been underlined in female patients.[20]

Although the Alvarado Score has several clinical advantages, such as its simplicity, its cheapness and its reproducibility, it has never enjoyed wide application in clinical practice

TABLE 6.1 The Alvarado Scoring System for Acute Appendicitis

			Alvarado Score
Symptoms		Migratory right iliac fossa pain	1
		Nausea/vomiting	1
		Anorexia	1
Signs		Right iliac fossa tenderness	2
		Elevation of temperature	1
		Rebound tenderness right iliac fossa	1
Laboratory		Leukocytosis	2
		Neutrophilic shift to the left (>75%)	1
Total Score			0–10
Sum	0–4	Not likely appendicitis	
	5–6	Equivocal	
	7–8	Probably appendicitis	
	9–10	Highly likely appendicitis	

probably because of its suboptimal discriminatory properties in difficult cases (children, women and the elderly). It was constructed from retrospective data, analyzed by univariate analysis, and uses dichotomous variables, some of which are unspecific and subjective (anorexia, nausea, relocation of pain).

To overcome the drawbacks of the Alvarado Score, Andersson introduced the **Appendicitis Inflammatory Response Score (AIR).**[21] This is constructed from seven indicators with independent prognostic value, which are graded according to the severity of symptoms and signs; laboratory variables are divided into intervals and C-reactive protein has been incorporated, as its discriminating power in the assessment of AA has been demonstrated in several studies.[16,18,22] According to the score, patients are divided into three groups. In Group 1 (score 0–4), patients have a very low probability of suffering from AA (sensitivity and specificity), in Group 2 (score 5–8) patients have a higher probability of having AA (sensibility and specificity) and in Group 3 (score 9–12) patients have a very high probability of suffering from AA (Table 6.2).

The AIR score was developed and validated on a prospective cohort of patients with suspected AA. Its intent is not to make a primary diagnosis of AA, but to discriminate

TABLE 6.2 The Andersson Scoring System for Acute Appendicitis

			AIR Score
Vomiting			1
Pain in right inferior fossa			1
Rebound tenderness or muscular defense		light	1
		medium	2
		strong	3
Body temperature >38°C			0
			1
WBC count		10.0–14.9 × 109/L	1
		15.0 × 109/L	2
Polymorphonuclear leukocytes		70–84%	1
		>85%	2
CRP concentration		10–49 g/L	1
		>50 g/L	2
Total score			0–12
Sum	0–4	Low probability. Outpatient follow up if unaltered general condition.	
	5–8	Indeterminate group. In hospital active observation with re-scoring/imaging or diagnostic laparoscopy according to local traditions.	
	9–12	High probability. Surgical exploration is proposed.	

objectively when there is uncertainty. Moreover, it can be a simple instrument to use for selecting patients for immediate surgery, observation or further investigation. The score can be shared and its results compared, facilitating clinical studies. It can be also used as a decisional support by less experienced surgeons. Finally, it can be repeated during observation in equivocal cases, improving the diagnostic accuracy.[23,24] Thanks to its high discriminating power, the AIR score outperforms the Alvarado Score and works better even among pediatric populations, because its clinical variables are easy to apply in children, who might have problems in identifying concepts like nausea, anorexia or migration of pain.

Instrumental Diagnosis

Radiological tests can be also used to aid the diagnosis of AA.

Advances in **ultrasonography (US)** quality and a novel technique called "graded compression" have improved its sensitivity for identifying AA[25]: steady pressure is applied to the right low quadrant in order to reduce bowel gas and collapse the normal bowel to assist in visualizing the appendix. AA is diagnosed from US when a non-compressible appendix with a diameter of more than 7 mm is visualized. Table 6.3 summarizes the US findings that help to diagnose AA.

The limits of US are operator's ability, obesity, increased intestinal gas and the retrocecal location of the appendix. If the appendix is not visualized, the surgeon cannot correctly exclude AA; the exam is nondiagnostic and has a poor predictive value. The reported sensitivity of US is 78–94% and its specificity reaches 81–94%[26]; it certainly is the most reliable technique in centers which perform large numbers of studies.

Computer tomography is a very effective tool for the diagnosis of AA at all stages. It is not operator dependent, can be interpreted by most radiologists and surgeons and it is not limited by body habitus. Its sensitivity and specificity are 90–100% and 91–99%, respectively; its positive predictive value is 92–98%.[26]

Diagnostic findings for AA in a CT scan include enlarged appendiceal diameter, appendiceal circumferential wall thickening of more than 2 mm with mural enhancement

TABLE 6.3 Ultrasonography Features to Diagnose Acute Appendicitis

Appendix more than 6 mm in anteroposterior diameter
Absence of peristatis
Lack of compressibility
Inflammation of the adjacent mesenteric or omental flat
Appendicolith
Loss of layer structure
Free fluid collection
Loculated collection (abscess)

TABLE 6.4 Computer Tomography Features to Diagnose Acute Appendicitis

Appendiceal diameter more than 6 mm
Appendicolith
Appendiceal wall thickness more than 2 mm
Periappendiceal inflammatory changes (adjacent or periappendiceal fat stranding, fluid collections, phlegmon or abscess formation)
Extraluminal air
Adjacent adenopathy
Adjacent bowel wall thickening
Focal cecal wall thickening

("target sign"), calcified appendicolith (seen in up to 30% of cases), signs of periappendiceal inflammation (Table 6.4).

Tip appendicitis, that is the inflammation limited to the distal tip of the appendix, is subtle to detect and may lead to a false negative interpretation. Ideally the appendix should be followed from its cecal base to its most distal portion.

In general if the appendix is not visible and there are no findings of inflammation in the right low quadrant, a diagnosis of AA can be excluded.[27,28]

Questions arise regarding the effect of CT on decreasing the numbers of unnecessary appendectomies. Although Flum suggests that imaging does not impact on negative appendectomy rates,[29] there are only a few studies and meta-analyses demonstrating the effectiveness of CT in reducing this, and resulting in cost savings.[30–36]

But one issue remains: should a patient with an unequivocal history and physical examination undergo advanced imaging? Anderson observed that the liberal use of CT leads to an increase in the number of cases of detected appendicitis.[4] Another critical concern is the effect of ionizing radiation on the future risk of malignancy. The radiation dose received in abdominal CT is approximately the same as 500 chest radiographs, or 4.5 years of natural background radiation.[28,37] Radiation exposure, as well as costs and availability of resources should be kept in mind and alternatives should be considered before its use. We think that US should be preferred over CT as the first line imaging test, in particular in children and pregnant women, even if there is no definitive evidence to support this concept.

TREATMENT

In the pre-antibiotic era, immediate appendectomy was strongly recommended in every case of suspected AA, because it was traditionally thought that AA progressed from uncomplicated to perforated cases due to fecalith obstruction. However the luminal obstruction theory has never been proven as the main cause of AA. In 1984, Arnbornsson and Benmark hypothesized that enteric bacteria play a predominant role in the pathogenesis of AA, while obstruction represents a contributing factor.[38] This theory has been corroborated by Luckmann's observations, as well as Andersson and Livingston's studies; these authors find that perforated and non-perforated appendicitis have different behaviors and epidemiologic trends, they act as separate disease entities and presumably have different pathophysiologies.[6,13,39]

Considering the infective etiology, AA can be managed with antibiotics, like other intra-abdominal inflammatory processes (e.g., colonic diverticulitis). Until now, antibiotic treatment has been considered a bridge to surgery[40] and laparoscopic appendectomy is still considered the main treatment. However, we can say that the current management of acute appendicitis is based mainly on tradition rather than evidence. Surgery has indeed some advantages; the first and most obvious is that it reduces risk of recurrence with an acceptable percentage of mortality and morbidity, even if stump appendicitis is an under-reported and poorly defined condition.[2,41]

The second benefit is that surgical intervention offers the opportunity to "have a look inside the abdomen". Although appendectomy is not an official diagnostic tool for malignancy, carcinoid is found in 3–7/1000 appendectomies and colon cancer in 0.85% of cases.[42,43] We should probably compare the risk of missing tumors with the 0.07–0.7% mortality risk of appendectomy performed for AA without perforation.

The third advantage arises from the reduced use of antibiotics. While antibiotics are generally stopped after the removal of a normal or phlegmonous appendix, patients treated by antibiotic therapy alone will receive a longer course of drugs. Thus, the increasing risk of antibiotics resistance is theoretically reduced by surgery.

Is there any reason to search for alternative therapeutic strategies to surgery? Is it always possible to substitute the traditional surgical approach to AA with a more conservative one? And how can we distinguish patients that might respond to antibiotics from those who are unlikely to do so?

The comparison of surgery and antibiotics in terms of efficacy in treating AA is intrinsically complex due to the huge disparity of treatments available. Hypothetically, surgery can treat acute appendicitis in 100% of cases with relatively low mortality and morbidity rates.

Mortality rates range from 0.07 to 0.7% and from 0.5 to 2.4% for appendicitis without or with perforation, respectively.[44,45] Although low, mortality associated with emergency appendectomy is age related and seven times higher than the general population matched per age and gender.[44] Furthermore, mortality is not strictly linked to the degree of inflammation/complication, but it is also related to the trauma of anesthesia and surgery.[46,47] Blomqvist published a very interesting epidemiological study of appendectomies performed in Sweden.[44] In this study, appendectomies performed for unspecific abdominal pain and for other diagnosis showed case fatality rates of 1.87% and 9.89%, respectively. For patients operated for suspected AA but discharged with no specific abdominal pain diagnosis, the cause of death was non-abdominal in 70% of cases. A more recent retrospective analysis conducted using data from the US National Inpatient Sample from 1998 to 2007 gave the rate of negative appendectomy as 11.83%, and among these patients mortality was 1.07%.[48] Age and co-morbidities increase the percentages of mortality following appendectomy to 0.1%, 2.6%, 6.8% and 16.4% in patients aged 40, 70, 80 and 90, respectively.[44]

Although some clinical trials have failed to define complications very well,[49–51] the morbidity scenery ranges from superficial wound infections to cardiac complications related to surgery and anesthesia. We divide the complications strictly related to surgery into short term and long term categories. Among the first are wound infections, deep infections, stump disruption and bleeding. Among the second are hernias, small bowel obstructions, stump appendicitis, infertility, and chronic abdominal pain.[52] Considering all factors, the post-appendectomy morbidity rate is around 10–19% for non-perforated AA and up to 30% when perforation occurs.[7,53,54] The perioperative complication rate for removing a normal appendix in patients suspected of having AA is the same as that for intervention in non-perforated AA.[55,56]

Finally, if we examine bowel obstruction due to postoperative peritoneal adhesion formation, almost 3% of patients undergoing appendectomy will be readmitted to hospital over the following 10 years,[57] and 1.5% of patients will require surgery within the following 30 years.[44,58,59] Moreover, extreme age accounts for a greater proportion of the morbidity than the percentage of the population who are treated for the disease. The possible causes for this are:

1) Failure to consider the diagnosis due to the atypical patient population and presentation;
2) The difficulties in collecting a history from many patients in the pediatric and geriatric population;
3) The comorbid conditions encountered in the geriatric population.

For all these reasons emergency appendectomy should be avoided in patients with uncomplicated appendicitis. Several studies, including five randomized trials, have suggested that antibiotics can be the first line of treatment.[49,50,57,58,60]

Historically, the link between appendicitis and surgery has necessarily been challenged by antibiotics in poor surgical scenarios, such as in nine US submarines[61] and in patients treated at sea aboard Soviet ships.[62] In the latter, Gurin, by reviewing the conservative treatment outcomes in 252 patients with AA on vessels of the Kalingrad fishing industry from 1975 to 1987, reported a recovery rate of 84.1% with only the use of antibiotics.[62] Thus, antibiotics offer the opportunity to treat AA when surgical resources are not easily available, such as in developing countries. Furthermore, health systems worldwide continually assess the cost effectiveness of all medical procedures carefully. A significant difference in hospital costs was reported by Hansson et al., with a reduction in expense of 25–50% in the antibiotic group compared to surgery.[58]

The antibiotic approach also offers the opportunity to avoid negative appendectomies, thus allowing a more correct use of health resources. This could also be useful in the context of patients for whom surgery would present a high risk.[63,64] Finally, there is no doubt that in successful cases (approximately 80%), antibiotic treatment can eliminate the mortality and morbidity risks related to surgery.

On the other hand, postoperative AA successfully treated by antibiotics still presents a potential source of recurrent appendicitis. The first year recurrence rate for AA successfully treated by antibiotics ranges between 10 and 15%.[51] Higher percentages were reported in the previous study of Eriksson and in a recent prospective study by Vons (36.8% and 26%, respectively).[49,57] Due to the rate of recurrence, according to the Ansaloni meta-analysis, clinical effectiveness favors surgery.[51] However, the Cochrane review showed that, due to the low to moderate quality of the studies, only minor complications in the two groups can be truly compared.[63]

Surprisingly, a different interpretation of the same studies has been published in the British Medical Journal, and the authors conclude that antibiotics are both effective and safe as primary treatment for patients with uncomplicated AA.[65]

The two main obstacles to the antibiotic treatment of AA are increasing antibiotic resistance and allergies. The frequent use and misuse of antibiotics in the treatment of intra-abdominal infections is leading to an increase of the number of bacteria which are resistant to different lines of antibiotics. The choice of the therapy regimen in each patient should be driven by the necessity to defeat the infection without creating a new resistant clone of microorganisms.[66]

The failure rate of antibiotics within 48 h varies from 5% to 47%, largely because primary defeat has been variously defined by different authors, leaving decisions to personal judgment even in large clinical trials.[51] When it does occur, all authors agree that surgery should be undertaken, and no mortality or significant morbidity is added even if operation is delayed.[65,67]

The Antibiotic Choice

Antimicrobial therapy plays an essential role in the management of intra-abdominal infections, particularly in critically ill patients who necessitate urgent empiric antibiotic therapy. An insufficient or otherwise inadequate antimicrobial regimen is one of the variables most strongly related with unfavorable outcomes.[68,69] Many different antibiotic regimens have been used to treat AA and many different studies have been undertaken to compare different antibiotic regimens in the treatment of localized or diffuse peritonitis.[70]

Definitive recommendations regarding the optimal antibiotic treatment for AA have not yet been established, but several types of antibiotics and their combinations have been proposed.[49,50] The possibility of treating AA and intra-abdominal infections in general effectively with a mono-therapy has made triple regimen antibiotic treatment obsolete.[64,71] Moreover, the possibility of administering oral antibiotics, which are absorbed so efficiently through the gastrointestinal tract, would allow many serious infections (acute diverticulitis and AA with phlegmon or abscess) to be treated on an outpatient basis with an efficacy equal to intravenous regimens.[71–73] Generally, in the majority of treatment protocols, antibiotics are administered intravenously for 24–48 h. Patients are then discharged with oral antibiotics for approximately 7–10 days.[49–51] In any case we would recommend the following recommendations by the Surgical Infection Society and the Infectious Diseases Society of America on intra-abdominal infections every time a conservative approach to AA is preferred[74]:

- Antibiotic therapy should be initiated once a patient receives a diagnosis of an intra-abdominal infection, or once such an infection is considered likely (Level of Evidence A-III).
- Antimicrobial therapy should be administered to all patients who receive a diagnosis of appendicitis (A-II).

- For those patients with intra-abdominal infection of mild-to-moderate severity, including acute diverticulitis and various forms of appendicitis, who will not undergo a source protocol procedure, regimens listed for treatment of mild-to-moderate severity infection are recommended, with a possibility of early oral therapy (B-III).
- Appropriate antimicrobial therapy includes an agent effective against facultative and aerobic gram-negative organisms and anaerobic organisms (A-I).
- For adults, antimicrobial therapy should be provided for a minimum of three days, until clinical symptoms and signs of infection resolve (B-III).
- Non-operative management of selected patients with acute, non-perforated appendicitis can be considered if there is a marked improvement in the patient's condition prior to operation (B-II).
- Non-operative management may also be considered as a part of a specific approach for patients who are hospitalized for 48 hours and show a substantial clinical improvement in clinical symptoms and signs within 24 hours while receiving antimicrobial therapy (A-II). The microbiology of patients managed non-operatively is likely similar to that of patients managed operatively, and regimens recommended for patients with complicated intra-abdominal infection are recommended.[75]
- Antibiotics used for the empirical treatment of community-acquired intra-abdominal infection should be active against enteric gram-negative aerobic and facultative bacilli and enteric gram-positive streptococci (A-I).
- Coverage for obligate anaerobic bacilli should be provided for distal small bowel, appendiceal and colon derived infection (A-I).
- Ampicillin-sulbactam is not required for use because of high rates of resistance to this agent among community-acquired *Escherichia coli* (B-II).

CONCLUSIONS AND FUTURE PROSPECTS

The surgical treatment of AA can be universally considered a milestone in modern medicine. Nevertheless, the literature has shown an unexpected worldwide interest in reconsidering this dogma as demonstrated by five randomized controlled trials in the last fifteen years and four meta-analysis presented in the last two years.[49–51,57,58,60,63,65,76]

The relationship between surgery and antibiotic therapy for the treatment of AA is still a challenging issue. Should we compare the rate of early failure in the antibiotic arm with the rate of negative appendicitis? Or should we compare the recurrence rate of antibiotic-treated appendicitis within 30 days and the long-term morbidity of surgery? Furthermore, how should we consider the tumors occasionally discovered in the appendicitis specimen? And finally, how should we analyze the risk of antibiotic resistance?

The comparison between antibiotics and surgery needs a homogenous and more objective patient selection and clear guidelines. Some advantages and some disadvantages are intrinsic in the treatment choice and cannot be compared.

In order to reduce the rate of negative appendicitis and misdiagnosis, there is a need to improve the power of our diagnostic tools. A clinical score can be used as the basis for a structured algorithm in the management of patients, including selective use of imaging depending on the probability of appendicitis. It could safely avoid hospitalization and unneeded investigations in patients in whom AA is unlikely.

US could represent an extension, but not a substitution, to a clinical score. CT scan should be used only in difficult cases to help pinpoint diagnosis in dedicated centers because of its drawbacks (radiation exposure, costs, lack of availability).

Appendectomy still remains the standard choice for AA, but it is considerd that this condition may be managed conservatively with antibiotics as a bridge to definitive surgery. Current evidence does not support the sole use of antibiotics as an alternative treatment modality to appendectomy in the management of AA.[77] However, many studies have demonstrated that uncomplicated appendicitis in adults can be safely and successfully treated by antibiotics in an in-hospital setting. On the other hand, even if conceivable, antibiotic treatment cannot be proposed as clinical standard practice in an outpatient setting. Obviously, if conservative treatment fails surgery is mandatory.

Hopefully, ongoing[78] or future multicenter, prospective, randomized studies comparing antibiotic therapy and surgery will provide a better insight into the current management of AA, and will help surgeons to select those patients for conservative treatment, if an evidence-based benefit from this approach can be definitely demonstrated.

References

1. Addiss DG, Shaffer N, Fowler BS, Tauxe RV. The epidemiology of appendicitis and appendectomy in the United States. *Am J Epidemiol* 1990;**132**:910–25.
2. Liang NK, Lo HG, Marks JL. Stump appendicitis: a comprehensive review of literature. *Am Surg* 2006;**72**: 162–6.
3. Mason RJ. Surgery for appendicitis: is it necessary? *Surg Infect (Larchmt)* 2008;**9**:481–8.
4. Andersson RE. The natural history and traditional management of appendicitis revisited: spontaneous resolution and predominance of prehospital perforations imply that a correct diagnosis is more important than an early diagnosis. *World J Surg* 2007;**31**:86–92.
5. Fitz RH. Perforating inflammation of the vermiform appendix. *Am J Med Sci* 1886;**92**:321–46.
6. Andersson RE, Hugander A, Thulin AJ. Diagnostic accuracy and perforation rate in appendicitis: association with age and sex of the patient and with appendicectomy rate. *Eur J Surg* 1992;**158**:37–47.
7. Hale DA, Molloy M, Pearl RH, Schutt DC, Jaques DP. Appendectomy: a contemporary appraisal. *Ann Surg* 1997;**225**:252–61.
8. Cuschieri J, Florence M, Flum DR, Jurkovich GJ, Lin P, Steele SR, et al. Negative appendectomy and imaging accuracy in the Washington state surgical care and outcomes assessment program. *Ann Surg* 2008;**248**:557–63.
9. Wagner PL, Eachempati SR, Soe K, Pieracci FM, Shou J, Barie PS. Defining the current negative appendectomy rate: for whom is preoperative computed tomography making an impact? *Surgery* 2008;**144**:276–82.
10. Poortman P, Oostvogel Hj, de Lange-de Klerk ES, Cuesta MA, Hamming JF. The use of imaging in the case of suspected acute appendicitis: option of Dutch surgeons. *Ned Tijdschr Geneeskd* 2009;**153**:B376.
11. Coldrey E. Five years of conservative treatment of acute appendicitis. *J Int Coll Surg* 1959;**32**:255–61.
12. Anonymous. Combined traditional Chinese and western medicine in acute appendicitis. *Chin Med J* 1977;**3**:266–9.
13. Livingston EH, Woodward WA, Sarosi GA, Haley RW. Disconnect between incidence of nonperforated and perforated appendicitis: implications for pathophysiology and management. *Ann Surg* 2007;**245**:886–92.
14. Soomro AG, Siddiqui FG, Abro S, Shaikh. NA, Memon AS. Diagnostic accuracy of Alvarado scoring system in acute appendicitis. *J Liaquat Univ Med Health Sci* 2008;**7**:93–6.
15. Khan I, Rehman AU. Application of Alvarado scoring system in diagnosis of acute appendicitis. *J Ayub Med Coll Abbotad* 2005;**17**:41–4.
16. Andersson RE. Meta-analysis of the clinical and laboratory diagnosis of appendicitis. *Br J Surg* 2004;**91**:28–37.

17. Teicher I, Landa B, Cohen M, Kabnick LS, Wise L. Scoring system to aid in diagnosis of appendicitis. *Ann Surg* 1983;**198**:753–9.
18. Graham DF. Computed-aided prediction of gangrenous and perforating appendicitis. *Br Med J* 1977;**2**:1375–7.
19. Alvarado A. A practical score for early diagnosis of acute appendicitis. *Ann Emerg Med* 1986;**15**:557–64.
20. Memon ZA, Irfan S, Fatima K, Iqbal MS, Sami W. Acute appendicitis: diagnostic accuracy of the Alvarado scoring system. *Asian J Surg* 2013;**36**:144–9.
21. Andersson M, Andersson RE. The appendicitis inflammatory response score: a tool for the diagnosis of acute appendicitis that outperforms the Alvarado score. *World J Surg* 2008;**32**:1843–9.
22. Andersson RE, Hugander AP, Ghazi SH, Ravn H, Offenbartl SK, Nyström PO, et al. Diagnostic value of disease history, clinical presentation, and inflammatory parameters of appendicitis. *World J Surg* 1999;**23**:133–40.
23. Andersson RE, Hugander A, Ravn H, Offenbartl K, Ghazi SH, Nystrom PO, et al. Repeated clinical and laboratory examinations in patients with an equivocal diagnosis of appendicitis. *World J Surg* 2000;**24**:479–85.
24. Graff L, Radford MJ, Werne C. Probability of appendicitis before and after observation. *Ann Emerg Med* 1991;**20**:503–7.
25. Prystowsky JB, Pugh CM, Nagle AP. Current problems in surgery. Appendicitis. *Curr Probl Surg* 2005;**42**:688–742.
26. Cole MA, Maldonado N. Evidence-based management of suspected appendicitis in the emergency department. *Emerg Med Pract* 2011;**13**:1–29.
27. Ganguli S, Raptopoulos V, Komlos F, et al. Right lower quadrant pain: value of the nonvisualized appendix in patients at multidetector CT. *Radiology* 2006;**241**:175–80.
28. Rosendahl K, Aukland SM, Fosse K. Imaging strategies in children with suspected appendicitis. *Eur Radiol* 2004;**14**:L138–45.
29. Flum DR, McClure TD, Morris A, Koepsell T. Misdiagnosis of appendicitis and the use of diagnostic imaging. *J Am Coll Surg* 2005;**201**:933–9.
30. Rao PM, Rhea JT, Novelline RA, Mostafavi AA, McCabe CJ. Effect of computed tomography of the appendix on treatment of patients and use of hospital resources. *N Engl J Med* 1998;**338**:141–6.
31. Wilson EB, Cole JC, Nipper ML, Cooney DR, Smith RW. Computed tomography and ultrasonography in the diagnosis of appendicitis: when are they indicated? *Arch Surg* 2001;**136**:670–5.
32. Kim K, Rhee JE, Lee CC, Kim KS, Shin JH, Kwak MJ, et al. Impact of helical computed tomography in clinically evident appendicitis. *Emerg Med J* 2008;**25**:477–81.
33. Krajewski S, Brown J, Phang PT, Raval M, Brown CJ. Impact of computed tomography of the abdomen on clinical outcomes in patients with acute right lower quadrant pain: a meta-analysis. *Can J Surg* 2011;**54**:43–53.
34. Santos DA, Manunga Jr J, Hohman D, Avik E, Taylor EW. How often does computed tomography change the management of acute appendicitis? *Am Surg* 2009;**75**:918–21.
35. Balthazar EJ, Rofsky NM, Zucker R. Appendicitis: the impact of computed tomography imaging on negative appendectomy and perforation rates. *Am J Gastroenterol* 1998;**93**:768–71.
36. Rao PM, Rhea JT, Rattner DW, Venus LG, Novelline RA. Introduction of appendiceal CT: impact on negative appendectomy and appendiceal perforation rates. *Ann Surg* 1999;**229**:344–9.
37. McCollough CH, Schueler BA, Atwell TD, Braun NN, Regner DM, Brown DL, et al. Radiation exposure and pregnancy: when should we be concerned? *Radiographics* 2007;**27**:909–18.
38. Armbjornssorv E, Bengmark S. Role of obstruction in the pathogenesis of acute appendicitis. *Am J Surg* 1984;**147**:390–2.
39. Luckmann R. Incidence and case fatality rates for acute appendicitis in California. A population-based study of the effect of age. *Am J Epidemiol* 1989;**129**:905–18.
40. Stahalfeld K, Hower J, Homitsky S, Madden J. Is acute appendicitis a surgical emergency? *Am Surg* 2007;**73**:626–9.
41. Subramanian A, Liang MK. A 60-year literature review of stump appendicitis: the need for a critical view. *Am J Surg* 2012;**203**:503–7.
42. Roggo A, Wood WC, Ottinger LW. Carcinoid tumors of the appendix. *Ann Surg* 1993;**217**:385–90.
43. Lai HW, Loong CC, Tai LC, Wu CW, Lui WY. Incidence and odds ratio of appendicitis as first manifestation of colon cancer: a retrospective analysis of 1873 patients. *J Gastroenterol Hepatol* 2006;**21**:1693–6.

44. Blomqvist P, Ljung H, Nyren O, Ekbom A. Appendectomy in Sweden 1989–1993 assessed by the inpatient registry. *J Clin Epidemiol* 1998;**51**:859–65.
45. Margenthaler JA, Longo WE, Virgo KS, Johnson FE, Oprian CA, Henderson WG, et al. Risk factors for adverse outcomes after the surgical treatment of appendicitis in adults. *Ann Surg* 2003;**238**:59–66.
46. Wen SW, Naylor CD. Diagnostic accuracy and short-term surgical outcomes in cases of suspected acute appendicitis. *Can Med Assoc J* 1995;**152**:1617–26.
47. McDonald PJ, Royle GT, Taylor I, Karran SJ. Mortality in a university surgical unit: what is an avoidable death? *J R Soc Med* 1991;**84**:213–6.
48. Seetahal SA, Bolorunduro OB, Sookdeo TC, Oyetunji TA, Greene WR, Frederick W, et al. Negative appendectomy: a 10-year review of a nationally representative sample. *Am J Surg* 2011;**201**:433–7.
49. Eriksson S, Granstrom L. Randomized controlled trial of appendectomy versus antibiotic therapy for acute appendicitis. *Br J Surg* 1995;**82**:166–9.
50. Styrud J, Eriksson S, Nilsson I, Ahlberg G, Haapaniemi S, Neovius G, et al. Appendectomy versus antibiotic treatment in acute appendicitis. A prospective multicenter randomized controlled trial. *World J Surg* 2006;**30**:1033–7.
51. Ansaloni L, Catena F, Coccolini F, Ercolani G, Gazzotti F, Pasqualini E, et al. Surgery versus conservative antibiotic treatment in acute appendicitis. Systematic review and meta-analysis of randomized controlled trials. *Dig Surg* 2011;**28**:210–21.
52. Ditzel M, van Ginhoven M, van der Wal JBC, Hop W, Coene PP, Lange JF, et al. What patients and surgeons should know about the consequences of appendectomy for acute appendicitis after long-term follow-up: factors influencing the incidence of chronic abdominal complaints. *J Gastrointest Surg* 2013;**17**:1471–6.
53. Colson M, Skinner KA, Dunnington G. High negative appendectomy rates are no longer acceptable. *Am J Surg* 1997;**174**:723–7.
54. Guller U, Jain N, Peterson ED, Muhlbaier LH, Eubanks S, Pietrobon R. Laparoscopic appendectomy in the elderly. *Surgery* 2004;**135**:479–88.
55. Bijnen CL, Van den Broek WT, Bijnen AB, De Ruiter P, Gouma DJ. Implications of removing a normal appendix. *Dig Surg* 2003;**20**:215–21.
56. Flum DR, Koepsell T. The clinical and economic correlates of misdiagnosed appendicitis: nationwide analysis. *Arch Surg* 2002;**137**:799–804.
57. Vons C, Barry C, Maitre S, Pautrat K, Leconte M, Costaglioli B, et al. Amoxicillin plus clavulanic acid versus appendectomy for treatment of acute uncomplicated appendicitis: an open-label, non-inferiority, randomized controlled trial. *Lancet* 2011;**7**:1573–9.
58. Hansson J, Körner U, Khorram-Manesh A, Solberg A, Lundholm K. Randomized clinical trial of antibiotic therapy versus appendectomy as primary treatment of acute appendicitis in unselected patients. *Br J Surg* 2009;**96**:473–81.
59. Andersson RE. Small bowel obstruction after appendicectomy. *Br J Surg* 2001;**88**:1387–91.
60. Malik AA, Bari SU. Conservative management of acute appendicitis. *J Gastrointest Surg* 2009;**13**:966–70.
61. Adams ML. The medical management of acute appendicitis in a nonsurgical environment: a retrospective case review. *Milit Med* 1990;**155**:345–7.
62. Gurin NN, Slobodchuk IUS, Gavrilov IUF. The efficacy of the conservative treatment of patients with acute appendicitis on board ships at sea. *Vestn Khir* 1992;**148**:144–50.
63. Wilms IM, de Hoog DE, de Visser DC, Janzing HM. Appendectomy versus antibiotic treatment for acute appendicitis. *Cochrane Database Syst Rev* 2011;**11**:CD008359.
64. Sakorafas GH, Mastoraki A, Lappas C, Sampanis D, Danias N, Smyrniotis V. Conservative treatment of acute appendicitis: heresy or an effective and acceptable alternative to surgery? *Eur J Gastroenterol Hepatol* 2011;**23**:121–7.
65. Varadhan KK, Neal KR, Lobo DN. Safety and efficacy of antibiotics compared with appendectomy for treatment of uncomplicated acute appendicitis: meta-analysis of randomized controlled trials. *BMJ* 2012;**344**:e2156.
66. Sartelli M, Catena F, Coccolini F, Pinna AD. Antimicrobial management of intra-abdominal infections: Literature's guidelines. *World J Gastroenterol* 2012;**18**:865–71.
67. Bowers WF, Hughes CW, Bonilla KB. The treatment of acute appendicitis under suboptimal conditions. *US Armed Forces Med J* 1958;**9**:1545–57.
68. Paul M, Shani V, Muchtar E, Kariv G, Robenshtok E, Leibovici L. Systematic review and meta-analysis of the efficacy of appropriate empiric antibiotic therapy for sepsis. *Antimicrob Agents Chemother* 2010;**54**:4851–63.

69. Dellinger RP, Levy MM, Carlet JM, Bion J, Parker MM, Jaeschke R, et al. Surviving sepsis campaign: international guidelines for management of severe sepsis and septic shock. *Crit Care Med* 2008;**36**:296–327.
70. Coccolini F, Catena F, Ansaloni L, Ercolani G, Di Saverio S, Gazzotti F, et al. A prospective, multi centre, randomized clinical study to compare the efficacy and safety of Ertapenem for three days versus Ampicillin-Sulbactam for three days in the treatment of localized community acquired intra-abdominal infection. (T.E.A. Study: Three days Ertapenem vs. three days Ampicillin-sulbactam). *BMC Gastroenterol* 2011;**11**:42.
71. Powell LL, Wilson SE. The role of b-lactam antimicrobials as single agents in treatment of intra-abdominal infection. *Surg Infect (Larchmt)* 2000;**1**:57–63.
72. Oliak D, Yamini D, Ydani VM, Lewis RJ, Arnell T, Vargas H, et al. Initial nonoperative management for periappendiceal abscess. *Dis Colon Rectum* 2001;**44**:936–41.
73. Brown CV, Abrishami M, Muller M, Velmahos GC. Appendiceal abscess: immediate operation or percutaneous drainage? *Am Surg* 2003;**69**:829–32.
74. Solomkin JS, Mazuski JE, Bradley JS, Rodvold KA, Goldstein EJ, Baron EJ, et al. Diagnosis and management of complicated intra-abdominal infection in adults and children: guidelines by the surgical infection society and the infectious diseases society of America. *Clin Infect Dis* 2010;**50**:133–64.
75. Liu K, Fogg L. Use of antibiotics alone for treatment of uncomplicated acute appendicitis: a systematic review and meta-analysis. *Surgery* 2011;**150**:673–83.
76. Brook I, Frazier EH. Aerobic and anaerobic microbiology in intra-abdominal infections associated with diverticulitis. *J Med Microbiol* 2000;**49**:827–30.
77. Fitzmaurice GJ, McWilliams B, Hurreiz H, Epanomeritakis E. Antibiotics versus appendectomy in the management of acute appendicitis: a review of the current evidence. *Can J Surg* 2011;**54**:304–7.
78. Pisano M, Ansaloni L. The ASAA-Study. Antibiotics vs. surgery in acute appendicitis: an intention to treat prospective randomized controlled study. *ClinicalTrials.gov Identifier*: NCT01421901.

CHAPTER

7

Assessment of Severity and Prognosis in Surgical Patients with Secondary Peritonitis

Rodolfo Leonel Bracho-Riquelme

Instituto de Investigación Científica de la Universidad Juárez del Estado de Durango, Durango, México

INTRODUCTION

Intra-abdominal infection (IAI) represents an important cause of complication and death among critical patients. Mortality rates range from 20 to 40%, in spite of numerous advances in antibiotic therapy and intensive care.[1] Secondary peritonitis is an acute peritoneal infection resulting from the loss of integrity of the gastrointestinal tract or from infected viscera. It is the most common form of complicated IAI treated by surgeons.[2] The frequency of occurrence of these conditions argues for a system of evaluation.

An on-demand strategy for relaparotomy or simply the need for outcome information are common situations in which an abdominal infection scoring system may be of use for surgeons.[3–4]

Using various measurement criteria, a severity of illness index determines the levels of severity of illness within a diagnostic group.[5] These scales are necessary for prognostic stratification of patients for prospective clinical trials in clinical research.[6] Unfortunately, to date the available scoring systems used to study patients with IAI have been unable to reconcile the needs of the individual surgeon and the demands of clinical researchers. In this chapter, a brief summary of the evolution of the scores used for IAI is given. The scores most frequently registered in MEDLINE and their historic forerunners are described. A succinct account of each score is given, underlining its aim, the medical scenarios and conditions for which it was devised, and the manner in which the variables of interest were selected.

Microbiology for Surgical Infections
DOI: http://dx.doi.org/10.1016/B978-0-12-411629-0.00007-6

The idea of the chapter is to provide the reader with the necessary information to choose the scoring criteria required for a particular situation, and to serve as a basis for further discussion on this topic, which is always in development. Comments on some of the conflicting results found in the literature will also be made.

THE PROGRESSION OF SEVERITY SCORES

In 1981 Knaus et al.[7] proposed the Acute Physiology and Chronic Health Evaluation (APACHE) system to classify patients treated in intensive care units (ICUs). Even though a severity of illness index is a scale that assesses the probability of mortality, APACHE is aimed at classifying groups of patients, not individuals, to control for case mix, compare outcomes, evaluate new therapies, and study the utilization of ICUs. It is composed of two parts: a physiology score and a preadmission health evaluation. In the first portion (Table 7.1), the clinical record is surveyed for abnormalities among 34 possible physiological measurements, recorded in the first 32 h of admission. Each parameter is evaluated within a range from 0 to 4. Preadmission health status is assigned by reviewing the patient's medical history over the three to six months before admission. On this basis, the patient is placed into one of four chronic health categories (A to D): A, if the patient was in prior good health; B, if he has mild to moderate limitations due to a chronic medical problem; C, if he has a chronic disease producing serious but no incapacitating restriction of activity; and D, severe restriction of activity due to disease. The patient's complete APACHE classification is the sum of the physiological measurements and the letter reflecting the chronic health evaluation.

Elebute and Stoner[8] developed a system for grading the severity of sepsis (SS), which was intended to be simple, in order to classify patients within the setting of a district hospital. It scored the attributes of sepsis under four headings: local effects of infection, pyrexia, secondary effects of sepsis and laboratory data (Table 7.2). Pyrexia and laboratory data have a numerical scale from 0 to 3 (mild to very severe sepsis). After an initial trial, it became obvious that the local effects of tissue infection were the major determinant of the aggregate score and were given more weight (0, 2, 4 and 6). Finally, the secondary effects could not be graded, and fell at or close to the "very severe" end of the scale. They were therefore treated as existence criteria, being given only one score if present. Of the 15 patients studied with SS, not all had IAI.

That same year, Stevens[9] devised a scoring system for surgeons, to quantify the severity of a septic process (SSS), based on the state of deterioration of functions of seven organ systems: lung, kidney, coagulation, cardiovascular, liver, gastrointestinal tract and neurological (Table 7.3). Surgical sepsis was defined as a source of infection improved by drainage or debridement. Since the failure of two organ systems, in sepsis, is a more serious threat to the patient than that expected from the sum of individual organ failure risks, the values given to each of the three organs with the most severe dysfunction were squared and then summed to obtain the final rating. The authors established the source of infection in the 30 patients studied at operation or autopsy. Using the working definitions, and lacking further information, it must be inferred that the source of infection in these cases was not exclusively IAI.

TABLE 7.1 Acute Physiology and Chronic Health Evaluation: Acute Physiology Score.

Points	+4	+3	+2	+1	0	+1	+2	+3	+4
Cardiovascular:									
Heart rate ventricular response	180 or >	141–179	111–140		70–110		56–69	41–55	40 or less
Mean blood pressure (mmHg)	160 or >	131–159	111–130		70–110		51–69		50 or less
R atrial pressure/CVP (mmHg)			26 or >	16–25	1–15	<1			
CPK-MB or ECG evidence of acute MI	Yes				No				
ECG-arrhythmias		Atrial arrhythmias + hemodynamic instability	Atrial arrhythmias alone					>6 PVCs/min	Ventricular tachycardia or fibrillation
Lactate mEq/L (serum)	>8	3.5–8			0–3.4				
PH (blood)	7.7 or >	7.6–7.69		7.51–7.59	7.33–7.5		7.25–7.32	7.15–7.24	<7.15
Respiratory:									
Respiratory rate total nonventilated	50 or >	35–49	26–34		12–25	10–11	7–9		6 or less
$P(A\text{-}a)O_2$ (100%) or [a]	>500	351–499		200–350	<200				
P_aCO_2	70 or >	61–69	50–60		30–49		25–29	20–24	<20

(Continued)

TABLE 7.1 (Continued)

Points	+4	+3	+2	+1	0	+1	+2	+3	+4
Renal:									
Urine output/day			5 L or >	3501–4999 ml	700–3500 ml		480–699 ml (20–29 ml/h)	120–479 ml (5–20 ml/h)	<120 ml/day (5 ml/h)
Serum BUN	>150	101–150	81–100	21–80	10–20		<10		
Serum creatinine	>7	3.6–7	2.1–3.5	1.6–2	0.6–1.5	<0.6			
Gastrointestinal:									
Serum amylase international units	2000 or >	500–1999			500 or >				
Serum albumin	>8				3.5–8	2.5–3.4	<2.5		
Bilirubin (total)		15 or >		5.1–14.9	0–5				
ALKP-alkaline phosphatase (serum) international units				>160	0–160				
SGOT			1500 or >	101–1499	0–100				
Anergy (skin test)[b]	Total		Relative		None				
Hematological:									
Hematocrit	>60		51–60	47–50	30–46		20–29		<20
WBC-white blood count (total)	>40,000		20,001–40,000	15,001–20,000	3,000–15,000		1,000–2,999		<1,000

(Continued)

TABLE 7.1 (Continued)

Points	+4	+3	+2	+1	0	+1	+2	+3	+4
Platelets			>1,000,000	600,001–1,000,000	80,000–600,000		20,000–79,999		<20,000
Protime (in sec >control) no anticoagulants	>12	5.1–12	3.1–5		0–3				
Septic:									
CSF-positive culture	Yes				No				
Blood positive culture	Yes				No				
Fungal positive culture	Blood and/or CSF	2 sites other than blood or CSF		1 site other than blood or CSF	None				
Temperature °C (rectal)	>41.0	39.1–41.0		38.6–39.0	36.0–38.5	34.0–35.9	32–33.9	30.0–31.9	29.9 or less
Metabolic:									
Serum calcium mg/100 ml	16 or >		14–15.9	11.1–13.9	8–11.0		5.0–7.9		<5
Serum glucose	>800	500–800		251–499	70–250		50–69	30–49	<30
Serum sodium	>180	161–180	156–160	151–155	130–150		120–129	110–119	<110
Serum potassium	>7	6.1–7		5.6–6	3.5–5.5	3–3.4	2.5–2.9		<2.5
Serum HCO_3^-		>40		31–40	20–30	10–19		5–9	<5
Serum osmolarity	>350	321–350		301–320	20–300		240–259	220–239	<220
Neurological:									
Glasgow coma score	3	4–6	7–9	10–12	13–15				

[a] *P(A-a)* O_2 *= [Fi* O_2 *(713) – Pa*CO_2 *– Pa*O_2*]*

[b] *Total anergy – no response to all provocative skin tests including mumps and fungal. Relative-reduced response to skin indicative of compromised cellular immunity.*

TABLE 7.2 Grading of Sepsis

I. Local effects of tissue infection	Score
Wound infection with purulent discharge/enterocutaneous fistula:	
Requiring only light dressing changed not more than once daily	2
Requiring to be dressed with a pack, dressing needing to be changed more than once daily, requiring application of a bag and/or requiring suction	4
Peritonitis:	
Localized peritonitis	2
Generalized peritonitis	6
Chest infection:	
Clinical or radiological signs of chest infection with productive cough	2
Clinical or radiological signs of chest infection with a cough producing purulent sputum	4
Full clinical manifestations of lobar/bronchopneumonia	6
Deep-seated infection (e.g. subphrenic abscess, pelvic abscess, empyema thoracis, acute or chronic osteomyelitis)	6
II. Pyrexia (oral temperature)	**Score**
Maximum daily temperature(°C) *	
36–37.4	0
37.5–38.4	1
38.5–39	2
>39	3
<36	3
	Add
Minimum daily temperature >37.5°C in 1 day	1
If 2 or more temperature peaks above 38.4°C in 1 day	1
If any rigours occur in a day	1
III. Secondary effects of sepsis	**Score**
Obvious jaundice (in the absence of established hepatobiliary disease)	2
Metabolic acidosis:	
Compensated	1
Uncompensated	2
Renal failure	3
Gross disturbance of mental orientation/level of consciousness (e.g., delirium, coma) and/or other focal neurological manifestations of pyrexia/septicemia (having excluded other causes)	3
Bleeding diathesis (from disseminated intravascular coagulation)	3

TABLE 7.2 (Continued)

IV. Laboratory data	Score
Blood culture:	
Single positive culture	1
Two or more positive cultures separated by 24 h	3
Single positive culture + history of invasive procedure	3
Single positive culture + cardiac murmur and/or tender enlarged spleen	3
Leucocyte count ($x10^9$/l):	
12–30	1
>30	2
<2.5	3
Hemoglobin level in the absence of obvious bleeding (g/dl):	
7–10	1
<7	2
Platelet count ($x10^9$/l):	
100–150	1
<100	2
Plasma albumin level (g/l):	
31–35	1
25–30	2
<25	3
Plasma total bilirubin level in the absence of clinically obvious jaundice:	
>25 μmol/L	1

**Temperature should be recorded at least 4 times in 24 h. The record for 24 h period is assessed as above and "pyrexia score" computed.*

Variations in the mean number of data points collected per patient in the APACHE index, which introduces systematic bias to in-patient scoring, motivated the development of the Simplified Acute Physiology Score (SAPS) by Le Gall.[10] A mandatory 13, easily measured, physiological variables combined with age are used (Table 7.4). The selection of the variables was determined by an APACHE survey using those found to be measured in 90% of the patients. Data is collected during the first 24 h after ICU admission, using only the most abnormal value, when measured repeatedly. A fixed value of 3 was assigned to ventilated patients, rather than the alveolar-arterial oxygen tension difference (P[A-a]O_2) value used in APACHE. All remaining variables were scored from 0 to 4. SAPS was

TABLE 7.3 Septic Severity Score: Levels of Organ Dysfunction

System	1	2	3	4	5
Lung	O_2 by mask	Intubated; no PEEP	PEEP, 0%–10%	PEEP, >10%; PO_2 > 50 mmHg	Maximal PEEP; PO_2 <50 mmHg
Kidney	CL, 1.5–2.5 mg/dl	CL, 2.6–3.5 mg/dl	CL, >3.6 mg/dl; adequate urine volume	CL, >3.6 mg/dL; urine volume, 20–50 mL/hr	CL, >3.6 mg/dL; urine volume, 20 mL/hr
Coagulation	Ecchymosis; PT, PTT, and platelet count, normal	PPT, 45–65 s; PT, 12–14 s	Platelets, 20,000–100,000/cu mm; PTT, >50 s; PT >14s	Platelets, 20,000/ cu mm; elevated PT and PTT	Increased FSP and euglobulin, bleeding
Cardio-vascular	Slight hypotension	Livedo; moderate hypotension	Vasopressors, moderate doses	Vasopressors, large doses	Profound BP decrease despite vasopressors
Liver	LDH and SGOT increased; bilirubin, normal	Bilirubin 1.5–2.5 mg/dl	Bilirubin, 2.6–4.0 mg/dL	Bilirubin, 4.9–8.0 mg/dl	Precoma bilirubin, >8.0 mg/dL
GL tract	Mild ileus	Moderate ileus	Severe ileus	Bleeding due to erosive gastritis	Mesenteric venous thrombosis
Neurologic	Obtunded	Disoriented	Irrational	Hyporeactive	Coma

CL, creatinine level; PT prothrombin time; GI, gastrointestinal; PTT, partial thromboplastin time; PEEP, positive end-expiratory pressure; LDH, lactate dehydrogenase; and FSP, fibrin split products.

evaluated in 679 consecutive patients admitted to eight multidisciplinary referral ICUs in France. It correctly classified patients into groups of increasing probability of death, irrespective of diagnosis. As with APACHE, the authors suggested it should not be used for individual prognosis or treatment decisions.

Due to the complexity of the APACHE index, Knaus et al. developed the simpler APACHE-II system.[11] The physiological part was reduced from 34 to 12 variables (Table 7.5). As in the original prototype version, a panel of experts decided which variables should be eliminated and they changed the thresholds and weights of three of the original physiological variables. Now the acute physiology score would be determined from the most severe physiological value during the initial 24 h after ICU admission. Unlike APACHE, measurement of all 12 physiological values is mandatory in APACHE-II. The chronic health evaluation was changed. Weights were assigned to age groups, and to specific chronic health problems. Additional points were assigned if the patient was non-operative or emergency postoperative (5 points) or elective postoperative (2 points). A maximum possible score of 71 was possible, adding all of the aforementioned scores. In addition to the original uses, the new system was also intended for individual outcome estimation and to be useful in good clinical decision making.

Based on the empirical weighted peritonitis index they previously introduced, Linder et al.[12] developed the Mannheim Peritonitis Index (MPI) in 1987. The purpose of this index is

TABLE 7.4 Simplified Acute Physiology Score

	4	3	2	1	0	1	2	3	4
Age (yr)					≤45	46–55	56–65	66–75	>75
Heart rate (beat/min)	≥180	140–179	110–139		70–109		55–69	40–54	<40
Systolic blood pressure (mmHg)	≥190		150–189		80–149		55–79		<55
Body temperature (°C)	≥41	39–40.9		38.5–38.9	36–38.4	34.0–35.9	32.0–33.9	30–31.9	<30
Spontaneous respiratory rate (breath/min)	≥50	35–49		25–34	12–24	10–11	6–9		<6
or									
Ventilation or CPAP								Yes	
Urinary output (L/24 h)			>5.00	3.50–4.99	0.70–3.49		0.50–0.69	0.20–0.49	<0.20
Blood urea (mmol/L)	≥55	36–54.9	29–35.9	7.5–28.9	3.5–7.4	<3.5			
Hematocrit (%)	≥60		50.0–59.9	46–49.9	30–45.9		20.0–29.9		<20.0
White blood cell count ($10^3/mm^3$)	≥40		20.0–39.9	15.0–19.9	3.0–14.9		1.0–2.9		<1.0
Serum glucose (mmol/L)	≥44.5	27.8–44.4		14.0–27.7	3.9–13.9		2.8–3.8	1.6–2.7	<1.6
Serum potassium (mEq/L)	≥7	6–6.9		5.5–5.9	3.5–5.4	3.0–3.4	2.5–2.9		<2.5
Serum sodium (mEq/L)	≥180	161–179	156–160	151–155	130–150		120–129	110–119	<110
Serum HCO_3 (mEq/L)		>40		30.0–39.9	20.0–29.9	10.0–19.9		5.0–9.9	<5.0
Glasgow coma score					13–15	10–12	7–9	4–6	3

to predict lethal outcome from IAI in the individual patient, using only clinical risk factors that are routinely documented pre- and intra-operatively. In a retrospective pilot study of 1,243 patients with purulent peritonitis, 17 risk factors were analyzed by logistic discriminant analysis. Only eight factors, found to be relevant, were included in the MPI. Seven variables were dichotomous and one trichotomous (Table 7.6). The possible scores ranged from 0 to 47.

TABLE 7.5 Acute Physiology and Chronic Health Evaluation II: Acute Physiology Score, Sum of the 12 Individual Variable Points

Physiological variable	+4	+3	+2	+1	+0	+1	+2	+3	+4
Temperature – rectal °C	≥41	39–40.9		38.5–38.9	36–38.4	34–35.9	32–33.9	30–31.9	≤29.9
Mean arterial pressure – mmHg	≥160	130–159	110–129		70–109		50–69		≤49
Heart rate (ventricular response)	≥180	140–179	110–139		70–109		55–69	40–54	≤39
Respiratory rate (non-ventilated or ventilated)	≥50	35–49		25–34	12–24	10–11	6–9		≤5
Oxygenation A • aDO_2 or PaO_2 (mmHg) a Fi O_2 ≥0.5 record A • aD O_2	≥500	350–499	200–349		<200				
b Fi O_2 <0.5 record only PaO_2					PO_2 >70	PO_2 61–70		PO_2 55–60	PO_2 <55
Arterial pH	≥7.7	7.6–7.69		7.5–7.59	7.33–7.49		7.25–7.32	7.15–7.24	<7.15
Serum sodium (mmol/L)	≥180	160–179	155–159	150–154	130–149		120–129	111–119	≤110
Serum potassium (mmol/L)	≥7	6–6.9		5.5–5.9	3.5–5.4	3–3.4	2.5–2.9		<2.5
Serum creatinine (mg/100mL) (Double point score for acute renal failure)	≥3.5	2–3.4	1.5–1.9		0.6–1.4		<0.6		
Hematocrit (%)	≥60		50–59.9	46–49.9	30–45.9		20–29.9		<20
White blood count (total/mm^3)1,000 s	≥40		20–39.9	15–19.9	3–14.9		1–2.9		<1
Glasgow coma score (GCS) Score = 15 minus actual GCS									
Serum HCO_2 (venous mmol/L) (Not preferred, use if no ABGs)	≥52	41–51.9		32–40.9	23–31.9		18–21.9	15–17.9	<15

(*Continued*)

TABLE 7.5 (Continued)

Acute Physiology and Chronic Health Evaluation II: Previous Health Status			
B Age Points	**C Chronic Health Points**		**Score**
Assign points to age as follows Age (yrs) — Points ≤44 — 0 45–54 — 2 55–64 — 3 65–74 — 5 ≥75 — 6	If the patient has a history of severe organ system insufficiency or is immune-compromised assign points as follows a: for nonoperative or emergency postoperative patients – 5 points **or** b: for elective postoperative patients – 2 points **Definitions** Organ insufficiency or immunocompromised state must have been evident prior to this hospital admission and conform to the following criteria: Liver – biopsy proven cirrhosis and documented portal hypertension; episodes of past upper GI bleeding attributed to portal hypertension; or prior episodes of hepatic failure/encephalopathy/coma Cardiovascular – New York Heart Association Class IV.	Respiratory – Chronic restrictive, obstructive, or vascular disease resulting in severe exercise restriction (i.e., unable to climb stairs or perform household duties); or documented chronic hypoxia, hypercapnia, secondary polycythemia, severe pulmonary hypertension (>40 mmHg), or respirator dependency. Renal – receiving chronic dialysis. Immunocompromised – the patient has received therapy that suppresses resistance to infection (e.g., immunosuppression, chemotherapy, radiation, long term or recent high dose steroids, or has a disease that is sufficiently advanced to suppress resistance to infection, e.g., leukemia, lymphoma, AIDS).	Sum of A + B + C A APS points B Age points C Chronic Health points Total APACHE II

Copeland et al.[13] presented a scoring system to be used across the general surgical spectrum, whose main use would be for surgical audit. This included surgical success, in addition to mortality and morbidity rates. Developed over two years using discriminant analysis of 62 individual factors, a 12-factor, four-grade physiological score was the result, termed the Physiological and Operative Severity Score for the enUmeration of Mortality and morbidity (POSSUM) system. Twelve physiological variables are scored at the time of surgery; the operative severity is determined in grades 1 to 4, and in addition complications are recorded in the established format (Table 7.7).

APACHE, APACHE-II and SAPS were built subjectively using a panel of experts to select variables and weights. Le Gall et al.[14] proposed a new Simplified Acute Physiology Score (SAPS-II) using a logistic regression modelling technique for the selection of variables and to assign their weight. The multicenter study group consisted of 12,997 patients (developmental sample 8,369 and validation sample 4,628). All were adult (≥18 years of age) ICU patients, and excluded burn, coronary care and cardiac surgery patients. Data collection included patient demographic information, variables from the original SAPS, new variables chosen by a team of coordinators, and vital status at discharge; 37 variables in all. Only 17 variables were included in the final SAPS-II: 12 physiological variables, age, type of admission, and three variables related to underlying disease (Table 7.8). The points assigned for each variable varied from 0 to 3 (for temperature) up to 0 to 26 (for Glasgow Coma Score). An important purpose of the new score was to convert its score to a probability of hospital mortality.

TABLE 7.6 Mannheim Peritonitis Index

		Presence	
Risk Factor	**Scores**	**Yes**	**No**
Age >50 years	5	( )	( )
Sex: female	5	( )	( )
Organ failure*	7	( )	( )
Neoplasm	4	( )	( )
Preoperative peritonitis of >24 hr	4	( )	( )
Non-colonic origin	4	( )	( )
Diffuse generalized peritonitis	6	( )	( )
Exudate (only one yes answer)		( )	( )
Clear	0	( )	( )
Purulent	6	( )	( )
Fecal	12	( )	( )

**Kidney failure = creatinine level >177 μmol/L or urea level >167 mmol/L or oliguria <20 ml/hour; pulmonary insufficiency = PO2 <50 mmHg or PCO2 >50 mmHg; intestinal obstruction/paralysis >24 hours or complete mechanical ileus, shock hypodynamic or hyperdynamic.*

ICU patients' mortality rate correlates with the number of failing organ systems, and the degree of dysfunction within a given system. On this basis, Marshal et al.[15] developed and evaluated a scoring system to quantify the severity of the multiple organ dysfunction syndrome as an outcome of illness. They reviewed the literature to determine how an organ system dysfunction or failure had been characterized in previous studies. Individual organ dysfunction variables were identified and, considering content validity and reproducibility, the ideal descriptor was searched for. Candidate variables were evaluated for their ability to predict ICU mortality in a dose-dependent fashion among 692 ICU patients (developmental sample 336 and validation sample 356), with a mortality rate of 9.4% (n = 65). Increasing abnormality of each variable was graded on a scale from 0 to 4, and calibrated so that a score of 0 correlated with a ICU mortality rate of less than 5%, while a score of 4 correlated with a mortality of ≥50% (Table 7.9).

Vincent et al.[16] presented the Sepsis-related Organ Failure Assessment (SOFA) score. Its aim is to improve our understanding of the natural history of organ dysfunction/failure and to assess the effects of new therapy upon its course. It was not designed to predict outcome, but rather to describe a sequence of complications in the critically ill. Hence, the SOFA score does not compete with the existing severity indexes, but complements them. The panel of experts who created the score limited the number of organs studied to six (Table 7.10). Even though the authors mentioned that the gut has been cited as a possible motor of multiple organ failure, its evaluation was too complex and it was omitted. A score from 0 (normal) to 4 (most abnormal) was assigned to each organ, recording the worst values on each day.

TABLE 7.7 Physiological and Operative Severity Score for the enumeration of Mortality and Morbidity

Physiological Severity Score	1	2	4	8
Age (years)	≤60	61–70	≥71	
Cardiac signs	No failure	Diuretic, digoxin, antianginal or hypertensive therapy	Peripheral oedema; warfarin therapy	Raised jugular venous pressure
Chest radiograph			Borderline cardiomegaly	Cardiomegaly
Respiratory history	No dyspnea	dyspnea on exertion	Limiting dyspnea (one flight)	Dyspnea at rest (rate ≥30 /min)
Chest radiograph		Mild COAD	Moderate COAD	Fibrosis or consolidation
Blood pressure (systolic) (mmHg)	110–130	131–170 100–109	≥171 90–99	– ≤89
Pulse (beats/min)	50–80	81–100 40–49	101–120	≥121 ≤39
Glasgow coma score	15	12–14	9–11	≤8
Hemoglobin (g/100 ml)	13–16	11.5–12.9 16.1–17.0	10.0–11.4 17.1–18.0	≤9.9 ≥18.1
White cell count ($\times 10^{12}$/l)	4–10	10.1–20.0 3.1–4.0	≥20.1 ≤3.0	
Urea (mmol/L)	≤7.5	7.6–10.0	10.1–15.0	≥15.1
Sodium (mmol/L)	≤136	131–135	126–130	≤125
Potassium (mmol/L)	3.5–5.0	3.2–3.4 5.1–5.3	2.9–3.1 5.4–5.9	≤2.8 ≥6.0
Electrocardiogram	Normal		Atrial fibrillation (rate 60–90)	Any other abnormal rhythm or ≥5 ectopics/min, Q waves or ST/T wave changes
Operative severity score	**1**	**2**	**4**	**8**
Operative severity*	Minor	Moderate	Major	Major+
Multiple provedures	1		2	>2
Total blood loss (ml)	≤100	101–500	501–999	≥1000

(Continued)

TABLE 7.7 (Continued)

Operative severity score	1	2	4	8
Peritoneal soiling	None	Minor (serous fluid)	Local pus	Free bowel content, pus or blood
Presence of malignancy	None	Primary only	Nodal metastases	Distant metastases
Mode of surgery	Elective		Emergency resuscitation >2 h possible[†]. Operation < 24 h after admission	Emergency (inmediate surgery <2 h needed)

Complications Record Sheet

Name
Hospital no.
Diagnosis
Operation
Date of admission
Date of operation
Date of discharge
Surgeon
Anesthetist
Outcome*
Hemorrhage
- Wound
- Deep
- Other

Infection**
- Chest
- Wound
- Urinary tract
- Deep
- Septicemia
- Pyrexia of unknown origin
- Other

Wound dehiscence
- Superficial
- Deep

Anastomotic leak
Thrombosis
- Deep vein thrombosis
- Pulmonary embolus
- Other
- Cerebrovascular accident
- Myocardial infarct

Cardiac failure
Impaired renal function (urea increase >5 mmol/L from preoperative level)
Hypotension (<90 mmHg for 2 h)
Respiratory failure
Any other complication
In the event of death give date
Post-mortem findings

Give dates for all complications;
***for infection give bacteria cultured if known;*
[†]indicates that resuscitation is possible even if this is not actually utilized.

Perforating lesions of the colon affect a heterogeneous group of patients, often elderly, and usually present as abdominal emergencies with high morbidity and mortality. Biondo et al.[17] decided to assess the prognostic value of various factors in patients with left colonic peritonitis in order to develop a scoring method that might allow groups of patients with different mortality risks to be defined, and so help in their surgical management. Data from 156 patients who underwent emergency operation for distal colonic peritonitis were studied (score building group 101 and validation group 55), with a mortality of 22.4% (n = 35). Significant factors in univariate and multivariate analysis were employed to develop the left colonic Peritonitis Severity Score (PSS). Factors that were significant only in univariate analysis scored 2 points if present and 1 point if not. When the factor considered in the scoring system was significant in multivariate logistic regression, it scored from 1 to 3 points (Table 7.11).

DISCUSSION

Prognosis in IAI is difficult to establish because it includes a variety of different pathological conditions.[18] The spectrum of IAI includes primary peritonitis, secondary peritonitis, tertiary peritonitis and intra-abdominal abscesses as well as focal infections such as diverticulitis, appendicitis and cholangitis-cholecystitis.[19] These conditions have differences in management and outcome. An uncomplicated-IAI may be managed non-operatively, while complicated IAI requires surgical intervention. Even within these defined categories there may be differences. A healthcare-acquired-IAI, for example, has a worse prognosis than a community-acquired one, due to the underlying health status and the likelihood of infection by multi-drug resistant organisms.[20–21]

The scoring indexes reviewed can be divided into disease-independent scores (APACHE, SAPS, APACHE-II, POSSUM, SAPS-II, MOF, SOFA)[7,10–11,13–14,16] or peritonitis-specific scores (MPI, PSS).[12,17] Two scores were sepsis-specific (SS and SSS),[8–9] but for infections not limited to the abdominal cavity.

The fact that a disease-independent system can be used to assess patients with IAI has an explanation. It has been shown that the outcome of peritonitis is often determined by the extent of the derangement of systemic physiology, appropriately addressed by these scores.[22]

The applicability of a score relates to the condition and scenario for which it was devised, along with the matter of its complexity. APACHE,[7] SAPS[10] and SOFA[16] were not intended for individual outcome estimation or decision making, but for classifying groups of patients to compare outcomes in clinical studies. MOF[15] offers only a gross prognostic evaluation. APACHE-II, POSSUM, SAPS-II, MPI, PSS[11–14,17] were designed to predict lethal outcome in IAI, and seem to be of use. PSS is limited to left colonic peritonitis.[17]

Even though the POSSUM index has been shown to be helpful for categorizing groups of patients and for prognosis, it is used in the literature mainly for surgical audit purposes.[23]

I believe that SS and SSS, of historical relevance, have been surpassed by the newer physiology and multi-organ dysfunction scores. So, for the purpose classifying groups of patients to compare outcomes in clinical studies and for the prediction of individual mortality in IAI, of the indexes discussed, the best estimators are APACHE-II, SAPS-II and MPI.

TABLE 7.8 Simplified Acute Physiology Score (SAPS II)*

Variable	Points: 26	13	12	11	9	7	6	5	4	3	2	0	1
Age, yrs												<40	
Heart rate, beats/min				<40							40–69	70–119	
Systolic BP, mm Hg		<70						70–99				100–199	
Body temperature, °C (°F)												<39° (<102.2°)	
Only if ventilated or continuous pulmonary artery pressure PaO_2, mm Hg/Fi O_2				<100	100–199		≥ 200						
PaO_2, kPa/Fi O_2				<13.3	13.3–26.5		≥ 26.6						
Urinary output, L/d				<0.500					0.500–0.999			≥ 1.000	
Serum urea level, mmol/L (g/L) or serum urea nitrogen level, mg/dL												<10.0 (<0.60) <28	
WBC count (10^3/cu mm)			<1.0									1.0–19.9	
Serum potassium, mmol/d										<3.0		3.0–4.9	
Serum sodium level, mmol/L								<125				125–144	≥ 145
Serum bicarbonate level, mEq/L							<15			15–19		≥ 20	
Bifirubin level, μmol/L (mg/dL)												<68.4 (<4.0)	
Glasgow Coma Score	<6	6–8				9–10		11–13				14–15	
Chronic diseases													
Type of admission												Scheduled surgical	
Sum of points													

**SAPS indicates Simplified Acute Physiology Score; BP blood pressure; Fi O_2, fraction of inspired oxygen; kPa, kilopascal; WBC, white blood cell; and AIDS, acquired immunodeficiency syndrome.*

2	3	4	6	7	8	9	10	12	15	16	17	18
				40–59				60–69	70–74	75–79		≥80
		120–159		≥160								
≥200												
	<39° (≥102.2°)											
			10.0–29.9 (6.0–1.79) 28–83				≥30.0 (≥1.80) ≥84					
	≥20.0											
	≥5.0											
		68.4–102.5 (4.0–5.9)				≥102.6 (≥6.0)						
						Metastatic cancer	Hematologic malignancy				AIDS	
			Medical		Unscheduled surgical							
						Total SAPS II Score		=				Points
						Risk of Hospital Death		=				%

TABLE 7.9 Multiple Organ Dysfunction Score

Organ System	0	1	2	3	4
Respiratory[a]					
($PO_2/Fi\ O_2$ ratio)	>300	226–300	151–225	76–150	≤75
Renal[b]					
(serum creatinine)	≤100	101–200	201–350	351–500	>500
Hepatic[c]					
(serum bilirubin)	≤20	21–60	61–120	121–240	>240
Cardio-vascular[d]					
(PAR)	≤10.0	10.1–15.0	15.1–20	20.1–30.0	>30.0
Hematologic[e]					
(Platelet count)	>120	81–120	51–80	21–50	≤20
Neurologic[f]					
(Glasgow coma score)	15	13–14	10–12	7–9	≤6

[a]*The $PO_2/Fi\ O_2$ ratio is calculated without reference to the use or mode of mechanical ventilation, and without reference to the use or level of positive end-expiratory pressure;*
[b]*The serum creatinine concentration is measured in μmol/L, without reference to the use of dialysis;*
[c]*The serum bilirubin concentration is measured in μmol/L;*
[d]*The pressure-adjusted heart rate (PAR) is calculated as the product of the heart rate (HR) multiplied by the ratio of the right atrial (central venous) pressure (RAP) to the mean arterial pressure (MAP): PAR = HR × RAP/mean B;*
[e]*The platelet count is measured in platelets/mL 10^{-3};*
[f]*The Glasgow Coma Score is preferably calculated by the patient's nurse, and is scored conservatively (for the patient receiving sedation or muscle relaxants, normal function is assumed, unless there is evidence of intrinsically altered mentation).*

It is also relevant that in 1987 the Surgical Infection Society proposed that the APACHE-II system be used for grading the severity of the infection and the stratification of patient risk of mortality.[22–24] The fact that the score had been used in 5,815 intensive care admissions from 13 hospitals appeared to be sufficient support for its proposal. But of these admissions, only 106 according to the data available in the article clearly had IAI due to gastrointestinal perforation or obstruction. This is only 1.8% of the study group.[11] This underlines uneasiness felt by some about this score. A patient can be evaluated by this system and have no infectious process at all, let alone an IAI. In spite of this potential wekaness, the APACHE-II score has outlasted earlier criticism of not being specific enough for grading IAI. This criticism is more a conceptual than a factual objection, as explained above.[22]

Another objection to the APACHE score, in all its versions, has been its complexity. It was intended that the APACHE-II version should overcome this handicap, but this apparently did not occur. During the validation process of this version, all 12 physiological measurements were available for only 87% of the 5,815 patients studied. This omission occurred in spite of the fact that there was a training process and supervision of the

TABLE 7.10 Sepsis-related Organ Failure Assessment

	1	2	3	4
Respiration				
PaO_2/FiO_2 mmHg	<400	<300	<200 with respiratory support	<100 with respiratory support
Coagulation				
Platelets x $10^3/mm^3$	<150	< 100	<50	<20
Liver				
Bilirubin, (mg/dL μmol/L)	1.2–1.9 (20–32)	2.0–5.9 (33–101)	6.0–11.9 (102–204)	>12.0 (<204)
Cardiovascular				
Hypotension	MAP <70 mmHg	Dopamine ≤5 or dobutamine (any dose)[a]	Dopamine >5 or epinephrine ≤0.1 or norepinephrine ≤0.1	Dopamine >15 or epinephrine >0.1 or norepinephrine >0.1
Central nervous system				
Glasgow coma score	13–14	10–12	6–9	<6
Renal				
Creatinine, mg/dL (μmol/L) or urine output	1.2–1.9 (110–170)	2.0–3.4 (171–299)	3.5–4.9 (300–440) or <500 mL/day	>5.0 (>440) or < 200 mL/day

[a]*Adrenergic agents administered for at least 1 h (doses given are in μg/kg • min).*

TABLE 7.11 Peritonitis Severity Score

Score	1	2	3
Age (y)	≤70	>70	–
ASA* grade	I–II	III	IV
Preoperative organ failure	None	–	One or more organ
Immuno-compromised status	No	Yes	–
Ischemic colitis	No	Yes	–
Peritonitis stage	1–2	3–4	–

**ASA, American Society of Anesthesiologists.*

participant centers in the validation study, and that it was "mandatory" to measure all study variables.[7,11]

In everyday surgical practice, this flaw in the score is very real. Not all patients operated on for IAI are sent to ICUs, so in this scenario, the measurement of all variables is improbable. Many third-world hospitals, with serious shortages of equipment and lack of staff, do not even have an ICU.[25] SAPS also requires ICU patient admission.[10,14]

The aforementioned problems are overcome by the MPI. It is one of the simplest scoring systems in use that allows the surgeon to easily determine outcome risk during initial surgery. The collection of retrospective data is possible and valid, because MPI only requires information routinely found in surgical registers.[26]

Using the MPI, the cytokine dynamics in secondary peritonitis during the early induced-response have been studied. The grade-response relationship between the severity of peritonitis and the pivotal cytokines was described.[27] The study of what occurs in IAI in a cellular or molecular milieu, using the APACHE-II or SAPS scores, would be desirable. Very little has been published in this field.[28–29]

Given that IAI constitutes a heterogeneous group of conditions, APACHE-II and SAPS-II, as stated, are disease-independent systems. So, at least theoretically, this diversity of conditions does not hamper the results obtained from these assessments. MPI was designed for secondary peritonitis in a "narrow sense", with the exclusion of peritonitis of positive outcome (due to appendicitis or iatrogenic) and of negative outcome (postoperative peritonitis, due to mesenteric occlusion, and secondary to pancreatitis).[12] There are publications that study patients with these conditions, and this does not appear to affect its outcome predictive capacity.[26,30–31] Nonetheless, MPI has more restrictions.

A multicenter study established that MPI, for a threshold score of 26 points, has an 86% sensitivity, 74% specificity and an accuracy of 83%.[26] Another multicenter study found APACHE-II to be superior to MPI in discriminatory ability and reliability, but MPI made sharper predictions, though none of the scores were of use for predicting the outcome for individual patients.[31]

Individual centers have found a MPI sensitivity, specificity and correct prognosis of 93%, with a higher accuracy than calculated for APACHE-II.[32–33] But the results are inconsistent. Another center found that the area under the curve for accuracy of APACHE-II was 0.86, while it was 0.72 for MPI.[34] Many explanations can be found for these discrepancies, especially the differences in the study groups, with dissimilar conditions that are not comparable.

While some authors found that high APACHE-II scores might be helpful in predicting the need for relaparotomies,[35] others found that neither APACHE-II nor MPI were of clinical use for the identification of patients with ongoing infection needing relaparotomy.[3] High scores in both APACHE-II and MPI have been suggestive as indicators of tertiary peritonitis and the presence of fungal infection of the peritoneum.[36–37]

Collectively, these discrepancies have caused us to reflect on the following. MPI and APACHE-II have in essence important differences. MPI is a peritonitis-specific score, which can be calculated immediately after surgery, in any hospital level. APACHE-II is a disease-independent score that must be measured 24 h after ICU admission. So the two are essentially different, complementary scores that need not be compared.

Another matter that caught our attention was their usefulness for individual prognosis and for decision making for relaparotomy. Here we find what we believe to be conceptual errors. When one classifies, the individuals studied are put into groups. So why would we expect a score intended to be used for classifying severity to predict an individual's outcome? This seems to be a postmodern, pseudoscientific form of divination. All we can expect is that the score gives us a prognostic estimate for that severity group, not a prediction of individual outcome. APACHE-II, SAPS, and MPI do this well.

As to the decision for relaparotomy, this is a medical act, based on experience and good medical judgment. These cannot be substituted by any score which, once again, categorizes patients, to clarify our ideas or for comparison.

CONCLUSIONS

Postoperatively, a surgeon who operated on a patient with secondary peritonitis should score his patient using the MPI, because of its simplicity and, at that moment, he has all the information to do so. If the patient is sent to UCI, he should be scored 24 h after admission, either by APACHE-II or SAPS. These are complementary scores, taken at different points in time, and will be of use for the surgeon to inform family members regarding which outcome group the patient falls into.

MPI, SAPS and APACHE-II scores are useful for helping the surgeon discern which of the patients operated on because of secondary peritonitis might require a relaparotomy or may evolve toward tertiary peritonitis. They do not substitute for good medical judgment, experience and a diligent follow-up.

If a clinical researcher requires a scoring system to classify groups of patients in order to compare outcomes or to evaluate a new therapy, the scoring system most practical for the purpose of the study should be chosen. This choice should consider whether a disease-specific score is required, or if underlying health conditions or physiological variables are the focus of interest.

In any case, a surgeon and/or clinical researcher studying IAI must always keep in mind that this is a heterogeneous group of conditions, so they may wish to select the study participants taking this into account. If studying secondary peritonitis, MPI may be preferred. If participant selection is not possible or not desirable, disease-independent scores are preferable.

References

1. Suarez ME. Infecciones Intra-abdominales: Peritonitis y Abscesos. *MEDICRIT* 2004;**1**:146–80.
2. Mazuski JE, Solomkin JS. Intra-abdominal infections. *Surg Clin North Am* 2009;**89**:421–37.
3. van Ruler O, Kiewiet JJ, Boer KR, Lamme B, Gouma DJ, Boermeester MA, et al. Failure of available scoring systems to predict ongoing infection in patients with abdominal sepsis after their initial emergency laparotomy. *BMC Surg* 2011;**11**:38.
4. Moreno RP. Metnitz PGH. Severity Scoring Systems: Tools for the Evaluation of Patients and Intensive Care Units. In: Parrillo PE, Dellinger RP, editors. *Critical Care Medicine*. 3rd ed. Philadelphia: Mosby; 2007. p. 1547–65.

5. <http://www.ncbi.nlm.nih.gov/mesh?Db=mesh&Term=%22Severity%20of%20Illness%20Index%20%22[MESH]>; [accessed 6.11.2012].
6. Wagner DP, Knaus WA, Draper EA. Statistical validation of a severity of illness measure. *Am J Public Health* 1983;**73**:878–84.
7. Knaus WA, Zimmerman JE, Wagner DP, Draper EA, Lawrence DE. APACHE-acute physiology and chronic health evaluation: a physiologically based classification system. *Crit Care Med* 1981;**9**:591–7.
8. Elebute EA, Stoner HB. The grading of sepsis. *Br J Surg* 1983;**70**:29–31.
9. Stevens LE. Gauging the severity of surgical sepsis. *Arch Surg* 1983;**118**:1190–2.
10. Le Gall JR, Loirat P, Alperovitch A, Glaser P, Granthil C, Mathieu D, et al. A simplified acute physiology score for ICU patients. *Crit Care Med* 1984;**12**:975–7.
11. Knaus WA, Draper EA, Wagner DP, Zimmerman JE. APACHE-II: a severity of disease classification system. *Crit Care Med* 1985;**13**:818–29.
12. Linder MM, Wacha H, Feldmann U, Wesch G, Streifensand RA, Gundlach E. The Mannheim peritonitis index. An instrument for the intraoperative prognosis of peritonitis. *Chirurg* 1987;**58**:84–92 [German].
13. Copeland GP, Jones D, Walters M. POSSUM: a scoring system for surgical audit. *Br J Surg* 1991;**78**:355–60.
14. Le Gall Jr, Lemeshow S, Saulnier F. A new Simplified Acute Physiology Score (SAPS-II) based on a European/North American multicenter study. *JAMA* 1993;**270**:2957–63.
15. Marshall JC, Cook DJ, Christou NV, Bernard GR, Sprung CL, Sibbald WJ. Multiple organ dysfunction score: a reliable descriptor of a complex clinical outcome. *Crit Care Med* 1995;**23**:1638–52.
16. Vincent JL, Moreno R, Takala J, Willatts S, De Mendonça A, Bruining H, et al. The SOFA (Sepsis-related Organ Failure Assessment) score to describe organ dysfunction/failure. On behalf of the Working Group on Sepsis-Related Problems of the European Society of Intensive Care Medicine. *Intensive Care Med* 1996;**22**:707–10.
17. Biondo S, Ramos E, Deiros M, Ragué JM, De Oca J, Moreno P, et al. Prognostic factors for mortality in left colonic peritonitis: a new scoring system. *J Am Coll Surg* 2000;**191**:635–42.
18. Wittmann DH, Schein M, Condon RE. Management of secondary peritonitis. *Ann Surg* 1996;**224**:10–8.
19. Peetermans WE, Hiele M. Abdominal sepsis. *Acta Clin Belg* 1998;**53**:251–4.
20. Sartelli M. A focus on intra-abdominal infections. *World J Emerg Surg* 2010;**5**:9.
21. González-Ojeda A, Velázquez-Ramírez GA. Peritonitis secundaria. *Rev Invest Clin* 2005;**57**:706–15.
22. Ukwenya AY, Muhammad H, Nmadu PT. Assessing the severity of intraabdominal infections; the value of APACHE-II Scoring System. *Nigerian J Sur Res* 2006;**8**:24–9.
23. Wang H, Chen T, Wang H, Song Y, Li X, Wang J. A systematic review of the Physiological and Operative Severity Score for the enUmeration of Mortality and morbidity and its Portsmouth modification as predictors of post-operative morbidity and mortality in patients undergoing pancreatic surgery. *Am J Surg* 2013;**205**:466–72.
24. Nyström PO, Bax R, Dellinger EP, Dominioni L, Knaus WA, Meakins JL, et al. Proposed definitions for diagnosis, severity scoring, stratification, and outcome for trials on intraabdominal infection. Joint Working Party of SIS North America and Europe. *World J Surg* 1990;**14**:148–58.
25. Bracho-Riquelme RL, Melero-Vela A, Torres-Ramírez A. Mannheim peritonitis index validation study at the Hospital General de Durango (México). *Cir Cir* 2002;**70**:217–25.
26. Billing A, Fröhlich D, Schildberg FW. Prediction of outcome using the Mannheim peritonitis index in 2003 patients. *Br J Surg* 1994;**81**:209–13.
27. Bracho-Riquelme RL, Reyes-Romero MA, Torres-Valenzuela A, Flores-García AI. The grade-response relation between severity of peritonitis and serum cytokine concentrations explains Mannheim Peritonitis Index threshold. *Surg Infect* 2010;**11**:379–86.
28. Sendt W, Amberg R, Hassan A, von Specht BU, Schöffel U. Secondary peritonitis: severity of disease and activation of peritoneal cells. *Eur J Surg* 2001;**167**:426–32.
29. Karlsson S, Pettilä V, Tenhunen J, Laru-Sompa R, Hynninen M, Ruokonen E. HMGB1 as a predictor of organ dysfunction and outcome in patients with severe sepsis. *Intensive Care Med* 2008;**34**:1046–53.
30. Seifert J, Schirrmacher E. Retrospective comparison of peritonitis scores on 195 patients with a diffuse purulent peritonitis. *Theor Surg* 1990;**5**:33–5.
31. Ohmann C, Wittmann DH, Wacha H. Prospective evaluation of prognostic scoring systems in peritonitis. *Eur J Surg* 1993;**159**:267–74.

32. Fugger R, Rogy M, Herbst F, Schemper M, Schuli F. Validierungsstudie zum Mannheimer Peritonitis-Index. *Chirurg* 1988;**59**:598–601.
33. Rogy M, Függer R, Schemper M, Koss G, Schulz F. The value of 2 distinct prognosis scores in patients with peritonitis. The Mannheim Peritonitis Index versus the Apache II score. *Chirurg* 1990;**61**:297–300.
34. Delibegovic S, Markovic D, Hodzic S, Nuhanovic A. Evaluation of prognostic scoring systems in the prediction of the outcome in critically ill patients with perforative peritonitis. *Acta Inform Med* 2010;**18**:191–5.
35. Viehl CT, Kraus R, Zürcher M, Ernst T, Oertli D, Kettelhack C. The Acute Physiology and Chronic Health Evaluation II score is helpful in predicting the need of relaparotomies in patients with secondary peritonitis of colorectal origin. *Swiss Med Wkly* 2012;**142**:w13640.
36. Panhofer P, Izay B, Riedl M, Ferenc V, Ploder M, Jakesz R, et al. microbiology and prognostic scores help to differentiate between secondary and tertiary peritonitis. *Langenbecks Arch Surg* 2009;**394**:265–71.
37. Shan YS, Hsu HP, Hsieh YH, Sy ED, Lee JC, Lin PW. Significance of intraoperative peritoneal culture of fungus in perforated peptic ulcer. *Br J Surg* 2003;**90**:1215–9.

CHAPTER

8

Pathogenesis and Management of Dialysis Access Infections

Hariharan Regunath, Venkatesh Kumar Ariyamuthu* and Kunal Chaudhary*,†*

*Department of Medicine, University of Missouri Columbia, Missouri, †Harry S Truman Veterans Hospital, Columbia, Missouri

INTRODUCTION

The prevalence of Chronic Kidney Disease (CKD) leading to end-stage renal disease (ESRD) requiring dialysis as renal replacement therapy (RRT) continues to increase.[1] Both hemodialysis (HD) and peritoneal dialysis (PD), the two major modalities of RRT require placement of vascular and peritoneal access devices. A well-functioning dialysis access is of paramount importance for delivery of optimal dialysis. Due to the complex immune-modulatory effects of ESRD, these patients are susceptible to infections; and specifically dialysis access related infections continue to prevail as an important cause of hospitalization, morbidity and mortality in patients receiving dialysis.[1,2] Thus it is imperative that precautions for preventing access-related infections are taken, and adequate management of such infections is instituted in a timely fashion to salvage the access as well as to prevent further complications.

DIALYSIS ACCESS

Hemodialysis accounts for a majority of ESRD patients on maintenance dialysis.[1] This requires a vascular access by placement of either a tunneled or non-tunneled central venous catheter (CVC), or surgical construction of either an arterio-venous fistula (AVF) or arterio-venous graft (AVG). Tunneled catheters have a dacron cuff that incites an

Microbiology for Surgical Infections
DOI: http://dx.doi.org/10.1016/B978-0-12-411629-0.00008-8

inflammatory fibrosis along the subcutaneous track, forming a mechanical barrier from the vascular site of entry to the exit site at the skin surface. They are the choice at present for interim short term dialysis of those awaiting a renal transplant, maturation of AVF, and in patients who are not candidates for AVF or AVG. Non-tunneled catheters are typically used mainly for short term dialysis. HD catheters are often a source of infections leading to septicemia in HD patients, but compared to catheters, AVF and AVG are less likely to get infected.

PD accounts for a smaller percentage of patients on dialysis.[3] This modality requires placement of a PD catheter, different types of which are available based on preferences and experience of the treatment centers. In simple terms they are flexible biocompatible tubes with one end placed intra-abdominally, a mid-portion tunneling through the abdominal wall and subcutaneous tissue, and the other end reaching the skin surface for external access, thus forming a fistulous tract from peritoneal cavity to skin surface. PD catheter-related peritonitis is associated not only with increased morbidity (catheter loss, switch to HD, ultrafiltration failure) but also mortality from infectious and non-infectious complications (e.g., cardiovascular or cerebrovascular events from underlying sepsis).[4]

Infectious complications in both these modalities range from exit site and tunnel infection to frank sepsis and septic shock. Catheter or graft related blood stream infection (CRBSI) in HD, and peritonitis in PD, threaten the continued use of the associated device and mandate removal.

EPIDEMIOLOGY

Although both HD and PD patients are at increased risk of infections, the overall incidence of blood stream infections is higher in HD patients and peritonitis in PD patients. The hospitalization rates related to bacteremia and sepsis are highest for HD at 116 per 1000 patient years. Even though there has been a decline in vascular access infection-related hospitalization rates since 2005, the rates in 2009–2010 were 0.11 per patient year with a thirty day re-hospitalization rate at 6 percent.[1,5] In general, gram-positive infections predominate followed by gram-negative infections, and fungi also share a minimal percentage.

Hemodialysis

With the increasing proportion of HD patients (>80%) starting dialysis on tunneled catheters, CRBSIs are on the rise.[6] The causative organism appears to vary with the type of access used.[7–10] The incidence and prevalence of blood stream infections are reported to be highest with CVCs (77%), followed by AVG and AVF (23%).[11,12] The incidence of CRBSIs from tunneled CVCs is less than half that from non-tunneled CVCs.[7]

Staphylococci are responsible for the majority of CRBSIs, with coagulase negative staphylococci being the most common cause in those with CVCs, and *Staphylococcus aureus* in those with AVF or AVG.[13,14] Staphylococcal CRBSIs are particularly concerning as they are associated with increased morbidity and mortality because of increased virulence, propensity for antimicrobial resistance and metastatic complications like endocarditis, osteomyelitis, septic arthritis, spinal epidural abscess and intracranial abscess. Methicillin resistant *S. aureus* (MRSA) alone accounts for 25% of all infections in HD patients, and those with their blood stream infections have a two-fold higher mortality than those with methicillin sensitive *S. aureus* (MSSA) blood stream infections.[15] The widespread use of vancomycin to treat MRSA has resulted in the emergence of vancomycin intermediate *S. aureus* (VISA – vancomycin minimal inhibitory concentration [MIC] 4–8 μg/mL) and vancomycin resistant *S. aureus* (VRSA – vancomycin MIC >16 μg/mL).[16,17] Of the reported cases of VRSA in the US, over one third occurred in patients on HD.[18] Heterogeneous vancomycin intermediate staphylococcus aureus (hVISA) is a subset of MRSA with vancomycin MIC ≥2 μg/mL and is associated with a two-fold increase in vancomycin treatment failures.[15]

Staphylococcus epidermidis and other coagulase negative staphylococci were responsible for up to 35% of CRBSIs. In this group also, most are methicillin resistant and vancomycin resistance is an emerging problem.[15] Enterococcal infections, especially vancomycin resistant enterococci (VRE) pose challenges because pharmacological options for treatment are very limited. Outpatient vancomycin use has been associated with increased rates of VRE colonization.[19] VRE has been isolated from inanimate surfaces and health personnel gowns, hands and other equipment in hemodialysis centers.[20–22]

Gram-negative bacilli have been implicated in 21–30% of blood stream infections. They are reported to be more common in CVCs than AVF and AVG.[12,23] Emerging multi-drug resistance in this group is a significant concern, and it should be noted that HD patients are at a 13.6-fold higher risk of infections with extended spectrum β lactamase (ESBL) producing gram-negative bacilli.[24]

Fungal CRBSIs are commonly caused by non-albicans candida (*C. glabrata* and *C. krusei*) and increasing azole resistance is a rising concern.[25,26]

Peritoneal Dialysis

Peritonitis is a major complication in PD patients, causing many to switch to HD.[27] Again, gram-positive organisms (61%) are the most common cause of peritonitis, followed by gram-negative (20.9%) and fungal infections (4%).[28,29] About 10–40% are culture negative and a prior antibiotic therapy is an important risk factor.[30] *S. epidermidis* alone accounts for 30% of peritonitis, which is usually mild and has a good response to antibiotic therapy.[28] Methicillin resistant *S. epidermidis* (MRSE) is associated with increased relapse rates.[31] *S. aureus* peritonitis is more severe, primarily due to its virulent nature and increasing antibiotic resistance. Those who are nasal carriers of this organism are at a three-fold increased risk of exit site infections, tunnel infections and peritonitis.[32] The formation of biofilms on the inner surface of PD catheters by either of these organisms is a frequent cause of antibiotic resistance and relapses and usually requires catheter removal.[33,34]

Gram-negative organisms account for 20–30% of peritonitis overall. Over the years, improvements in connection techniques for PD have decreased the incidence of gram positives but not gram negatives.[35] The ability of gram-negative organisms to develop resistance through plasmid mediated and chromosomal beta-lactamases (especially ESBLs) confers a higher morbidity and poor outcomes.[36–38] *Escherichia coli*, *Klebsiella* and *Pseudomonas* contribute equally, but *Pseudomonas* is notorious for forming tough biofilms and is hence associated with increased rates of relapses, antimicrobial resistance and catheter loss. Exit site infections due to *Pseudomonas* often need a prolonged dual antibiotic course.

Fungal peritonitis is usually very severe and hence carries a high mortality rate. A preceding antibiotic use is usually present in those with fungal peritonitis.[39,40] Both *C. albicans* and non-albicans cause 70% of fungal peritonitis and once detected, treatment coupled with catheter removal improves outcomes.[39]

PATHOGENESIS AND RISK FACTORS

The three interplaying determinants in the pathogenesis of bacteremia are as follows[41]:

1. Defective host immune system.
2. Adherence and virulence characteristics of the infecting organisms.
3. Dialysis access devices and the dialysis apparatus itself.

Uremia and related metabolic complications (anemia, hypoalbuminemia, uremic solutes, hyperparathyroidism, iron overload, malnutrition, etc.) are known to cause impaired phagocytosis and degranulation of polymorphonuclear neutrophils.[42,43] Apart from the type of HD access device, the type of catheter material and possibly surface irregularity and thrombogenecity also play a role. *In vitro* studies have demonstrated that resistance to bacterial adherence was less with polyvinylchloride and polyethylene catheters than silicone elastomer or Teflon catheters.[10,44] The production of extracellular polysaccharides (slime or biofilm), by certain organisms such as staphylococci, *Pseudomonas* and *Candida*, favor increased virulence, adherence to catheter surface and resistance to antibiotics.[45–47] Other catheter-related risk factors that increase the risk of infections include a failure to maintain sterile precautions while placing or accessing CVCs, a longer duration of catheter use, history of CRBSI and prior catheter salvage.

Hemodialysis

The possible routes of contamination of catheters in decreasing order of occurence are as follows:

1. Migration of skin flora along the external (extra luminal) surface of catheters to colonize the tip or a contiguous extension of an exit site or tunnel infection. This is the most common mechanism for non-tunneled CVCs.[48–51]
2. Intraluminal colonization from contamination of catheter hub. This is a common route of contamination for tunneled catheters because the peri-catheter fibrosis along the subcutaneous tract limits extra-luminal migration.[48–50,52]
3. Intraluminal colonization of the catheter with a contaminated solvent/infusate.[53]
4. Hematogenous metastatic seeding of the catheter from distant sources of infection.[54]

Presence of diabetes mellitus, hepatitis C virus infection and intake of immunosuppressive medications also confer increased risk.[55,56] Inadequate water treatment, dialyzer reuse, dialysate or dialysis equipment contamination are some HD procedure related factors that increase the risk of infections.[56] In recent years a "center effect" has been proposed as a risk factor for increased incidence of blood stream infections based on a couple of multicenter studies in the US, where certain dialysis centers were identified as independent risk factors.[9,57] However, a large cohort study in Canada failed to show such an association.[58]

AVGs are usually made of Teflon (polytetrafluoroethylene or PTFE) or polyurethane, which, being foreign bodies, are associated with increased infection rates. Although AVFs are least likely to be infected, this can occur as a post-operative complication from improper aseptic technique or due to non-sterile technique during access.[59] Patients with any of the following factors are considered to be at increased risk for arterio-venous access infections.[60–62]

1. Peri-graft or peri-fistula fluid collections or hematoma.
2. Pseudo-aneurysm.
3. Prolonged post-dialysis bleeding.
4. Surgical manipulation of AVG.
5. Scratching over the needle sites of access.
6. Use of AVF or AVG for non-dialysis access or injection drug abuse.
7. Femoral site for AVG.

Peritoneal Dialysis

Peritonitis is the ultimate complication that results in PD failure and necessitates a switch to HD.[27] The key factors in pathogenesis that are unique to PD modality and different from HD are as follows:

1. Patients handle the access device in PD, whereas in HD trained persons (nurses or technicians) handle them.
2. The frequency of handling the access device is higher (at least 3–4 times daily) compared to HD (twice or thrice weekly).
3. Non-physiological PD fluid that is high in dextrose dwells in the peritoneal cavity and impairs host defense.[63] Also, lactate in the dialysate has been shown to cause stunning of leucocytes. A lower pH in conventional PD fluids and the use of heat sterilization through increased amounts of glucose degradation products (GDPs) causes mesothelial cell and neutrophil apoptosis.[64,65]

Similar to HD, the usual source of infection is either touch contamination during access (intraluminal) or an extension from exit site infection and tunnel infection (periluminal spread). Other less common, but possible, sources of infection include hematogenous seeding from a different source of infection, gastrointestinal tract from transvisceral migration, diverticulitis, cholecystistis, and procedures like diagnostic and therapeutic enemas, colonoscopy or dental extraction. These can allow transient bacteremia to seed the peritoneum. In addition, infection can occur very rarely from vaginal leak due to the anatomical continuity of the female genital tract with the intra-peritoneal lining at the fimbriae of the fallopian tubes.[66–68]

Increasing age, being a current smoker at commencement of dialysis, female sex and use of a pre-twin bag connecting system have been reported as independent risk factors for PD-related peritonitis.[69] Exit sites placed in the abdomen are more mobile, resulting in more trauma and pull on the exit site; both of which increase the risk of exit site infections.[70,71] For this reason when compared to PD catheters with sternal exit sites, those with abdominal exit sites may be associated with increased incidence of infections.[63,70] Exit site infections are associated with increased incidence of PD related peritonitis even when completely resolved. In addition to the increased risk, tunnel infections usually need catheter removal.[72] Biofilm formation by staphylococci and pseudomonas confer similar risk of relapse and antibiotic resistance as in HD.[73] Nasal carriage of *S. aureus* becomes an even more important risk factor due to the organism's virulence and the fact that patients handle PD catheters.[32,74]

Only a small fraction (5%) of PD patients develop peritonitis secondary to intra-abdominal infections like appendicitis, or cholecystitis. However managing such patients can be clinically challenging because of diagnostic issues and they tend to have an increased mortality.[75]

CLINICAL MANIFESTATIONS, DIAGNOSIS, TREATMENT AND PREVENTION

Hemodialysis

A vascular access related infection is suspected when there are signs of inflammation around the device with positive blood cultures and no other localizing signs or apparent source of infection. Clinical manifestations include fever, pain, swelling, tenderness, induration and redness at the site of insertion and along the tract in case of tunneled catheter; and purulent discharge from the exit site. Chills and fever on dialysis are also an indication. Infections of AVF or AVG usually present with warmth, redness and pain at the access site, but not necessarily in every case.[62] Difficult access, poor flow, prolonged post-dialysis bleeding and formation of pseudo-aneurysms, apart from being risk factors, can be the manifestations of arterio-venous (AVF/AVG) access infections. Depending on the inflammatory response, the spectrum of clinical presentation ranges from a bacteremia or systemic inflammatory response syndrome (SIRS) to frank septic shock.

Blood culture is the first and most important step towards diagnosis of HD access related infection. The following definitions aid in proper diagnosis, methods of collection of blood sample and categorization of access related infections.

Exit Site Infection and Tunnel Infection

Exit site infection is defined as signs of inflammation confined to the area surrounding the catheter and the presence of an exudate that proves to be positive on culture. The inflammation must not extend >2 cm from the exit site in the case of non-tunneled catheter or superiorly beyond the cuff in case of a tunneled catheter; which if present is indicative of a tunnel infection.[11,13] This may or may not be associated with pain and tenderness along the subcutaneous tract and a culture-positive exudate at the exit site.[13]

Catheter Related Bacteremia

The National Kidney Foundation – Kidney Diseases Outcomes and Quality Initiate (NKF-KDOQI) Guidelines have adopted the definitions provided by the Centers for Disease Control (CDC), who define a Central Line Associated Blood Stream Infection (CLABSI) as positive blood cultures in a symptomatic (fever, leukocytosis and/or hypotension, in any combination) patient who had a CVC in place at the time of blood draw and that was placed within the preceding 48 hours. Subgroups are also defined as follows:[13,52]

1. Definite: Isolation of same organism >15 colony-forming units (CFU) per catheter segment (semi-quantitative method) simultaneously from the catheter tip and a peripheral or CVC sample in a patient with symptoms and no other identified source of infection (either community acquired or hospital acquired infection).
2. Probable: Catheter tip culture is negative but blood cultures are positive or vice versa, in a symptomatic patient (with no apparent other source of infection) who responds well to antibiotic therapy with symptom defervescence, with or without catheter removal.
3. Possible: Blood cultures positive and symptoms improve with antibiotic therapy or catheter removal.

CRBSI, the term preferred by the Infectious Diseases Society of America (IDSA), is precisely not synonymous with CLABSI, as this latter term was primarily used for surveillance purposes by the CDC. A recent update from IDSA required either of the following to be present for a diagnosis of CRBSI:[76]

1. Quantitative culture results from CVC blood sample demonstrating at least a five-fold increase in the bacterial colony count from that of a peripheral venous blood sample.[77]
2. Automated radiometric blood culture detection systems identifying culture positivity from a CVC blood sample at least 60 minutes prior to culture positivity from a peripheral venous blood sample (this is called "differential time to positivity" or DTP).[78]

In order to improve the yield of blood cultures, it is recommended to draw the same volume of blood for each sample, preferably by a phlebotomy team with proper technique (clean hub with alcohol or alcoholic chlorhexidine or tincture iodine) and always before starting antibiotic therapy, failing which would provide a false negative result.[76] In everyday clinical practice, CRBSI is a presumed diagnosis for the following reasons:

1. The CVC (for catheter tip culture) cannot be removed in all cases soon after diagnosis is suspected (e.g., in patients with poor peripheral access or no other options for dialysis).
2. The IDSA criteria could not be met in all cases due to limitations in expertise and resources.

To demonstrate that a bacteremia is a result of an exit site or tunnel infection, the exudate culture from the exit site and the blood cultures must grow the same organism.[76,79]

For AVF or AVG infections, supportive radiological investigations may be needed for diagnosis. A duplex ultrasound may demonstrate a peri-graft or peri-fistula fluid collection and likewise, a tagged white blood cell scan (nuclear scan using indium 111-labeled white blood cells) may localize infection by showing increased uptake at the graft site. The latter study has a good sensitivity but lacks specificity, as inflammation from any cause can result in increased uptake.[62,80]

Treatment

Systemic antimicrobial therapy and catheter or access device management (removal, exchange or salvage) are the two broad aspects of treatment. Catheter removal is not always feasible in HD patients, who depend on it for regular dialysis needs. Initial work-up should also focus on identifying an alternative source of infection when signs and symptoms are non-specific (e.g., one positive blood culture and fever, leukocytosis and/or hypotension).

ANTIMICROBIAL THERAPY

Empirical antibiotics are usually started soon after suspicion of access device related infections, but it is important that blood cultures are drawn prior to their initiation. The local prevalence rates of bacterial isolates and their cumulative resistance patterns per institutional antibiogram should be taken into account. In general, initial empirical therapy includes a reasonably good coverage for common gram-positive and gram-negative organisms. Subsequent therapy must be guided by the susceptibility pattern of the isolated organism.

Given the fact that gram-positive infections are the most prevalent, vancomycin is the usual empirical antibiotic of choice, especially in areas with increased prevalence of MRSA isolates.[76] When isolates are MRSA-exhibiting, a minimum inhibitory concentration of >2 mg/ml for vancomcyin or VRE, then daptomycin is the drug of choice.[81] Linezolid alone is usually not favored for empirical therapy. When isolates show methicillin-sensitive *S. aureus* or *S. epidermidis*, therapy should be scaled down to penicillin or cephalosporin as per sensitivity pattern.[76]

Gram-negative coverage is usually indicated for critically ill patients and in areas where local surveillance data indicates increased prevalence. The usual choice is a beta-lactam – beta-lactamase inhibitor combination, fourth generation cephalosporin, carbapenem, with or without an aminoglycoside. Combination anti-pseudomonal therapy is indicated for patients with a history of prior infection, severe sepsis, immuno-suppression and neutropenic patients.[76]

Empiric antifungal therapy may be indicated in the following situations:

1. Candida infection/colonization at multiple sites.
2. Prolonged use of broad spectrum antibiotics.
3. Critically ill patients with use of total parenteral nutrition (TPN).
4. Immuno-suppressed patients (bone marrow or solid organ transplants, hematological malignancy).
5. Presence of femoral access.

The recommended drugs for antifungal therapy are either an echinocandin (e.g., caspofungin or micafungin) or sometimes fluconazole (in those patients with no prior azole use and in an area with very low prevalence of non-albicans candida).[76]

Pharmacokinetics of antibiotics is altered in ESRD patients who are dependent on dialysis. The type of dialysis filter, body weight (volume of distribution), inter-dialysis period, variable removal during dialysis, and residual renal functions are a few of the known factors that affect the ideal dosing of antibiotics, especially vancomycin.[15,82,83] The duration of antibiotic therapy is dependent upon the causative organism, clinical picture and presence or absence of metastatic infectious complications or other sites of infection.[15] More complete details in regard to the causative organism and specific dosage are beyond the scope of this chapter.

MANAGEMENT OF VASCULAR ACCESS DEVICE

NON-TUNNELED CATHETERS All non-tunneled catheters associated with bacteremia are recommended to be removed.

TUNNELED CATHETERS For tunneled catheters, the options for management include catheter removal; change over guide-wire and catheter salvage or antibiotic lock.

CATHETER REMOVAL Interruption of HD is itself a barrier towards catheter removal due to ongoing dialysis needs, especially in patients with such catheters as the only available option for dialysis (e.g., failed AVF/AVG). Nevertheless some situations mandate catheter removal and are enumerated in Table 8.1.

CATHETER SALVAGE Catheter salvage, which includes systemic antibiotic therapy coupled with antibiotic lock, has emerged as an alternative viable option for selected patients without a metastatic focus of infection. An antibiotic lock solution is a mix of heparin and an antibiotic (per sensitivity pattern and at very high concentration) that is administered into each lumen soon after dialysis and allowed to remain there in the inter-dialytic period

TABLE 8.1 Indications for Dialysis Access Device Removal

Indications for Hemodialysis Catheter Removal[13,76,90]	Indications for Peritoneal Dialysis Catheter Removal
Persistence of symptoms >36 hours or the presence of severe sepsis, hemodynamic instability and metastatic infection.	Relapsing peritonitis – defined as an episode of peritonitis with the same genus/species that caused a preceding episode of peritonitis within 4 weeks of completion of a course of antibiotics or one sterile episode.
Blood cultures that continue to remain positive >72 hours of antibiotic therapy.	Refractory peritonitis – defined as failure to respond to appropriate antibiotics ≤5 days (lack of improvement in cell count or clinical symptoms).
If isolates grow difficult to clear organisms (biofilm producing) like *S. aureus*, *Pseudomonas* or fungi.	Refractory exit site and tunnel infections.
Associated tunnel infection (with any organism) and exit site infections (except non-severe infection with *S. epidermidis*).	Fungal peritonitis.
Arterio-venous graft with infection, fluid collection or abscess formation.	Peritonitis associated with intra-abdominal pathology or abscess.

for the duration of systemic antibiotic therapy. This serves to pharmacologically eradicate biofilms on the intraluminal aspect of the catheter.[84] Gentamicin, vancomcyin, cephalosporins, citrate and taurolidine are the various agents used, either alone or in selected combinations.[85,86] Meta-analyses of studies on antibiotic lock therapy have reported a success rate of 60–70%, but also noted several limitations; a few worth mentioning include publication bias, paucity of large trials and absence of data on antimicrobial resistance.[87] The success rate for cure was highest for gram-negative infections, followed by *S. epidermidis* and enterococci, making this modality relatively organism specific.[88] It is not recommended as an option for *S. aureus* (especially MRSA) infections which are associated with very high relapse rates, complications and metastatic spread. Similarly, salvage is not a good choice for recurrent CRBSIs, except in very rare situations with no other options for HD access and infection recurrence, even with other sterile precautions or preventive measures, as discussed later.[89,90]

EXCHANGE OVER A GUIDE-WIRE When clinical improvement occurs within 48–72 hours of initiation of systemic antibiotic therapy and blood cultures turn negative, exchange over a guide-wire is an available option. However, this method is preferred only for those without tunnel infection and CRBSIs with few non-virulent organisms.[91] The NKF-KDOQI guidelines recommend systemic antibiotic therapy for at least three weeks, with periodic monitoring of blood cultures during and after antibiotic therapy. This approach did result in a reduction in unnecessary short term catheter insertions for routine dialysis needs while awaiting completion of antibiotic therapy, cure of infection and subsequent establishment of a different long term access for dialysis.[92]

ARTERIO-VENOUS ACCESS (AVF/AVG) Infections of AVF, unless associated with abscess, fluid collection or septic emboli, do not warrant surgery and usually resolve completely in six weeks with systemic antibiotic therapy.[13] AVG infections require surgical intervention, the type of which is determined by age of graft, and the extent and severity of infection. Graft infections occurring within a month (before graft incorporation by fibrous tissue) of placement mandate removal regardless of the extent of infection.[13] Similarly, if abscess or infected fluid collections embed the whole graft, removal is needed regardless of the age of the graft.[93] Graft preservation techniques, such as limited partial excision of infected graft or subtotal excision (that retains only the arterial segment maintaining patency), may be an option for those patients with minimal infection of the exposed graft and without signs of severe sepsis.[94,95]

Prevention

The incidence of infections is lowest with AVF. This observation, coupled with the benefits of reduction in morbidity and mortality in the long term, led to the formation of Fistula First Breakthrough Initiative in 2003 by Centers for Medicare and Medicaid services (CMS) in association with the Institute of Health Care Improvement in the USA to promote preferential and early placement of AVF for HD access. Since then, the prevalence of AVF has increased from 33% in 2003 to 60% in 2011, but without a reasonable decline in CVC use. Hence these organizations are now targeting measures to minimize the need for CVC for HD.[96]

The following is a list of suggested interventions for the prevention of hemodialysis-related bloodstream infections.[87] It is not all-inclusive, but contains only those factors for which conclusive evidence exists, as adopted from CDC[10,97] and NKF-KDOQI[13]:

1. Avoid use of HD access for non-dialysis access.[10,13,97]
2. Conduct surveillance of blood stream infections.[13,97]
3. Put in place staff and patient education about infection control measures.[13,97]
4. Use hand hygiene measures before accessing.[13,97]
5. Patient and staff members should wear a surgical mask prior to access.[13]
6. Apply topical antimicrobials, either povidone iodine or polyantibiotic (bacitracin/gramicidin/polymixin B) ointments, at exit sites.[10,13,85,97]
7. Use chlorhexidine for skin antisepsis.[13,97]

Although the nasal carriage of *S. aureus* has been associated with increased incidence of CRBSIs, nasal decolonization is not routinely recommended by experts for widespread use, due to the concerns about the rapid development of resistance.[98,99] However, this may still serve as a viable option for individual patients with recurrent blood stream infections.[100] The following strategies were reported by two different study groups:

1. Topical mupirocin applied twice or thrice weekly for two weeks followed by weekly application for up to nine months.[101]
2. Oral rifamipin for five days.[99]

Antibiotic coating of catheters had not been shown to reduce the incidence of CRBSI and hence is not recommended at this time. Antimicrobial locks (vancomycin, gentamicin, minocycline) have been associated with decreased rates of both CRBSI and exit site infections, but due to the concern over selecting out resistant organisms and drug related toxicities (e.g., ototoxicity with gentamicin) this method is not recommended for routine long term prevention.[102] Despite this, antimicrobial locks may still be an option for patients with recurrent infections despite maintaining preventive interventions mentioned above.[10]

Peritoneal Dialysis

Exit Site Infection

ES infection is defined by the presence of purulent drainage from the exit site and/or erythema. Preceding trauma, an upward facing exit site and lack of prophylactic antibiotic use are predisposing factors for exit site infections. They are usually managed in the outpatient setting, but assume significance because up to one fourth of exit site infections can lead to peritonitis. The exudate should be sent for gram stain and culture for identification and sensitivity.

Isolated tunnel infections are rare, as they usually occur in conjunction with exit site infection. This presents as tenderness, erythema and swelling along the subcutaneous tract, but can be occult requiring an ultrasound for diagnosis. For this reason, exit site and tunnel infections of PD catheters are usually considered collectively, as the etiology and management are usually indifferent in a majority of cases.

Peritonitis

The clinical signs and symptoms vary depending on the causative organism. PD peritonitis secondary to gram-negative organisms, *S. aureus*, fungi and *Streptococcus* are usually more painful than that caused by coagulase negative staphylococci.[27,103,104] Abdominal pain and cloudy effluent are the two most common presenting features of peritonitis. The effluent becomes cloudy due to the presence of leucocytes in the fluid.[105] Other common symptoms include nausea, fever and hypotension.

The usual diagnostic criteria include:

1. Cloudy effluent.
2. Cell count more than 100 leucocytes/ml.
3. Percentage of polymorphonuclear leukocytes >50%.
4. Positive culture.

However, if the fluid has been in the peritoneum for less than 2 hours, or if the patient cannot mount a good inflammatory response, the white blood cell count may be less than 100/ml. In such cases, the percentage of polymorphs exceeding 50% may be an important clue. A predominance of lymphocytes should suggest fungal or mycobacterial peritonitis. Eosinophil predominance may point toward a drug-induced etiology for peritonitis. Very high levels of amylase and lipase in PD effluent should prompt an aggressive search for secondary causes of peritonitis.[106] PD fluid culture will reveal the organism in the majority of patients provided a proper culture technique is utilized. However, the culture can be negative in up to one fifth of patients with peritonitis. An ideal technique should include obtaining PD effluent after a dwell time of more than an hour and the specimen should be from a centrifuge of a 50 ml sample of this PD effluent that is then suspended in 3–5 ml of saline and plated into both solid culture media and standard blood culture medium. This culture method is considered to be most sensitive and decreases the chances of culture negativity.[37]

Treatment

Exit Site and Tunnel Infection

Antimicrobial prophylaxis as a part of exit site care has decreased the incidence of exit site infections. Topical mupirocin, gentamicin, medical grade honey have all been used.[107,108] In one study the use of polysporin for prophylaxis resulted in a higher incidence of fungal peritonitis and is discouraged. Mild infection of the exit site is treated with topical antibacterial agents like hydrogen peroxide or chlorhexidine. Empirical treatment for *Pseudomonas* and *S. aureus* should be included in a patient with prior history of infections with these organisms. Generally, oral antibiotic therapy is sufficient except in patients with MRSA infections, for whom intravenous therapy is needed. Typically, gram-positive organisms are treated with an oral penicillin or cephalosporin. If *Pseudomonas* infection is suspected, an oral fluoroquinolone is the drug of choice. The usual duration of treatment is 14 days or until the exit site appears clean and non-inflamed, whichever is longer.

Peritonitis

PD peritonitis is typically treated with intra-peritoneal antibiotics. The dosing and frequency of antibiotics vary according the modality of PD (continuous ambulatory peritoneal dialysis [CAPD] vs. continuous cycling peritoneal dialysis [CCPD]). The initial antibiotic regimen should include both gram-positive (with vancomycin when there is a high incidence of MRSA infection, or otherwise a first generation cephalosporin) and gram-negative (third generation cephalosporin or an aminoglycoside) coverage.[37] Intra-peritoneal antibiotics have shown to achieve therapeutic serum concentrations, especially when the peritoneum is inflamed, when it allows transport of antibiotics into the bloodstream. It is recommended that continuous antibiotic (added to each bag) be used for the first 24–48 hours, and after that switch to intermittent dosing. The intermittent dose should be added to the cycle with the long dwell (night-time for CAPD and day-time for automated peritoneal dialysis [APD]). Short term use (less than two weeks) of intra-peritoneal aminoglycosides may not impair residual renal function.[109] However, repeated or prolonged courses of antibiotics, especially for the regimens with aminoglycosides are generally not advised. Apart from antibiotic use, heparin should be instilled in PD fluid to prevent fibrin formation and PD catheter malfunction. Indications for PD catheter removal in patients with peritonitis are enumerated in Table 8.1.

Prevention

Elimination of nasal carriage of *S. aureus* with intranasal mupirocin, a downward directed tunnel, avoidance of trauma and use of intravenous antibiotic during placement, providing effective training for handling, adequate instruction on thorough hand washing and drying prior to accessing the catheter and prophylactic topical antibiotics (e.g., gentamicin, mupirocin) at the exit site are a few evidence-based preventive measures for PD catheter-related infections. Continuous quality improvement programs are essential for PD centers for root cause analysis, continuous monitoring and surveillance of infections. Procedures that disrupt the mucosal barrier (colonoscopy, hysteroscopy, dental extraction, cholecystectomy) need antibiotic prophylaxis to prevent peritonitis.[110]

CONCLUSIONS AND FUTURE PROSPECTS

The recent improvements in diagnosis and surveillance of CRBSI, and adherence to novel preventive strategies have resulted in better identification of infective organisms and a decline in the incidence and recurrence of dialysis access related infections. Even so, such infections continue to pose a significant burden due to rising antimicrobial resistance. Successful management incorporates decisions about catheter removal and the institution of appropriate antibiotics with optimal dose and duration. Although catheter or graft removal is indicated in proven cases of related bloodstream infections, the salvage methods serve as a potential rescue option in selected patients without alternative access options or infected with relatively less virulent organisms. A future focus on developing sensitive and specific novel diagnostic methods to detect access device associated infections, and producing accesses which are less prone to infection would help to improve the longevity of catheters and grafts. This would result in better dialysis adequacy, avoid morbidity and mortality in dialysis patients, and so would reduce health care costs.

References

1. Collins AJ, Foley RN, Chavers B, Gilbertson D, Herzog C, Johansen K, et al. United States renal data system 2011 annual data report: atlas of chronic kidney disease & end-stage renal disease in the United States. *Am J Kidney Dis* 2012;**59**:A7, e1–420.
2. Sarnak MJ, Jaber BL. Mortality caused by sepsis in patients with end-stage renal disease compared with the general population. *Kidney Int* 2000;**58**:1758–64.
3. Jain AK, Blake P, Cordy P, Garg AX. Global trends in rates of peritoneal dialysis. *J Am Soc Nephrol* 2012;**23**:533–44.
4. Boudville N, Kemp A, Clayton P, Lim W, Badve SV, Hawley CM, et al. Recent peritonitis associates with mortality among patients treated with peritoneal dialysis. *J Am Soc Nephrol* 2012;**23**:1398–405.
5. Aslam N, Bernardini J, Fried L, Burr R, Piraino B. Comparison of infectious complications between incident hemodialysis and peritoneal dialysis patients. *Clin J Am Soc Nephrol* 2006;**1**:1226–33.
6. Collins AJ, Foley RN, Gilbertson DT, Chen SC. The state of chronic kidney disease, ESRD, and morbidity and mortality in the first year of dialysis. *Clin J Am Soc Nephrol* 2009;**4**:S5–11.
7. Stevenson KB, Hannah EL, Lowder CA, Adcox MJ, Davidson RL, Mallea MC, et al. Epidemiology of hemodialysis vascular access infections from longitudinal infection surveillance data: predicting the impact of NKF-DOQI clinical practice guidelines for vascular access. *Am J Kidney Dis* 2002;**39**:549–55.
8. Lafrance JP, Rahme E, Lelorier J, Iqbal S. Vascular access-related infections: definitions, incidence rates, and risk factors. *Am J Kidney Dis* 2008;**52**:982–93.
9. Tokars JI, Miller ER, Stein G. New national surveillance system for hemodialysis-associated infections: initial results. *Am J Infect Control* 2002;**30**:288–95.
10. O'Grady NP, Alexander M, Burns LA, Dellinger EP, Garland J, Heard SO, et al. Guidelines for the prevention of intravascular catheter-related infections. *Clin Inf Dis* 2011;**52**:e162–93.
11. Akoh J. Vascular access infections: epidemiology, diagnosis, and management. *Cur Inf Dis Rep* 2011;**13**:324–32.
12. Klevens RM, Edwards JR, Andrus ML, Peterson KD, Dudeck MA, Horan TC. Dialysis Surveillance Report: National Healthcare Safety Network (NHSN)-data summary for 2006. *Semin Dial* 2008;**21**:24–8.
13. Vascular Access Work Group. Clinical practice guidelines for vascular access. *Am J Kidney Dis* 2006;**48**: S248–73.
14. Alexandraki I, Sullivan R, Zaiden R, Bailey C, McCarter Y, Khan A, et al. Blood culture isolates in hemodialysis vascular catheter-related bacteremia. *Am J Med Sci* 2008;**336**:297–302.
15. Fitzgibbons LN, Puls DL, Mackay K, Forrest GN. Management of gram-positive coccal bacteremia and hemodialysis. Am J Kidney Dis 2011;**57**:624–40.
16. Sader HS, Jones RN, Rossi KL, Rybak MJ. Occurrence of vancomycin-tolerant and heterogeneous vancomycin-intermediate strains (hVISA) among *Staphylococcus aureus* causing bloodstream infections in nine USA hospitals. *J Antimicrob Chemother* 2009;**64**:1024–8.
17. Sievert DM, Rudrik JT, Patel JB, McDonald LC, Wilkins MJ, Hageman JC. Vancomycin-resistant *Staphylococcus aureus* in the United States, 2002–2006. *Clin Infect Dis* 2008;**46**:668–74.
18. Kallen AJ, Arduino MJ, Patel PR. Preventing infections in patients undergoing hemodialysis. *Expert Rev Anti Infect Ther* 2010;**8**:643–55.
19. Atta MG, Eustace JA, Song X, Perl TM, Scheel Jr PJ. Outpatient vancomycin use and vancomycin-resistant enterococcal colonization in maintenance dialysis patients. *Kidney Int* 2001;**59**:718–24.
20. Grabsch EA, Burrell LJ, Padiglione A, O'Keeffe JM, Ballard S, Grayson ML. Risk of environmental and healthcare worker contamination with vancomycin-resistant enterococci during outpatient procedures and hemodialysis. *Infect Control Hosp Epidemiol* 2006;**27**:287–93.
21. Centers for Disease Control and Prevention (CDC). Invasive methicillin-resistant *Staphylococcus aureus* infections among dialysis patients – United States, 2005. *MMWR Morb Mortal Wkly Rep* 2007;**56**:197–9.

22. Servais A, Mercadal L, Brossier F, Venditto M, Issad B, Isnard-Bagnis C, et al. Rapid curbing of a vancomycin-resistant *Enterococcus faecium* outbreak in a nephrology department. *Clin J Am Soc Nephrol* 2009;**4**:1559–64.
23. Dopirak M, Hill C, Oleksiw M, Dumigan D, Arvai J, English E, et al. Surveillance of hemodialysis-associated primary bloodstream infections: the experience of ten hospital-based centers. *Infect Control Hosp Epidemiol* 2002;**23**:721–4.
24. Saely S, Kaye KS, Fairfax MR, Chopra T, Pogue JM. Investigating the impact of the definition of previous antibiotic exposure related to isolation of extended spectrum beta-lactamase-producing *Klebsiella pneumoniae*. *Am J Infect Control* 2011;**39**:390–5.
25. Pyrgos V, Ratanavanich K, Donegan N, Veis J, Walsh TJ, Shoham S. *Candida* bloodstream infections in hemodialysis recipients. *Med Mycol* 2009;**47**:463–7.
26. Barberino MG, Silva N, Rebouças C, Barreiro K, Alcântara AP, Netto EM, et al. Evaluation of blood stream infections by *Candida* in three tertiary hospitals in Salvador, Brazil: a case-control study. *Braz J Inf Dis* 2006;**10**:36–40.
27. Voinescu CG, Khanna R. Peritonitis in peritoneal dialysis. *Int J Artif Organs* 2002;**25**:249–60.
28. Mujais S. Microbiology and outcomes of peritonitis in North America. *Kidney Int Suppl* 2006;**103**:S55–62.
29. Davenport A. Peritonitis remains the major clinical complication of peritoneal dialysis: the London, UK, peritonitis audit 2002–2003. *Perit Dial Int* 2009;**29**:297–302.
30. Fahim M, Hawley CM, McDonald SP, Brown FG, Rosman JB, Wiggins KJ, et al. Culture-negative peritonitis in peritoneal dialysis patients in Australia: predictors, treatment, and outcomes in 435 cases. *Am J Kidney Dis* 2010;**55**:690–7.
31. Szeto CC, Kwan BC, Chow KM, Lau MF, Law MC, Chung KY, et al. Coagulase negative staphylococcal peritonitis in peritoneal dialysis patients: review of 232 consecutive cases. *Clin J Am Soc Nephrol* 2008;**3**:91–7.
32. Nouwen J, Schouten J, Schneebergen P, Snijders S, Maaskant J, Koolen M, et al. *Staphylococcus aureus* carriage patterns and the risk of infections associated with continuous peritoneal dialysis. *J Clin Microbiol* 2006;**44**:2233–6.
33. Anwar H, Dasgupta MK, Costerton JW. Testing the susceptibility of bacteria in biofilms to antibacterial agents. *Antimicrob Agents Chemother* 1990;**34**:2043–6.
34. Finkelstein ES, Jekel J, Troidle L, Gorban-Brennan N, Finkelstein FO, Bia FJ. Patterns of infection in patients maintained on long-term peritoneal dialysis therapy with multiple episodes of peritonitis. *Am J Kidney Dis* 2002;**39**:1278–86.
35. Szeto CC, Chow KM. Gram-negative peritonitis – the Achilles heel of peritoneal dialysis? *Perit Dial Int* 2007;**27**:S267–71.
36. Szeto CC, Chow VC, Chow KM, Lai RW, Chung KY, Leung CB, et al. Enterobacteriaceae peritonitis complicating peritoneal dialysis: a review of 210 consecutive cases. *Kidney Int* 2006;**69**:1245–52.
37. Li PK, Szeto CC, Piraino B, Bernardini J, Figueiredo AE, Gupta A, et al. Peritoneal dialysis-related infections recommendations: 2010 update. *Perit Dial Int* 2010;**30**:393–423.
38. Jarvis EM, Hawley CM, McDonald SP, Brown FG, Rosman JB, Wiggins KJ, et al. Predictors, treatment, and outcomes of non-Pseudomonas gram-negative peritonitis. *Kidney Int* 2010;**78**:408–14.
39. Miles R, Hawley CM, McDonald SP, Brown FG, Rosman JB, Wiggins KJ, et al. Predictors and outcomes of fungal peritonitis in peritoneal dialysis patients. *Kidney Int* 2009;**76**:622–8.
40. Michel C, Courdavault L, al Khayat R, Viron B, Roux P, Mignon F. Fungal peritonitis in patients on peritoneal dialysis. *Am J Nephrol* 1994;**14**:113–20.
41. Jaber BL. Bacterial infections in hemodialysis patients: pathogenesis and prevention. *Kidney Int* 2005;**67**:2508–19.
42. Vanholder R, Ringoir S. Polymorphonuclear cell function and infection in dialysis. *Kidney Int Suppl* 1992;**38**: S91–5.
43. Pesanti EL. Immunologic defects and vaccination in patients with chronic renal failure. *Infect Dis Clin North Am* 2001;**15**:813–32.
44. Ashkenazi S, Weiss E, Drucker MM. Bacterial adherence to intravenous catheters and needles and its influence by cannula type and bacterial surface hydrophobicity. *J Lab Clin Med* 1986;**107**:136–40.
45. Mack D, Davies AP, Harris LG, Rohde H, Horstkotte MA, Knobloch JK. Microbial interactions in *Staphylococcus epidermidis* biofilms. *Anal Bioanal Chem* 2007;**387**:399–408.

46. Murga R, Miller JM, Donlan RM. Biofilm formation by gram-negative bacteria on central venous catheter connectors: effect of conditioning films in a laboratory model. *J Clin Microbiol* 2001;**39**:2294–7.
47. Douglas LJ. *Candida* biofilms and their role in infection. *Trends Microbiol* 2003;**11**:30–6.
48. Cooper GL, Hopkins CC. Rapid diagnosis of intravascular catheter-associated infection by direct gram staining of catheter segments. *N Engl J Med* 1985;**312**:1142–7.
49. Cheesbrough JS, Finch RG, Burden RP. A prospective study of the mechanisms of infection associated with hemodialysis catheters. *J Infect Dis* 1986;**154**:579–89.
50. Raad I, Costerton W, Sabharwal U, Sacilowski M, Anaissie E, Bodey GP. Ultrastructural analysis of indwelling vascular catheters: a quantitative relationship between luminal colonization and duration of placement. *J Infect Dis* 1993;**168**:400–7.
51. Safdar N, Maki D. The pathogenesis of catheter-related bloodstream infection with noncuffed short-term central venous catheters. *Intensive Care Med* 2004;**30**:62–7.
52. Horan TC, Andrus M, Dudeck MA. CDC/NHSN surveillance definition of health care-associated infection and criteria for specific types of infections in the acute care setting. *Am J Infect Control* 2008;**36**:309–32.
53. Raad I, Hanna HA, Awad A, Alrahwan A, Bivins C, Khan A, et al. Optimal Frequency of changing intravenous administration sets: is it safe to prolong use beyond 72 hours? *Inf Control Hosp Epidemiol* 2001;**22**:136–9.
54. Anaissie E, Samonis G, Kontoyiannis D, Costerton J, Sabharwal U, Bodey G, et al. Role of catheter colonization and infrequent hematogenous seeding in catheter-related infections. *Eur J Clin Microbiol Inf Dis* 1995;**14**:134–7.
55. Hoen B, Paul-Dauphin A, Hestin D, Kessler M. EPIBACDIAL: a multicenter prospective study of risk factors for bacteremia in chronic hemodialysis patients. *J Am Soc Nephrol* 1998;**9**:869–76.
56. Katneni R, Hedayati SS. Central venous catheter-related bacteremia in chronic hemodialysis patients: epidemiology and evidence-based management. *Nat Clin Pract Nephrol* 2007;**3**:256–66.
57. Tokars JI, Light P, Anderson J, Miller ER, Parrish J, Armistead N, et al. A prospective study of vascular access infections at seven outpatient hemodialysis centers. *Am J Kidney Dis* 2001;**37**:1232–40.
58. Lafrance JP, Iqbal S, Lelorier J, Dasgupta K, Ritchie J, Ward L, et al. Vascular access-related bloodstream infections in First Nations, community and teaching Canadian dialysis units, and other center-level predictors. *Nephron Clin Pract* 2010;**114**:c204–12.
59. Nassar GM, Ayus JC. Infectious complications of the hemodialysis access. *Kidney Int* 2001;**60**:1–13.
60. Anderson JE, Chang AS, Anstadt MP. Polytetrafluoroethylene hemoaccess site infections. *ASAIO J* 2000;**46**:S18–21.
61. Bonomo RA, Rice D, Whalen C, Linn D, Eckstein E, Shlaes DM. Risk factors associated with permanent access-site infections in chronic hemodialysis patients. *Infect Control Hosp Epidemiol* 1997;**18**:757–61.
62. Padberg Jr FT, Calligaro KD, Sidawy AN. Complications of arteriovenous hemodialysis access: Recognition and management. *J Vasc Surg* 2008;**48**:S55–80.
63. Teitelbaum I, Burkart J. Peritoneal dialysis. *Am J Kidney Dis* 2003;**42**:1082–96.
64. Lee DH, Choi SY, Ryu HM, Kim CD, Park SH, Chung HY, et al. 3,4-dideoxyglucosone-3-ene induces apoptosis in human peritoneal mesothelial cells. *Perit Dial Int* 2009;**29**:44–51.
65. Perl J, Nessim SJ, Bargman JM. The biocompatibility of neutral pH, low-GDP peritoneal dialysis solutions: benefit at bench, bedside, or both[quest]. *Kidney Int* 2011;**79**:814–24.
66. Fried L, Bernardini J, Piraino B. Iatrogenic peritonitis: the need for prophylaxis. *Perit Dial Int* 2000;**20**:343–5.
67. Coward RA, Gokal R, Wise M, Mallick NP, Warrell D. Peritonitis associated with vaginal leakage of dialysis fluid in continuous ambulatory peritoneal dialysis. *Br Med J (Clin Res Ed)* 1982;**284**:1529.
68. Scanziani R, Dozio B, Baragetti I, Grillo P, Colombo L, De Liso S, et al. Vaginal colonization with group B Streptococcus (*Streptococcus agalactiae*) and peritonitis in a woman on CAPD. *Nephrol Dial Transplant* 1999;**14**:2222–4.
69. Kotsanas D, Polkinghorne KR, Korman TM, Atkins RC, Brown F. Risk factors for peritoneal dialysis-related peritonitis: can we reduce the incidence and improve patient selection? *Nephrology (Carlton)* 2007;**12**:239–45.
70. Zimmerman DG. Presternal catheter design – an opportunity to capitalize on catheter immobilization. *Adv Perit Dial* 2010;**26**:91–5.
71. Piraino B, Bailie GR, Bernardini J, Boeschoten E, Gupta A, Holmes C, et al. ISPD guidelines/recommendations. *Periton Dial Int* 2005;**25**:107–31.

72. van Diepen AT, Tomlinson GA, Jassal SV. The association between exit site infection and subsequent peritonitis among peritoneal dialysis patients. *Clin J Am Soc Nephrol* 2012;**7**:1266–71.
73. Finkelstein ES, Jekel J, Troidle L, Gorban-Brennan N, Finkelstein FO, Bia FJ. Patterns of infection in patients maintained on long-term peritoneal dialysis therapy with multiple episodes of peritonitis. *Am J Kidney Dis* 2002;**39**:1278–86.
74. Luzar MA, Coles GA, Faller B, Slingeneyer A, Dah GD, Briat C, et al. *Staphylococcus aureus* nasal carriage and infection in patients on continuous ambulatory peritoneal dialysis. *N Engl J Med* 1990;**322**:505–9.
75. Tzamaloukas AH, Obermiller LE, Gibel LJ, Murata GH, Wood B, Simon D, et al. Peritonitis associated with intra-abdominal pathology in continuous ambulatory peritoneal dialysis patients. *Perit Dial Int* 1993;**13**: S335–7.
76. Mermel LA, Allon M, Bouza E, Craven DE, Flynn P, O'Grady NP, et al. Clinical Practice guidelines for the diagnosis and management of intravascular catheter-related infection: 2009 update by the Infectious Diseases Society of America. *Clin Inf Dis* 2009;**49**:1–45.
77. Safdar N, Fine JP, Maki DG. Meta-analysis: methods for diagnosing intravascular device-related bloodstream infection. *Ann Intern Med* 2005;**142**:451–66.
78. Raad I, Hanna HA, Alakech B, Chatzinikolaou I, Johnson MM, Tarrand J. Differential time to positivity: a useful method for diagnosing catheter-related bloodstream infections. *Ann Intern Med* 2004;**140**:18–25.
79. Bagdasarian N, Heung M, Malani PN. Infectious complications of dialysis access devices. *Infect Dis Clin North Am* 2012;**26**:127–41.
80. Williamson MR, Boyd CM, Read RC. 111 In-labeled leukocytes in the detection of prosthetic vascular graft infections. *Am J Roentgenol* 1986;**147**:173–6.
81. Boucher HW, Sakoulas G. Perspectives on Daptomycin resistance, with emphasis on resistance in *Staphylococcus aureus*. *Clin Infect Dis* 2007;**45**:601–8.
82. Pai AB, Pai MP. Optimizing antimicrobial therapy for gram-positive bloodstream infections in patients on hemodialysis. *Adv Chronic Kidney Dis* 2006;**13**:259–70.
83. Vandecasteele SJ, De Vriese AS. Vancomycin dosing in patients on intermittent hemodialysis. *Semin Dial* 2011;**24**:50–5.
84. Andris DA, Krzywda EA, Edmiston CE, Krepel CJ, Gohr CM. Elimination of intraluminal colonization by antibiotic lock in silicone vascular catheters. *Nutrition* 1998;**14**:427–32.
85. James MT, Conley J, Tonelli M, Manns BJ, MacRae J, Hemmelgarn BR. Meta-analysis: antibiotics for prophylaxis against hemodialysis catheter-related infections. *Ann Intern Med* 2008;**148**:596–605.
86. Snaterse M, Ruger W, Scholte Op Reimer WJ, Lucas C. Antibiotic-based catheter lock solutions for prevention of catheter-related bloodstream infection: a systematic review of randomised controlled trials. *J Hosp Infect* 2010;**75**:1–11.
87. Patel PR, Kallen AJ, Arduino MJ. Epidemiology, surveillance, and prevention of bloodstream infections in hemodialysis patients. *Am J Kidney Dis* 2010;**56**:566–77.
88. Poole CV, Carlton D, Bimbo L, Allon M. Treatment of catheter-related bacteraemia with an antibiotic lock protocol: effect of bacterial pathogen. *Nephrol Dial Transplant* 2004;**19**:1237–44.
89. Ashby DR, Power A, Singh S, Choi P, Taube DH, Duncan ND, et al. Bacteremia associated with tunneled hemodialysis catheters: outcome after attempted salvage. *Clin J Am Soc Nephrol* 2009;**4**:1601–5.
90. Sychev D, Maya ID, Allon M. Clinical management of dialysis catheter-related bacteremia with concurrent exit-site infection. *Semin Dial* 2011;**24**:239–41.
91. Robinson D, Suhocki P, Schwab SJ. Treatment of infected tunneled venous access hemodialysis catheters with guidewire exchange. *Kidney Int* 1998;**53**:1792–4.
92. Tanriover B, Carlton D, Saddekni S, Hamrick K, Oser R, Westfall AO, et al. Bacteremia associated with tunneled dialysis catheters: comparison of two treatment strategies. *Kidney Int* 2000;**57**:2151–5.
93. Akoh JA. Prosthetic arteriovenous grafts for hemodialysis. *J Vasc Access* 2009;**10**:137–47.
94. Raju S. PTFE grafts for hemodialysis access. Techniques for insertion and management of complications. *Ann Surg* 1987;**206**:666–73.
95. Ryan SV, Calligaro KD, Scharff J, Dougherty MJ. Management of infected prosthetic dialysis arteriovenous grafts. *J Vasc Surg* 2004;**39**:73–8.
96. Vassalotti JA, Jennings WC, Beathard GA, Neumann M, Caponi S, Fox CH, et al. Fistula first breakthrough initiative: targeting catheter last in fistula first. *Semin Dial* 2012;**25**:303–10.

97. O'Grady NP, Alexander M, Dellinger EP, Gerberding JL, Heard SO, Maki DG, et al. Guidelines for the prevention of intravascular catheter-related infections. *Am J Infect Control* 2002;**30**:476–89.
98. Lederer SR, Riedelsdorf G, Schiffl H. Nasal carriage of meticillin resistant Staphylococcus aureus: the prevalence, patients at risk and the effect of elimination on outcomes among outclinic haemodialysis patients. *Eur J Med Res* 2007;**12**:284–8.
99. Yu VL, Goetz A, Wagener M, Smith PB, Rihs JD, Hanchett J, et al. *Staphylococcus aureus* nasal carriage and infection in patients on hemodialysis. Efficacy of antibiotic prophylaxis. *N Engl J Med* 1986;**315**:91–6.
100. Vandecasteele SJ, Boelaert JR, De Vriese AS. *Staphylococcus aureus* infections in hemodialysis: what a nephrologist should know. *Clin J Am Soc Nephrol* 2009;**4**:1388–400.
101. Boelaert JR, Van Landuyt HW, Gordts BZ, De Baere YA, Messer SA, Herwaldt LA. Nasal and cutaneous carriage of *Staphylococcus aureus* in hemodialysis patients: the effect of nasal mupirocin. *Infect Control Hosp Epidemiol* 1996;**17**:809–11.
102. Rabindranath KS, Bansal T, Adams J, Das R, Shail R, MacLeod AM, et al. Systematic review of antimicrobials for the prevention of haemodialysis catheter-related infections. *Nephrol Dial Transplant* 2009;**24**:3763–74.
103. Kim D, Tapson J, Wu G, Khanna R, Vas SI, Oreopoulos DG. *Staph aureus* peritonitis in patients on continuous ambulatory peritoneal dialysis. *Trans Am Soc Artif Intern Organs* 1984;**30**:494–7.
104. Munoz de Bustillo E, Aguilera A, Jimenez C, Bajo MA, Sanchez C, Selgas R. Streptococcal versus *Staphylococcus epidermidis* peritonitis in CAPD. A comparative study. *Perit Dial Int* 1997;**17**:392–5.
105. Oliveira LG, Luengo J, Caramori JC, Montelli AC, Cunha Mde L, Barretti P. Peritonitis in recent years: clinical findings and predictors of treatment response of 170 episodes at a single Brazilian center. *Int Urol Nephrol* 2012;**44**:1529–37.
106. Caruana RJ, Burkart J, Segraves D, Smallwood S, Haymore J, Disher B. Serum and peritoneal fluid amylase levels in CAPD. Normal values and clinical usefulness. *Am J Nephrol* 1987;**7**:169–72.
107. Bernardini J, Bender F, Florio T, Sloand J, Palmmontalbano L, Fried L, et al. Randomized, double-blind trial of antibiotic exit site cream for prevention of exit site infection in peritoneal dialysis patients. *J Am Soc Nephrol* 2005;**16**:539–45.
108. Johnson DW, Clark C, Isbel NM, Hawley CM, Beller E, Cass A, et al. The honeypot study protocol: a randomized controlled trial of exit-site application of medihoney antibacterial wound gel for the prevention of catheter-associated infections in peritoneal dialysis patients. *Perit Dial Int* 2009;**29**:303–9.
109. Baker RJ, Senior H, Clemenger M, Brown EA. Empirical aminoglycosides for peritonitis do not affect residual renal function. *Am J Kidney Dis* 2003;**41**:670–5.
110. Piraino B, Bernardini J, Brown E, Figueiredo A, Johnson DW, Lye W-C, et al. ISPD Position statement on reducing the risks of peritoneal dialysis-related infections. *Perit Dial Int* 2011;**31**:614–30.

CHAPTER

9

Microbiological Diagnosis and Antimicrobial Therapy in Acute Pancreatitis

Lipika Singhal, Amber Prasad and Vikas Gautam

Department of Medical Microbiology, Postgraduate Institute of Medical Education and Research, Chandigarh, India

INTRODUCTION

Acute pancreatitis (AP) is defined as an acute inflammatory disease of the pancreas with variable involvement of peri-pancreatic tissues and/or remote organ systems. The commonest cause is cholelithiasis, with alcohol being the next common cause.[1] Disease severity in patients with acute pancreatitis varies from mild disease with minimal morbidity to severe disease. Clinically, AP comprises a mild edematous interstitial inflammation, which is self-limiting and resolves spontaneously without complications. Fifteen to 20% of patients develop severe pancreatitis. According to the Atlanta classification, severe acute pancreatitis (SAP) is defined as an AP associated with local and/or systemic complications.[1,2] There are two peaks of lethality in severe acute pancreatitis. In the first phase, extensive pancreatic inflammation and necrosis are followed by a systemic inflammatory response syndrome (SIRS) secondary to cytokine release that may lead to multiple organ dysfunction syndrome (MODS) within the first week and early mortality, without any apparent bacterial infection. Inappropriate intracellular activation of proteases occurs, along with conversion of trypsinogen to trypsin leading to pancreatic tissue damage in the presence of intracellular lysozyme enzymes. Trypsin in turn activates the up-regulation of pro-inflammatory chemokines and cytokines as interleukin-1 (IL-1), interleukin-6 (IL-6), interleukin-8 (IL-8), platelet activating factor (PAF) and tumor necrosis factor (TNF) which result in inflammation and cellular injury. The degree to which these mediators escape

Microbiology for Surgical Infections
DOI: http://dx.doi.org/10.1016/B978-0-12-411629-0.00009-X

into the circulation determines the course and severity of systemic inflammatory response. In mild AP, local inflammation is controlled by the host's inflammatory response with localization of pro-inflammatory mediators in the affected area. These inflammatory mediators may affect the intestinal microcirculation, leading to damage to the microvilli allowing bacterial translocation (BT) into the systemic circulation, which increases the risk of pancreatic infection.[3,4]

Unless this process is arrested and reversed by natural defense or therapeutic intervention, the second phase usually ensues and is related to the formation of infected pancreatic necrosis with possible progression to overt sepsis, MODS and death. Gloor et al. found that early deaths rarely occur in patients with severe acute pancreatitis. They reported that 90% of the deaths in patients with infected pancreatic necrosis occurred more than three weeks after disease onset, which may be attributed to modern intensive care treatment.[5] Similarly other researchers have also found MODS, the extent of pancreatic necrosis, infection and sepsis to be the major determinants of mortality in AP.[6,7]

INFECTED PANCREATIC NECROSIS (IPN)

Pancreatic necrosis may either be sterile or infected. Infectious complications are observed in 40–70% of all patients with severe acute pancreatitis and are characterized by diffuse bacterial inflammation of necrotic pancreatic and peri-pancreatic tissue, but without any significant pus collection.[8,9] Even if the extent of necrosis is substantial, mortality rates remain low as long as the necrotic process remains sterile. Pancreatic necrosis even in the presence of severe or prolonged systemic complications, if sterile, has a mortality rate of up to 20%, while in the case of infected necrosis it increases to over 50%.[10] Bacterial infections are the most often recognized, and their occurrence correlates with the extent of necrosis. The risk is about 20% if necrosis is less than 50% and increases up to 70% when pancreatic necrosis exceeds 50%.[11]

In contrast, pancreatic abscess is a localized collection of pus surrounded by a more or less distinct capsule, does not develop before the fifth week after onset of symptoms and after subsidence of the acute phase of pancreatitis.[9]

The most common mechanism involved in the pathogenesis of pancreatic infections is the translocation of bacteria from the gut. Bacterial translocation is the process whereby luminal bacteria migrate across the gastrointestinal mucosal barrier to extra-intestinal sites.[12,13] The gastrointestinal mucosal barrier may become damaged by ischemia during the hypovolemic phase of AP and by pro-inflammatory cytokines, and is an important key factor in pancreatic contamination.[14] The bacterial flora of the gastrointestinal tract varies longitudinally. The stomach ($<10^2$ CFU/ml) (CFU – colony-forming units), and both the lumen and mucosa of the upper regions of the small intestine (duodenum, 10^2 and jejunum, 10^4) are generally sparsely populated (Figure 9.1). Organisms likely to survive will be aciduric species such as some streptococci and lactobacilli (derived from the oral and nasal cavities). That is the reason why perforation of the upper gastrointestinal tract does not normally result in sepsis. The bacterial content increases distally, with approximately 10^8 CFU/ml of ileal content and up to 10^{10-12} CFU/ml of colonic content.[15,16] It includes both aerobes and anaerobes. In their study, Xue P et al. showed that gram-negative

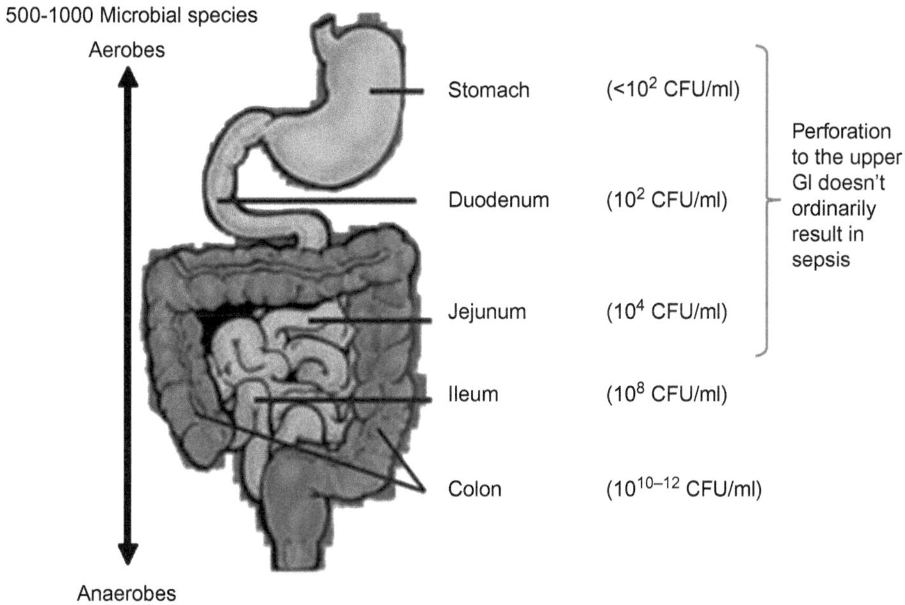

FIGURE 9.1 **The gastrointestinal flora of humans.**

bacteria were preponderantly found in SAP, comprising 56.6% of the isolated strains, whereas gram-positive bacteria and fungi accounted for 22.1% and 21.2% of the isolated strains, respectively.[17]

About three decades ago when antibiotic prophylaxis was not widely practiced, Beger et al. reported that organisms of gastrointestinal origin were predominantly cultured from infected pancreatic necrosis,[18] which is why *Escherichia coli* is the most common organism in this setting.[10,19] However, Gerzof et al. reported that *Klebsiella* was the most common isolate.[20] The infection may further extend to involve extra-pancreatic tissues with the blood being the most common site. The presence of bacteremia should raise the level of suspicion of infected necrosis in a patient already diagnosed with pancreatic necrosis.[19]

Pancreatic infections are most often monomicrobial, whereas extra-pancreatic infections are mostly polymicrobial.[19] Both gram-negative (most commonly *Escherichia coli* and *Klebsiella* species, less commonly *Enterobacter*, *Pseudomonas*, *Proteus*, and others) and gram-positive (enterococcal, streptococcal, and staphylococcal species) bacteria participate. In patients with pancreatic infections there occurs a change in the spectrum of the microorganisms from gram-negative to gram-positive and fungi as the disease progresses. This might be explained by the prolonged hospitalization and use of prophylactic antibiotic therapy against gram-negative bacteria.[19,21]

Fungal infections are often nosocomial and a strong correlation between fungal positivity and the duration of antibiotic exposure has also been reported.[8,22] The overall frequency of fungal contamination was about 7% of all pancreatic infections in the period before antibiotics were used prophylactically,[18] whereas at present the reported prevalence of fungal infection in patients with SAP varies from 11% to 68.5%.[8,23–26]

In recent years, candidial infection has been reported increasingly, which used to be considered as extremely unusual. Most researchers report *Candida albicans* as the commonest fungus isolated in patients, followed by non-albicans species such as *C. tropicalis* and *C. glabrata*.[25–27]

A candida infection may occur either as a part of a mixed infection or by itself. In a mixed infection, the role played by each organism is difficult to discern. Candida isolation in mixed cultures is considered to be pathogenic if there is evidence of systemic clinical deterioration and signs of systemic infection along with lack of improvement with antibacterial therapy alone. Waele et al. found the incidence of candida infection in patients receiving early fluconazole to be 16% in comparison to 58% in patients who did not receive fluconazole prophylactically.[24]

Animal studies indicate that even during acute pancreatic inflammation, adequate tissue penetration by fluconazole occurs into the inflamed pancreas. It is suggested by these experimental findings that fluconazole penetration is likely to be adequate to protect and treat fungal pancreatic infection in humans,[28] however, the selection of non-albicans *Candida* may be an important implication of such prophylactic administration.

The risk factors recognized for fungal infection by various studies are prolonged antibiotic therapy >4 weeks, prolonged hospital stay, presence of organ failure at admission, surgical intervention, requirement for mechanical ventilation and the use of total parental nutrition (TPN).[8,27,29] The use of broad-spectrum antibiotics alters the intestinal flora and facilitates transmigration of microorganisms, and finally, the infection of the pancreatic necrosis. One way of reducing the risk of fungal infection would be to limit the use of prophylactic antibiotics, as has been suggested by many researchers.[22,30] Other preventative measures could include enteral nutrition over TPN, timely change of central venous catheters, and maintenance of asepsis during interventional procedures.[22,30]

Patients with fungal infections who are identified early and treated adequately have better survival probabilities than the untreated group. Fungal infection *per se* is not a risk factor for mortality, but denial of antifungal therapy does increase mortality.[27]

Of all forms of pancreatic infection, infected necrosis carries the worst prognosis and determines the course of the disease and patient outcome.[2,7] It is associated with a high incidence of organ failure irrespective of its extent or whether it occurs early or late in the disease course.[7,31] However, there are no specific clinical signs or symptoms to differentiate sterile from infected necrosis. Sterile pancreatic necrosis is typically managed conservatively without drainage whereas infection of pancreatic necrosis is an indication for surgical intervention.[1,2] The challenge lies in the timely recognition of patients who eventually develop a severe form of the disease and are likely to benefit from early management, intensive care and prophylactic antibiotics. Although not specific, signs and symptoms such as fever, leukocytosis and aggravation of abdominal pain may indicate infection of pancreatic necrosis. However, these may not help clinicians to accurately assess the severity. Also, reliable exclusion of patients with severe acute pancreatitis helps to avoid unnecessary use of invasive and costly procedures in mild cases, facilitating optimal use of health care resources. Thus it is mandatory to distinguish between sterile and infected pancreatic necrosis, especially in those with >30% necrosis on computed tomography (CT) and persistent symptoms, or those with any degree of necrosis and signs of sepsis. This is achieved by fine needle aspiration

(FNA) for bacteriology of pancreatic or peri-pancreatic necrosis or the presence of retroperitoneal gas on CT.[32]

Early diagnosis and prognostic evaluation are extremely important in SAP. The management of these patients becomes challenging due to late hospitalization after onset of an acute attack, and due to difficulty in distinguishing mild from severe disease in the first 48 to 72 hours. A number of infections in patients with SAP are unsuspected, inadequately investigated, and remain untreated, resulting in high mortality. This is further emphasized in a study showing that patients with severe acute pancreatitis who were admitted to the intensive care unit after a delay of more than 24 hours have a greater risk of dying.[33]

Several scoring systems, incorporating various criteria for severity assessment, have been used for some decades. These include the Ranson (11 criteria), the Glasgow score (eight criteria) and the acute physiology and chronic health evaluation (APACHE II) score (14 criteria). The limitations of these scoring systems have either been their inability to obtain a complete score until at least 48 hours into the illness (Ranson and Glasgow scores), which is too late for therapeutic decision-making, or their complexity (APACHE II).[34]

Contrast-enhanced CT is presently the most accurate non-invasive single method to diagnose and to evaluate the severity of the disease.[35] It cannot, however, be performed in all patients with acute pancreatitis due to the risk of renal complications, and it is not always available. A definite microbiological diagnosis of the pathogen can be established only by ultrasound (US) or CT-guided FNA of the necrotic area.[36] Moreover, until the exact pathogen is identified, the treatment given is empirical using several antimicrobial drugs for a prolonged period of time and with doubtful results. There should be limited antibiotic use against infections of bacterial etiology, as their widespread use for nonbacterial etiology has led to the development of antibiotic resistance. Correct determination of the etiological agent is essential for management and depends on a combination of clinical acumen and laboratory support. Currently, FNA followed by gram staining and culture of the aspirate are recommended to diagnose infection, identify the pathogen and institute appropriate therapy.[36,37] The other commonly used samples include preoperative or perioperative aspirated/necrotic tissue from the pancreas, postoperative drain effluent and blood. A culture of drain effluents is less reliable and it may reflect contamination. The drain effluents are centrifuged (at 3000 g for 10 min) to collect the deposit. The deposit, aspirated samples or necrotic tissue are processed as such for direct microscopy (10% KOH wet mount and gram stain) and for culture on blood agar and Saburaud dextrose agar. Tissue samples from patients undergoing necrosectomy are homogenized in a mechanical homogenizer before culture. Any yeast or bacterial growth in these media is identified by the standard protocol.[19,27,36] The knowledge of the exact pathogen and the pattern of resistance may lead to a rational antibiotic treatment.

Early diagnosis of infection in a necrotizing pancreatitis is extremely important. Compared with the current "gold-standard" guided FNA, a non-invasive, rapid, low cost, accurate biochemical parameter available on a 24 hour basis for identifying patients with infected necrosis, indicating the severity and prognosis of the disease, would definitely contribute to an easier and safer diagnosis.[38] Serum amylase and lipase, the standard tests for acute pancreatitis diagnosis, are poor predictors of these parameters.

- **C-reactive protein (CRP):** This is the most widely used predictor of severity in acute pancreatitis, though it is not specific. There is a 48 to 72 hour delay until peak values are reached, which limits its use for early evaluation of the severity of the condition. At the time of admission to hospital, the sensitivity of CRP is 47% – which is similar to that of clinical examination. At 48 hours post-symptom onset, if a cut-off level of 150 mg/l is used, it has been shown to be an accurate severity predictor (sensitivity and specificity above 80%). It is useful in the follow-up of the disease. The advantages of CRP are that it is easy to measure, cheap and widely available.[39,40]
- **Interleukin-6:** The pro-inflammatory cytokine IL-6 is the key inducer of CRP synthesis in the liver. Peak concentrations of IL-6 are reached 24 to 48 hours before those of CRP. It distinguishes between severe and mild acute pancreatitis with 86% to 100% sensitivity and 71% to 100% specificity. The sensitivity of IL-6 for detecting severe acute pancreatitis was 100%, with a specificity of 86% at a cut-off concentration of 2.7 ng/L. Serum concentrations of IL-6 decrease swiftly and its use in clinical practice has also been limited by the complexity and cost of the assay.[39,41]
- **Procalcitonin (PCT):** This is an acute phase reactant, considered to be one of the earliest and most specific markers of bacterial infections in immunocompromised and immunodeficient patients.[42] Normally, the plasma PCT level in healthy individuals is very low (<0.5 ng/mL). During severe bacterial infections and septic conditions its level rises, reaching up to 1000 ng/mL.[43,44] It provides a rapid diagnostic test, available at the patient's bedside, and its half-life is suitable for daily monitoring of disease progress and treatment. It rises earlier than CRP in the course of severe infections and falls promptly with response to appropriate antibiotics, so serial determination may therefore be useful in monitoring severe infections for their occurrence and response to treatment.[42–44] A decrease in PCT levels of >30% of the initial value after 24 hours of antibacterial therapy indicates the appropriateness of the treatment and control of infection.[45] PCT >1.4 ng/mL has a diagnostic accuracy of 70% for infection of necrosis; and PCT >3.8 ng/mL predicts MODS with a diagnostic accuracy of 92%.[46] Postoperatively, PCT has proved to be a sensitive indicator of whether infected necrosis had been successfully evacuated, but surgery itself does not result in a postoperative PCT increase as long as infection is absent.[46] In contrast, acute phase parameters like CRP or IL-6 are elevated in response to both infectious and non-infectious (for example, postoperative) stimuli, so an estimation of the presence or absence of infection is not possible.[42] PCT can reduce the number of days of antibacterial treatment.[47] Yvette et al. observed that none of the serum samples of patients with sterile pancreatic necrosis exhibited PCT concentrations higher than 1.2 ng/mL, whereas in patients with infected pancreatic necrosis, PCT was found in relatively high concentrations (8.5 ± 4.8 ng/mL), and positive values (>1 ng/mL) were obtained prior to positive blood culture or surgical specimens showing bacterial growth. In contrast, levels of IL-6 were high in both infected and sterile pancreatic necrosis, and remained high for several days, even after surgical removal of the infected focus. Therefore PCT is a readily available, helpful, accurate marker that enables the discrimination of IPN facilitating a decision concerning surgical therapy.[48]
- **Interleukin-8:** IL-8 level estimation is inferior in the prediction of infected necrosis when compared to PCT and CRP with an overall sensitivity of 72% in patients with

infected necrosis. Only 20% of patients with primary infected necrosis were correctly identified. A constant rise in IL-8 concentrations to at least 140 pg/mL, usually during the second week of the disease, is indicative of septic multiple organ failure but it is not superior to PCT and even CRP level estimation.[40,41]

- **Interleukin-1** (IL-1) and **Tumor necrosis factor alpha** (TNF-α): IL-1 is produced during acute and chronic inflammation, but showed no correlation with the severity of acute pancreatitis. Release of tumor necrosis factor alpha (TNF-α), a macrophage-derived cytokine believed to be important during injury and sepsis, was more strongly elevated in severe than mild acute pancreatitis.[37] However, other studies suggest that measurement of TNF-α is of no clinical value in assessing the severity of acute pancreatitis.[39]

Thus, early prediction of the severity of acute pancreatitis can be made by well-validated scoring systems at 48 hours, but the serum markers procalcitonin and IL-6 allow earlier prediction (12 to 24 hours after admission).

ANTIMICROBIAL THERAPY

In clinical practice, antibiotics are used to cure both infected pancreatic necrosis and extra-pancreatic infections which appear during the course of acute pancreatitis and also as part of prophylaxis in patients with pancreatic necrosis in order to prevent possible infection from the necrosis. Carbapenems, cephalosporins and quinolones are the antibiotics which achieve high pancreatic tissue concentration.[49,50] The broad-spectrum antibiotic imipenem, which is effective against gram-negative organisms of gastrointestinal origin and penetrates well into pancreatic secretions, is currently the recommended antibiotic for those patients with documented pancreatic necrosis.[32] Another study also suggests the use of imipenem in reducing pancreatic and extra-pancreatic sepsis, although it neither reduces the need for surgical intervention nor the incidence of organ failure and death.[51] Patients with proven infected pancreatic necrosis should receive carbapenems, since fluoroquinolones or cephalosporins with metronidazole have been shown to be less effective.[13,52,53] If gram-positive bacteria (enterococcus, staphylococcus) are the etiological agents, vancomycin should be used, whereas for candidial infection anti-fungals such as fluconazole or amphotericin B should be instituted until results of culture and sensitivity are obtained.[2] The susceptibility can be obtained by the disk diffusion method, or more accurately by determining the minimum inhibitory concentration for a specific antibacterial agent against the identified pathogen by various conventional and automated methods.

In spite of many years of research, the role of prophylactic antibiotics in acute pancreatitis is highly debatable. Some researchers believe that no prophylactic antibiotic should be used in necrotizing pancreatitis; it shows clinical improvement with continued supportive care. If the condition deteriorates, FNA for bacteriology of pancreatic or peri-pancreatic necrosis appears to be a safe method,[54,55] and if it is sterile then supportive care should be continued. The other school of thought believes in instituting antibiotics for the prevention and early treatment to reduce morbidity and mortality from the beginning. During the past few decades, a number of studies have been undertaken in order to determine whether it is necessary to administer antibiotics for Acute Necrotizing Pancreatitis (ANP), with conflicting results (Table 9.1).[51,53,56–61]

These randomized controlled trials (RCTs) have evaluated different antibiotics for prophylaxis, including carbapenems,[21,44,51,53] third generation cephalosporins (60) and fluoroquinolones.[53,62] However, there are differences in the antimicrobial activity, penetration rate, persistence and therapeutic concentrations of these antibiotics in the necrotic pancreatic tissue. In fact, the choice of antibiotic for preventing infected necrosis during necrotizing pancreatitis should be based on all these requisites, and further detailed research is required on all these aspects. Further, the validity of most of the trials is affected by the small size of the sample and the absence of a blinding method. Whereas most of the studies do not recommend the administration of antibiotic prophylaxis, a few have detected a beneficial role of early prophylactic treatment in acute pancreatitis.[58,59] In their meta-analysis, Yao et al. suggested that prophylactic antibiotics reduce IPN but not mortality or the need for surgical intervention in patients with IPN.[61] Another piece of research showed a greater benefit for prophylactic early antibiotic treatment compared with on-demand therapy concerning super infection of the pancreas, mortality and sepsis. At 24 hours after induction of acute pancreatitis in a standardized experimental model, bacterial infection of lymph nodes of the colon could be significantly reduced from 83.3% to 0% by early antibiotic prophylaxis.[13]

Villatoro et al. reported that antibiotic prophylaxis appeared to be associated with significantly decreased mortality but not infected pancreatic necrosis. Due to these conflicting results, the Cochrane review concluded that more high quality clinical trials must be performed to confirm the benefits of antibiotic prophylaxis.[63] An individual patient data analysis would be more meaningful to identify potential candidates who can gain a survival benefit from antibiotic prophylaxis rather than the administration of prophylactic antibiotics as a routine treatment for SAP.

CONCLUSION AND FUTURE PROSPECTS

- Severe acute pancreatitis (SAP) is defined as an acute pancreatitis associated with local and/or systemic complications.
- In patients with SAP, even if the extent of necrosis is substantial, mortality rates remain low as long as the necrotic process remains sterile.
- Infected pancreatic necrosis carries the worst prognosis with up to 50% mortality.
- Bacterial infections are most common, with translocation of microorganisms from the gut being the mechanism.
- Both gram-negative (most commonly *E. coli* and *Klebsiella* species) and gram-positive (enterococcal, streptococcal, and staphylococcal species) bacteria participate.
- There occurs a change in the spectrum of the infecting microorganisms from gram-negative to gram-positive and fungi as the disease progresses.
- Fungal infections are often nosocomial, and a strong correlation between fungal positivity and the duration of antibiotic exposure has been reported.

TABLE 9.1 Antibiotics Used for Acute Pancreatitis in Various Controlled Trials and Meta-Analyses

Researchers (Ref No.) (Year of Study)	Antibiotic Tested	Comments
Pederzoli, et al. [51] (1993)	Imipenem (500 mg IV tid, 14 days)	Non-pancreatic infections also reduced (15% vs. 49%, $P < 0.01$). No change in multi-organ failure rate, need for surgery, or survival compared with no early antibiotic treatment
Sainio, et al. [60] (1995)	2 groups: first group received cefuroxime (1.5 g IV tid) and second non-antibiotic receiving group	Reduction in total infections, infections per patient, and operations also statistically significant in antibiotic receiving group
Bassi, et al. [53] (1998)	Pefloxacin (400 mg bid) vs. imipenem (500 mg tid), 14 days	Pefloxacin is inferior to imipenem in the prevention of infections associated with severe pancreatitis
Nordback, et al. [58] (2001)	Single-center randomized study, compared early vs. delayed imipenem treatment for acute necrotizing pancreatitis	Early imipenem therapy appears to significantly reduce the need for surgery and the overall number of major organ complications in acute necrotizing pancreatitis, and reduces by half the mortality rate
Isenmann, et al. [57] (2004)	Ciprofloxacin (400 mg bid) and metronidazole (500 mg bid) IV	No benefit of antibiotic prophylaxis with respect to the risk of developing infected pancreatic necrosis
Manes, et al. [59] (2006)	Compared early antibiotic treatment with treatment started after the demonstration of pancreatic necrosis	Early antibiotic treatment associated with a significant improvement in the prognosis of necrotizing acute pancreatitis, because of a reduction in the occurrence of septic complications
Villatoro, et al. [63] (2006)	Cochrane Database review including randomized controlled trials comparing antibiotics with placebo in AP with CT-proven necrosis	Antibiotic prophylaxis appeared to be associated with significantly decreased mortality but not IPN. Beta lactams were associated with significantly decreased mortality and infected pancreatic necrosis. No significant differences with quinolone plus imidazole treatment.
Dellinger EP, et al. [56] (2007)	A multicenter, prospective, double-blind, placebo-controlled randomized study comparing meropenem with placebo in 100 patients with clinically severe, confirmed necrotizing pancreatitis	No statistically significant difference between the treatment groups and didn't support early prophylactic antimicrobial use in patients with severe acute necrotizing pancreatitis
Yao L et al, [61] (2010)	A meta-analysis of randomized trials to compare prophylactic antibiotic treatment of patients with ANP versus placebo	Prophylactic antibiotic treatment reduced occurrence of IPN, but did not affect mortality or surgical intervention in patients with ANP

- Candida isolation in mixed cultures is considered to be pathogenic if there is evidence of systemic clinical deterioration and signs of systemic infection along with lack of improvement with antibacterial therapy alone.
- The infection may further extend to involve extra-pancreatic tissues with the blood being the most common site.
- The presence of bacteremia should raise the level of suspicion for infected necrosis in a patient already diagnosed with pancreatic necrosis.
- Early prediction of the severity can be made by scoring systems at 48 hours.
- Serum markers procalcitonin and IL-6 allow earlier prediction (12 to 24 hours).
- PCT rises earlier than CRP in the course of severe infections and falls promptly with response to appropriate antibiotics.
- PCT provides a rapid diagnostic test, available at the patient's bedside, and its half-life is suitable for daily monitoring of disease progress and treatment.
- Antibiotics are used to cure both infected pancreatic necrosis and extra-pancreatic infections. They are also used as a part of prophylaxis to prevent possible infection.
- In spite of many years of research the role of prophylactic antibiotics in acute pancreatitis is highly debatable.
- An individual patient data analysis would be more meaningful, to identify potential candidates who would gain survival benefit from antibiotic prophylaxis, rather than this being administered as a routine treatment.
- Correct determination of the etiological agent is essential for management and depends on a combination of clinical acumen and laboratory support.
- Currently, ultrasound (US) or computed tomography (CT) guided FNA of the necrotic area, followed by gram staining and culture, is recommended to diagnose infection and identify the pathogen.
- The antimicrobial susceptibility can be obtained by the disk diffusion method or more accurately by determining the minimum inhibitory concentration for a specific antibacterial agent against the identified pathogen.
- Carbapenems, cephalosporins and quinolones are the antibiotics which achieve high pancreatic tissue concentration. Patients with proven infected pancreatic necrosis should receive carbapenems, since a combination of fluoroquinolones or cephalosporins with metronidazole has been shown to be less effective.
- High-quality clinical trials must be performed to confirm the benefits of antibiotic prophylaxis.
- New therapeutic strategies attempting to prevent the activity of pro-inflammatory mediators or to block their synthesis should be evaluated through clinical trials as therapeutic options.

References

1. Werner J, Feuerbach S, Uhl W, Buchler MW. Management of acute pancreatitis: from surgery to interventional intensive care. *Gut* 2005;**54**:426–36.
2. Banks PA, Freeman ML. Practice Parameters Committee of the American College of G. Practice guidelines in acute pancreatitis. *Am J Gastroenterol* 2006;**101**:2379–400.

3. Kylanpaa L, Rakonczay Jr Z, O'Reilly DA. The clinical course of acute pancreatitis and the inflammatory mediators that drive it. *Int J Inflam* 2012;**2012**:360685.
4. Runzi M, Niebel W, Goebell H, Gerken G, Layer P. Severe acute pancreatitis: nonsurgical treatment of infected necroses. *Pancreas* 2005;**30**:195–9.
5. Gloor B, Muller CA, Worni M, Martignoni ME, Uhl W, Buchler MW. Late mortality in patients with severe acute pancreatitis. *Br J Surg* 2001;**88**:975–9.
6. Johnson CD, Abu-Hilal M. Persistent organ failure during the first week as a marker of fatal outcome in acute pancreatitis. *Gut* 2004;**53**:1340–4.
7. Isenmann R, Rau B, Beger HG. Bacterial infection and extent of necrosis are determinants of organ failure in patients with acute necrotizing pancreatitis. *Br J Surg* 1999;**86**:1020–4.
8. Gloor B, Muller CA, Worni M, Stahel PF, Redaelli C, Uhl W, et al. Pancreatic infection in severe pancreatitis: the role of fungus and multiresistant organisms. *Arch Surg* 2001;**136**:592–6.
9. Bittner R, Block S, Buchler M, Beger HG. Pancreatic abscess and infected pancreatic necrosis. Different local septic complications in acute pancreatitis. *Dig Dis Sci* 1987;**32**:1082–7.
10. Garg PK, Madan K, Pande GK, Khanna S, Sathyanarayan G, Bohidar NP, et al. Association of extent and infection of pancreatic necrosis with organ failure and death in acute necrotizing pancreatitis. *Clin Gastroenterol Hepatol* 2005;**3**:159–66.
11. Isenmann R, Beger HG. Natural history of acute pancreatitis and the role of infection. *Baillieres Best Pract Res Clin Gastroenterol* 1999;**13**:291–301.
12. Liu H, Li W, Wang X, Li J, Yu W. Early gut mucosal dysfunction in patients with acute pancreatitis. *Pancreas* 2008;**36**:192–6.
13. Fritz S, Hartwig W, Lehmann R, Will-Schweiger K, Kommerell M, Hackert T, et al. Prophylactic antibiotic treatment is superior to therapy on-demand in experimental necrotising pancreatitis. *Crit Care* 2008;**12**:R141.
14. Fritz S, Hackert T, Hartwig W, Rossmanith F, Strobel O, Schneider L, et al. Bacterial translocation and infected pancreatic necrosis in acute necrotizing pancreatitis derives from small bowel rather than from colon. *Am J Surg* 2010;**200**:111–7.
15. Berg RD. The indigenous gastrointestinal microflora. *Trends Microbiol* 1996;**4**:430–5.
16. Cunha BA. Sepsis and septic shock: selection of empiric antimicrobial therapy. *Crit Care Clin* 2008;**24**:313–34.
17. Xue P, Deng LH, Zhang ZD, Yang XN, Wan MH, Song B, et al. Infectious complications in patients with severe acute pancreatitis. *Dig Dis Sci* 2009;**54**:2748–53.
18. Beger HG, Bittner R, Block S, Buchler M. Bacterial contamination of pancreatic necrosis. A prospective clinical study. *Gastroenterology* 1986;**91**:433–8.
19. Noor MT, Radhakrishna Y, Kochhar R, Ray P, Wig JD, Sinha SK, et al. Bacteriology of infection in severe acute pancreatitis. *JOP* 2011;**12**:19–25.
20. Gerzof SG, Banks PA, Robbins AH, Johnson WC, Spechler SJ, Wetzner SM, et al. Early diagnosis of pancreatic infection by computed tomography-guided aspiration. *Gastroenterology* 1987;**93**:1315–20.
21. Buchler P, Reber HA. Surgical approach in patients with acute pancreatitis. Is infected or sterile necrosis an indication – in whom should this be done, when, and why? *Gastroenterol Clin North Am* 1999;**28**:661–71.
22. Kochhar R, Ahammed SK, Chakrabarti A, Ray P, Sinha SK, Dutta U, et al. Prevalence and outcome of fungal infection in patients with severe acute pancreatitis. *J Gastroenterol Hepatol* 2009;**24**:743–7.
23. Isenmann R, Schwarz M, Rau B, Trautmann M, Schober W, Beger HG. Characteristics of infection with Candida species in patients with necrotizing pancreatitis. *World J Surg* 2002;**26**:372–6.
24. De Waele JJ, Vogelaers D, Blot S, Colardyn F. Fungal infections in patients with severe acute pancreatitis and the use of prophylactic therapy. *Clin Infect Dis* 2003;**37**:208–13.
25. Hoerauf A, Hammer S, Muller-Myhsok B, Rupprecht H. Intra-abdominal Candida infection during acute necrotizing pancreatitis has a high prevalence and is associated with increased mortality. *Crit Care Med* 1998;**26**:2010–5.
26. Kujath P, Rosenfeldt M, Esnaashari H, Scheele J, Bouchard R. Fungal infections in patients with necrotizing pancreatitis: risk-factors, incidence, therapy. *Mycoses* 2005;**48**:36–40.
27. Chakrabarti A, Rao P, Tarai B, Shivaprakash MR, Wig J. Candida in acute pancreatitis. *Surg Today* 2007;**37**:207–11.

28. Shrikhande S, Friess H, Issenegger C, Martignoni ME, Yong H, Gloor B, et al. Fluconazole penetration into the pancreas. *Antimicrob Agents Chemother* 2000;**44**:2569–71.
29. Villatoro E, Mulla M, Larvin M. Antibiotic therapy for prophylaxis against infection of pancreatic necrosis in acute pancreatitis. *Cochrane Database Syst Rev* 2010;**5**:CD002941.
30. Berzin TM, Rocha FG, Whang EE, Mortele KJ, Ashley SW, Banks PA. Prevalence of primary fungal infections in necrotizing pancreatitis. *Pancreatology* 2007;**7**:63–6.
31. Petrov MS, Chong V, Windsor JA. Infected pancreatic necrosis: not necessarily a late event in acute pancreatitis. *World J Gastroenterol* 2011;**17**:3173–6.
32. Toouli J, Brooke-Smith M, Bassi C, Carr-Locke D, Telford J, Freeny P, et al. Working Party of the Program Commitee of the Bangkok World Congress of Gastroenterology 2002. Guidelines for the management of acute pancreatitis. *J Gastroenterol Hepatol* 2002;**17**:S15–39.
33. Brivet FG, Emilie D, Galanaud P. Pro- and anti-inflammatory cytokines during acute severe pancreatitis: an early and sustained response, although unpredictable of death. Parisian Study Group on Acute Pancreatitis. *Crit Care Med* 1999;**27**:749–55.
34. Ueda T, Takeyama Y, Yasuda T, Matsumura N, Sawa H, Nakajima T, et al. Simple scoring system for the prediction of the prognosis of severe acute pancreatitis. *Surgery* 2007;**141**:51–8.
35. Balthazar EJ. Acute pancreatitis: assessment of severity with clinical and CT evaluation. *Radiology* 2002;**223**:603–13.
36. Rau B, Pralle U, Mayer JM, Beger HG. Role of ultrasonographically guided fine-needle aspiration cytology in the diagnosis of infected pancreatic necrosis. *Br J Surg* 1998;**85**:179–84.
37. Dervenis C, Johnson CD, Bassi C, Bradley E, Imrie CW, McMahon MJ, et al. Diagnosis, objective assessment of severity, and management of acute pancreatitis. Santorini consensus conference. *Int J Pancreatol* 1999;**25**:195–210.
38. Lempinen M, Puolakkainen P, Kemppainen E. Clinical value of severity markers in acute pancreatitis. *Scand J Surg* 2005;**94**:118–23.
39. Dambrauskas Z, Gulbinas A, Pundzius J, Barauskas G. Value of routine clinical tests in predicting the development of infected pancreatic necrosis in severe acute pancreatitis. *Scand J Gastroenterol* 2007;**42**:1256–64.
40. Buchler M, Malfertheiner P, Schoetensack C, Uhl W, Scherbaum W, Beger HG. Value of biochemical and imaging procedures for the diagnosis and prognosis of acute pancreatitis – results of a prospective clinical study. *Z Gastroenterol* 1986;**24**:100–9.
41. Chen CC, Wang SS, Lee FY, Chang FY, Lee SD. Proinflammatory cytokines in early assessment of the prognosis of acute pancreatitis. *Am J Gastroenterol* 1999;**94**:213–8.
42. Hatzistilianou M. Diagnostic and prognostic role of procalcitonin in infections. *ScientificWorldJournal* 2010;**10**:1941–6.
43. Becker KL, Snider R, Nylen ES. Procalcitonin assay in systemic inflammation, infection, and sepsis: clinical utility and limitations. *Crit Care Med* 2008;**36**:941–52.
44. Schneider HG, Lam QT. Procalcitonin for the clinical laboratory: a review. *Pathology* 2007;**39**:383–90.
45. Hatzistilianou M, Rekleity A, Athanassiadou F, DeLutiis MA, Conti P, Catriu D. Serial procalcitonin responses in infection of children with secondary immunodeficiency. *Clin Invest Med* 2007;**30**:E75–85.
46. Rau BM, Kemppainen EA, Gumbs AA, Büchler MW, Wegscheider K, Bassi C, et al. Early assessment of pancreatic infections and overall prognosis in severe acute pancreatitis by procalcitonin (PCT): a prospective international multicenter study. *Ann Surg* 2007;**245**:745–54.
47. Christ-Crain M, Stolz D, Bingisser R, Müller C, Miedinger D, Huber PR, et al. Procalcitonin guidance of antibiotic therapy in community-acquired pneumonia: a randomized trial. *Am J Respir Crit Care Med* 2006;**174**:84–93.
48. Mandi Y, Farkas G, Takacs T, Boda K, Lonovics J. Diagnostic relevance of procalcitonin, IL-6, and sICAM-1 in the prediction of infected necrosis in acute pancreatitis. *Int J Pancreatol* 2000;**28**:41–9.
49. Spicak J, Martinek J, Zavada F, Moravek J, Melenovsky V. Penetration of antibiotics into the pancreas in rats: an effect of acute necrotizing pancreatitis. *Scand J Gastroenterol* 1999;**34**:92–7.
50. Papagoras D, Giamarellos-Bourboulis EJ, Kanara M, Douridas G, Paraskevopoulos I, Antzaklis G, et al. Pancreatic concentrations of cefepime in experimental necrotizing pancreatitis. *J Chemother* 2003;**15**:43–6.
51. Pederzoli P, Bassi C, Vesentini S, Campedelli AA. randomized multicenter clinical trial of antibiotic prophylaxis of septic complications in acute necrotizing pancreatitis with imipenem. *Surg Gynecol Obstet* 1993;**176**:480–3.

52. Kondratenko PG, Shirshov IV. Antibacterial therapy in acute nonbiliary infectious necrotizing pancreatitis. *Klin Khir* 2012;**10**:14–8.
53. Bassi C, Falconi M, Talamini G, Uomo G, Papaccio G, Dervenis C, et al. Controlled clinical trial of pefloxacin versus imipenem in severe acute pancreatitis. *Gastroenterology* 1998;**115**:1513–7.
54. Banks PA, Gerzof SG, Langevin RE, Silverman SG, Sica GT, Hughes MD. CT-guided aspiration of suspected pancreatic infection: bacteriology and clinical outcome. *Int J Pancreatol* 1995;**18**:265–70.
55. Hiatt JR, Fink AS, King III W, Pitt HA. Percutaneous aspiration of peripancreatic fluid collections: a safe method to detect infection. *Surgery* 1987;**101**:523–30.
56. Dellinger EP, Tellado JM, Soto NE, Ashley SW, Barie PS, Dugernier T, et al. Early antibiotic treatment for severe acute necrotizing pancreatitis: a randomized, double-blind, placebo-controlled study. *Ann Surg* 2007;**245**:674–83.
57. Isenmann R, Rünzi M, Kron M, Kahl S, Kraus D, Jung N, et al. German Antibiotics in Severe Acute Pancreatitis Study Group. Prophylactic antibiotic treatment in patients with predicted severe acute pancreatitis: a placebo-controlled, double-blind trial. *Gastroenterology* 2004;**126**:997–1004.
58. Nordback I, Sand J, Saaristo R, Paajanen H. Early treatment with antibiotics reduces the need for surgery in acute necrotizing pancreatitis – a single-center randomized study. *J Gastrointest Surg* 2001;**5**:113–8 [discussion 8–20].
59. Manes G, Uomo I, Menchise A, Rabitti PG, Ferrara EC, Uomo G. Timing of antibiotic prophylaxis in acute pancreatitis: a controlled randomized study with meropenem. *Am J Gastroenterol* 2006;**101**:1348–53.
60. Sainio V, Kemppainen E, Puolakkainen P, Taavitsainen M, Kivisaari L, Valtonen V, et al. Early antibiotic treatment in acute necrotising pancreatitis. *Lancet* 1995;**346**:663–7.
61. Yao L, Huang X, Li Y, Shi R, Zhang G. Prophylactic antibiotics reduce pancreatic necrosis in acute necrotizing pancreatitis: a meta-analysis of randomized trials. *Dig Surg* 2010;**27**:442–9.
62. García-Barrasa A, Borobia FG, Pallares R, Jorba R, Poves I, Busquets J, et al. A double-blind, placebo-controlled trial of ciprofloxacin prophylaxis in patients with acute necrotizing pancreatitis. *J Gastrointest Surg* 2009;**13**:768–74.
63. Villatoro E, Bassi C, Larvin M. Antibiotic therapy for prophylaxis against infection of pancreatic necrosis in acute pancreatitis. *Cochrane Database Syst Rev* 2006;**4**:CD002941.

SECTION III

CARDIOVASCULAR INFECTIONS

CHAPTER

10

Infective Endocarditis

Ali Alizzi, John-Paul Tantiongco† and Sarah Shepard***

***Cardiothoracic Surgical Unit, Ashford Hospital, Adelaide, Australia, †Cardiothoracic Surgical Unit, Flinders Medical Center, Adelaide, Australia, **SA Pathology, Adelaide, Australia**

INTRODUCTION

In the late 19th century, William Osler once said, *"few diseases present with greater difficulties in the way of diagnosis than malignant endocarditis"*.[1] This statement is as true now as it was then. Endocarditis is defined as inflammation of the endocardium, and typically refers to infections of the heart valve surfaces. The term "sub-acute bacterial endocarditis" reflects the frequent chronic course of infective endocarditis (IE), which may occur for a significant time before diagnosis is made and appropriate treatment is given. Because no definition exists for "sub-acute," the term infective endocarditis is generally accepted. The clinical and morphologic diagnosis of IE is often a challenging one, and if unrecognized and untreated it can be life-threatening. In the case of valvular endocarditis, many patients have an underlying congenital or acquired alteration, and valves with regurgitant alterations are particularly at risk, as are patients with prosthetic heart valves, immunosuppressed patients and IV drug users.[2]

Endocarditis is grouped into two broad categories; infectious and non-infectious.[2] Non-infectious endocarditis encompasses two entities: nonbacterial thrombotic endocarditis (also known as marantic endocarditis) and autoimmune endocarditis. For the purposes of this textbook we will focus primarily on infectious endocarditis.

RISK FACTORS

Infectious and non-infectious endocarditis occurs in two settings; structurally normal valves and structurally abnormal valves, which are prone to surface disruption. In people with normal valves, there are usually multiple episodes of transient bacteremia, such as

Microbiology for Surgical Infections
DOI: http://dx.doi.org/10.1016/B978-0-12-411629-0.00010-6

occurs in IV drug use, alcoholism, intravenous lines, poor oral hygiene, or patients with compromised immune systems or malignancy.[4] In both clinical and autopsy published material, approximately 25% of patients have neither structural valve abnormalities nor predisposing conditions to cause IE. The pathogenesis of infectious endocarditis begins as sterile thrombi, which then become secondarily infected, with the proliferation of organisms, enlargement of the thrombus and recruitment of neutrophils and proteases that destroy the valve.

It has been well documented that abnormal flow around the valve which increases the risk of infection is caused by congenital valve abnormalities, acquired valve disease, prostheses and prior cardiac surgery; these form the basis for antibiotic prophylaxis for dental procedures and other interventions.[2] This has been widely supported by the American College of Cardiology (ACC) and the American Heart Association (AHA) in their practice guidelines published over the years.

Hospitalization increases the risk of endocarditis due to the numerous procedures patients are subjected to; in nosocomial endocarditis, nearly 40% of patients have indwelling catheters or other intravascular device, such as a pacemaker, and a large proportion will have metastatic malignancies. Patients who are on hemodialysis are especially at risk with mortality rates of methicillin resistant *S. aureus* (MRSA) endocarditis in these patients being as high as 90%.[2] Factors increasing the morbidity and mortality of patients with infective endocarditis include advanced patient age, MRSA infection, left-sided cardiac disease and chronic renal failure.

INFECTIOUS AGENTS

A variety of gram-positive and gram-negative bacterial, mycobacterial, fungal, rickettsial and chlamydial organisms are responsible for infective endocarditis (Table 10.1), but far and away the most common agents are streptococci, and increasingly, staphylococci. Staphylococci cause 20 to 30% of bacterial endocarditis, and 80 to 90% of these are caused

TABLE 10.1 Causative Organisms for Infective Endocarditis

Community Acquired	Nosocomial
Staphylococcus aureus (30–50%) (minority MRSA)	*Staphylococcus aureus* (60–80%) (majority MRSA)
Alpha haemolytis (viridans) streptococci (10–35%)	Alpha haemolytic streptococci (<5%)
Culture negative (5–30%)	Culture negative (5%)
Enterococcus (5–10%)	Enterococcus (5%)
Fungi (<5%)	Fungi (10%)
Staphylococcus epidermidis (coagulase negative) (<5%)	*Staphylococcus epidermidis* (coagulase negative) (<5%)
Others (*Escherischia coli, Klebsiella, Corynebacterium* etc.) (<5%)	Others (*E. coli, Klebsiella, Corynebacterium* etc.) (5–10%)

by coagulase positive *S. aureus*.[4] Nosocomial infections are most frequently caused by staphylococci, followed by alpha hemolytic streptococci, enterococci and other organisms.[2] Community acquired infections are most often streptococcal (one third represented by enterococcus faecalis) with staphylococci being a very close second.[2] Staphylococcal endocarditis characteristically produces bulky vegetations that are very destructive to the valve tissue. Haemodynamic "jet lesions" are important in pathogenesis.

In children, more than 75% of endocarditis patients have pre-existing congenital disease such as Tetralogy of Fallot, ventricular septal defect, transposition of the great arteries and often have a history of prior surgery.[2]

CULTURE NEGATIVE ENDOCARDITIS

The rate of culture negative endocarditis varies from 7% to 33% and is higher in community acquired infections due to antibiotic treatment prior to diagnosis.[2]

Mylonakis et al. published in the New England Journal of Medicine that new diagnostic approaches, including culture and microbiological assessment of vegetations, have yielded a better understanding of blood-culture negative endocarditis in particular.[8] Compared with culture positive cases, the diagnosis of culture negative endocarditis is delayed, which results in an increased risk of valve destruction, septic emboli and higher mortality. Traditional techniques such as histology, electron microscopy and serology have been applied to identify difficult-to-culture agents of infective endocarditis. Immunohistochemistry has also significantly enhanced our ability to detect microorganisms. The most common organisms that are not detected routinely are *Coxiella burnetii* and *Bartonella* species.

GROSS PATHOLOGY OF INFECTIVE ENDOCARDITIS

Careful examination of the specimen with knowledge of the clinical history is paramount for reaching a diagnosis of infective endocarditis. This allows for a distinction to be made between a rheumatic or congenital origin in chronic valvulopathy. Microscopically, both will show evidence of calcification, fibrosis, the presence of an inflammatory cell infiltrate and occasionally foci of dystrophic amyloid deposition.[3] Left-sided valves are more frequently affected than right-sided valves (aortic > mitral > tricuspid > pulmonic). Contrary to popular belief, this holds true even in the case of IV drug use, although infections of the tricuspid and pulmonary valves should always raise the suspicion of drug abuse. The gross and microscopic findings are attributed in part to the virulence of the organism, and microbiologic cultures should be requested and performed routinely on all excised tissue.

Grossly, the lesions can vary in size and shape. Extension of these lesions can proceed to the valve leaflets, chordae and annular regions to form abscesses.[5,6] The appearance of active vegetation varies from soft gray-pink to firm yellow brown[7] (Figure 10.1). The finding of a true perforation is nearly pathognomonic for endocarditis, although the distinction between perforation and fenestration is necessary, as fenestrations are a normal finding in

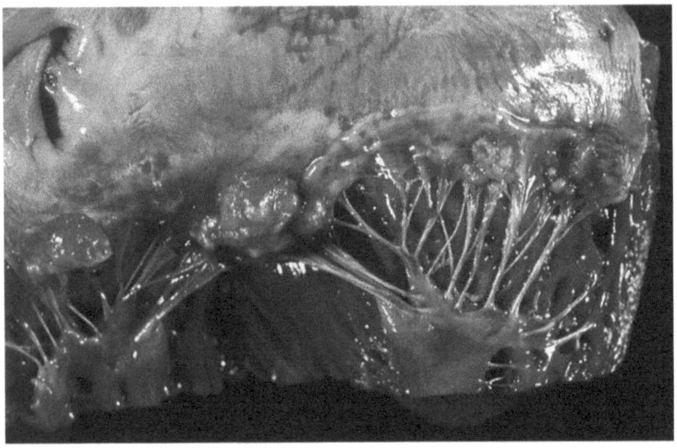

FIGURE 10.1 Gross appearance of an infected mitral valve with prominent vegetations.

FIGURE 10.2 Infective endocarditis of valve leaflets.

older patients. They are present near the commissure of the aortic or pulmonic valve, are not associated with vegetations and are characterized by a smooth, thin slit near the free edge of the valve without irregularities.

HISTOPATHOLOGY OF INFECTIVE ENDOCARDITIS

It must be noted that the most accurate diagnosis of infective endocarditis is made from the gross appearance of the valve by the pathologist and less so from examination under the microscope. However, proper histological examination of excised valvular tissue remains crucial, keeping in mind that histology cannot usually specify the causative agent. Haematoxylin and eosin (H&E) is the standard stain employed, and is used to identify the pattern of inflammation. In addition to this, a gram stain can allow the differentiation between gram-positive and gram-negative bacteria. Histologically, the vegetations consist of platelets, fibrin and acute and chronic inflammatory cells (Figure 10.2). The organisms responsible are generally present within the inflammatory infiltrate and are often intracellular within neutrophils or macrophages, and may not be evident, even with the application of special stains.

In the hands of a skilled observer, a rapid preliminary identification of the organism is sometimes possible from morphology.[9] This technique also has its limits, as some bacterial species (such as mycoplasma) lack a cell wall and are therefore not detected. The periodic acid-Schiff stain (PAS) may be useful in detecting foamy histiocytes and fungi, while a Warthin-Starry silver impregnation stain is among the most sensitive methods of detecting bacteria. Specific stains may also be employed to detect certain culture negative organisms, such as the Ziehl-Nielsen stain for detecting mycobacteria.

In a summary of techniques, Houpikan in 2002 highlighted the development of many highly specific antibodies for immuno-histochemical detection of pathogenic organisms in tissues. Excised tissue can be sampled fresh or after formalin fixation. Several techniques are now widely available, including immunoperoxidase stains and enzyme-linked immunosorbent (ELISA) or immunofluorescent (ELIFA) assays.[9] Direct immunofluorescence using fluoroscein conjugated monoclonal antibodies has also been employed for antigen detection in paraffin-embedded tissues.

Houpikan et al. also alluded in their paper to the role of electron microscopy (EM), which, although it may identify taxonomically significant features of organisms, has its limitations and is infrequently used in the diagnosis of IE as it is time consuming, tedious, expensive and requires experienced staff. This technique is usually only used when all other mechanisms of detection have failed. Growth-based diagnostic methods have benefited from the development of new media and also from the use of novel inoculation, incubation and detection techniques.[9]

The polymerase chain reaction can now be used to identify previously unculturable species from excised vegetations or from systemic emboli.[8,9] Its main advantages are that it is both easy and reliable, and it is culture independent. It can be used to detect any organism in a single reaction. However, microbial DNA contamination may still occur, even when rigorous technical precautions are taken, and therefore caution must be exercised in the interpretation when the organism has not been observed in stained valve tissues.[9]

Organizing vegetations contain chronic inflammatory cells (lymphocytes, macrophages and giant cells) and fewer neutrophils.

The findings of chronic inflammation with neovascularity and tissue destruction are sufficient for a diagnosis of IE, and the identification of the organisms is not required for confirmation. There is typically destruction of the underlying valve structure, both in acute as well as healed lesions. It has also been demonstrated that subclinical infections may be present in prosthetic and native valves without gross evidence of vegetations or perforations.

PROSTHETIC VALVE ENDOCARDITIS

Bacterial endocarditis is an important complication of heart valve replacement (and indeed all cardiac surgery). Infection of the heart valves may occur intra-operatively or may be the result of a subsequent bacteremia. Microscopic study of artificial heart valve prostheses has shown that, following insertion, a neoendocardium develops at the junction with the heart wall, and from there it grows centripetally over the sewing cloth toward the valve lumen.[3] The "ring abscess" is a distinctive and destructive, purulent, concentric lesion that results from infection of the annulus of native valves or by spread of infection from the sewing ring of prosthetic valves. Approximately two thirds are streptococcal and one third is staphylococcal. Besides being a source of bacteremia and septic emboli, ring abscesses have been reported to cause valvular dysfunction (more so in mitral valves) or conduction defects (more so in aortic valves).[4]

NON-INFECTIOUS ENDOCARDITIS (NONBACTERIAL THROMBOTIC ENDOCARDITIS)

The term "nonbacterial endocarditis" is a misnomer, as there is little, if any inflammation, only sterile thrombus, and it is defined as platelet-fibrin vegetations that are devoid of inflammatory cells or bacteria.[2] This type of endocarditis usually occurs in the setting of an underlying chronic inflammatory disease, malignant tumor, or coagulopathy, and is often identified incidentally at autopsy. The vegetations are arranged as linear or continuous aggregates along the lines of closure of the left-sided valves,[5] although tricuspid or pulmonic vegetations may occur secondary to central venous catheter-induced trauma. Embolization is common, and this is the usual indication for surgery.

Histologically, marantic vegetations consist predominantly of fibrin with a few trapped erythrocytes and rare acute and chronic inflammatory cells. The underlying valve is usually normal.

Autoimmune valvulitis overlaps with nonbacterial endocarditis, due to the common occurrence of a major risk factor (antiphospholipid syndrome) in lupus, which can be associated with sterile thrombi on the valve surfaces, usually without destruction of the leaflets.[2] Autoimmune valvulitis is poorly defined and generally refers to non-infectious Libman–Sacks vegetations. They often mimic infective endocarditis due to the histological appearance of fibrin, cores of fibrinous necrosis of the valve, inflammatory cell infiltrates, haematoxylin bodies, cellular debris and degenerating valve tissue.[7]

CLINICAL PRESENTATION

The typical classic presentation of infective endocarditis is rarely seen in the majority of cases in clinical practice. The variable presentations of different patients with IE and the evolving pathogens involved make diagnosis challenging. Due to the heterogeneous nature of clinical presentation, a high index of suspicion is adamant during a patient's history taking and examination. The history should include enquiries regarding constitutional symptoms of sepsis but also the complications of IE including heart failure, conduction disturbances and embolism. A specific inquiry about known risk factors for IE should also be made, such as invasive procedures, intravenous drug use, structural heart disease, previous endocarditis and prosthetic heart valves.

The clinical presentation is highly variable, partly depending on the causative microorganism, but also on the presence of pre-existing cardiac disease and the presence of co-morbidities. The disease may occur as acute, with a rapidly progressive course where patients may present with valvular infections or cerebral or pulmonary emboli. The acute setting is usually caused by more virulent organisms such as *S. aureus* and commonly effects normal native valves. Antibiotics alone seldom cure this type of infection and surgical intervention is often required. It is thought that acute IE may evolve too quickly for the development of immunologic vascular phenomena, which are more characteristic of sub-acute IE. In this setting, less virulent microorganisms are usually responsible, such as *S. viridans*, and antibiotic therapy alone is usually successful. The latter condition manifests with low-grade fever and non-specific symptoms that may elude initial assessment.

In patients with classical manifestations, the diagnosis of IE is fairly straightforward. The main clinical features include generalized malaise and weakness, fever and signs and symptoms of active valvular involvement. These may include symptoms of heart failure and a new heart murmur. Other manifestations of IE may include embolic phenomena occurring in approximately 30% of cases[10] which may cause cerebral complications such as transient ischemic attacks (TIAs) or variable severity of strokes. Systemic and pulmonary embolization may also occur, leading to single or multiple lung abscesses, splenic and hepatic embolization and coronary embolization leading to acute myocardial infarction. Some patients present with immunological vascular phenomena such as Osler nodes, petechial rash, splinter hemorrhages and digital clubbing. These classic presentations are seen more in sub-acute rather than acute IE, and patients are usually clinically stable. Two physical signs seen only in the setting of acute IE include Roth spots, which are pale areas surrounded by local hemorrhage in the region of the optic disc, and Janeway lesions (a local, painless, hemorrhagic area, a few millimeters in diameter on the palms and soles of the feet). Patients with acute IE are usually hemodynamically unstable and clinically unwell.

However, despite the classic textbook signs still seen in developing countries, peripheral stigmata of IE are increasingly uncommon elsewhere due to an earlier presentation of patients in the more developed parts of the world.

The physical examination must be performed diligently with special attention being paid to examination of the precordium, as heart murmurs are found in up to 85% of cases of IE.[11] Focus on dentition, all sites of invasive devices and recent procedures and potential embolic manifestations should also be detailed. However, fever remains the most common sign when patients present. This occurs in up to 90% of patients and may be with associated rigors, poor appetite and weight loss. New electrocardiographic (ECG) findings such as conduction abnormalities are present in up to 23% of patients with native valve endocarditis (NVE) and 47% of patients with prosthetic valve endocarditis (PVE). This usually indicates extension of the infection and formation of a para-valvular abscess.[12]

The pathogens involved in diagnosed patients with IE have evolved. In 2002, Cabell looked at the changing face of causative organisms and the burgeoning resistance in common bacterial pathogens, coupled with *S. aureus* surpassing the *S. viridans* group as the leading cause of IE in recent case series and how this has changed the clinical course of IE.[13] Hoen, also in 2002, published a one year, population-based survey conducted from January through December 1999 of hospitals in six French regions, representing 26% of the population, equivalent to 16 million inhabitants. He included a total of 390 adult inpatients diagnosed with IE according to the Duke Criteria. He looked at updating the epidemiology of IE in France based on a study from 1991. The team concluded that although the incidence of IE had not changed, there were important changes in disease characteristics, treatment, and outcomes.[14]

Incidence rates are more commonly associated with increased invasive medical procedures or device insertion and advancing age, and less commonly associated with rheumatic heart disease and poor dentition.[14,15] It is thought these have resulted in an overall worsening of the average clinical course of patients with endocarditis and have been associated with an increased number of serious complications and higher mortality rates.[16]

DIAGNOSIS

A thorough history, physical exam and clinical judgment may raise suspicion of IE, however the diagnosis is mainly based on two tests – blood culture and echocardiography. The Duke Criteria incorporate information from the patient's history, physical examination, echocardiography, microbiology and pathology to diagnose IE.[17] Initially proposed in 1994, these are the most widely accepted diagnostic criteria, proving to be superior to other diagnostic classifications. They were revised in 2000 into the Modified Duke Criteria.[18]

Major Criteria

1. Blood culture positive meeting the following criteria:
   a. Bacterial growth from two separate blood cultures showing common strains such as *Streptococcus viridans*, *S. aureus*, or community acquired enterococci, in the absence of a primary focus, or:
   b. Bacterial growth consistent with infective endocarditis specified as:
      - At least two positive cultures of blood drawn >12 hours apart, or
      - All of three or a majority of >4 separate cultures of blood (with the first and last samples drawn at least 1 hour apart).
   c. Positive blood culture for atypical organisms such as *Coxiella burnetii* or phase I IgG antibody titer to *C. burnetii* >1:800.
2. Positive endocardial involvement as seen on echocardiography for infective endocarditis. Trans-esophageal echo (TEE) is recommended in patients with prosthetic valves with suspected endocarditis on clinical grounds, or patients with complicated endocarditis, such as those with para-valvular abscess. Transthoracic echo (TTE) is the first test in other patients as follows:
   a. Patients with proven echo imaging of vegetations (Figure 10.3).
   b. Annular or root abscess.
   c. New partial dehiscence of a prosthetic valve.
   d. Evidence of a newly regurgitant valve (worsening or changing of pre-existing murmur not sufficient).

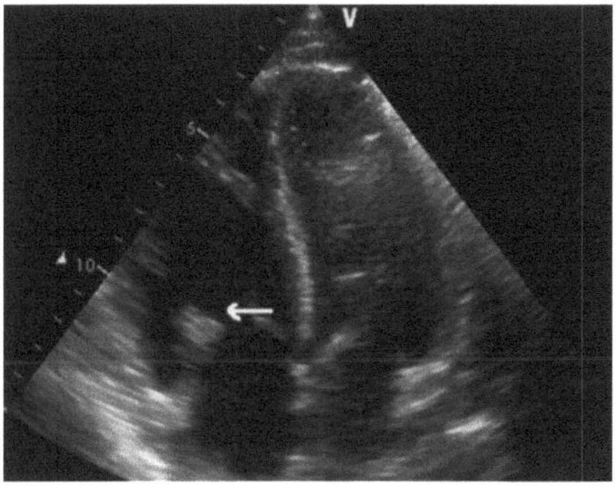

FIGURE 10.3 **Transesophageal echocardiogram showing vegetation of the tricuspid valve (arrow).**

Minor Criteria

1. Fever over 38°C.
2. Positive blood cultures but not meeting those in the major criterion.
3. Vascular phenomena such as major arterial emboli, septic pulmonary infarcts, and Janeway lesions.
4. Immunological phenomena such as glomerulonephritis, Osler nodes, Roth spots, and rheumatoid factor.
5. Predisposing factors to IE such as a history of intravenous drug use or previous valvular pathology.

Interpretation of these criteria for diagnosis includes:

Definite endocarditis = 2 major criteria,
or 1 major + 3 minor criteria,
or 5 minor criteria
Possible endocarditis = 1 major + 1 minor,
or 3 minor criteria
Rejected diagnosis = Alternative diagnosis established.
or resolution of symptoms within four days of antibiotic treatment.
or no proof of IE during surgical procedures or autopsy samples within four days of antibiotic treatment.
or the above mentioned major and minor criteria not met in any form.

Echocardiography must be performed in all cases of suspected IE. The sensitivity of TTE ranges from 40 to 63% and that of TEE from 90 to 100%.[19] TTE is usually the first mode of investigation, while TEE can be of further assisstance in cases of poor quality images, prosthetic heart valves or initial negative TTE scan.[20] However, the AHA 2010 guideline update on pacemaker-related infections recommends TEE to evaluate the left-sided heart valves for all adults suspected of having endocarditis, even if TTE has shown lead-adherent masses (Figure 10.3).

Blood culture is the best method for identifying the microorganisms causing IE. Cultures are positive in the majority of cases (up to 90%), but may be negative in some patients such as those on antibiotic therapy, Bartonella species infections, fungal IE and Q-fever IE. Ideally, blood cultures from a number of samples taken at different times should be analyzed prior to any antibiotic treatment when IE is suspected.

MANAGEMENT

Medical Treatment

Medical treatment is initiated with supportive management starting from admission to hospital and placing the patient under continuous monitoring for any change in general condition. The mainstay of IE management is empirical antibiotic therapy in accordance with sensitivity results. Intravenous (IV) antibiotic administration is preferred over the oral route, as it achieves more reliable therapeutic levels. Orally administered antibiotics have been used as suppressive therapy for incurable valvular infections (i.e., inoperable PVE).

The AHA guidelines for treating IE caused by the most frequently encountered micro-organisms are detailed and complex. The following is a summary of the antibiotic treatment for IE:

1. Adult NVE caused by penicillin-susceptible *S. viridans*, and other susceptible streptococci:
   - Intravenous Penicillin G at 12–18 million U/d for 4 weeks, or
   - Intravenous Ceftriaxone at 2 g/d 4 weeks. Intramuscularly (IM) for short periods may be an alternative if access problems develop.
   - Intravenous Gentamicin at 3 mg/kg/d may be added to Penicillin G or Ceftriaxone as a short-course therapy for 2 weeks in selected patients such as those with uncomplicated NVE caused by sensitive *S. viridans* and of less than 3 months' duration
   - Intravenous vancomycin at 30 mg/kg/d for 4 weeks is an alternative antibiotic in patients who are allergic to penicillin. Gentamycin may be added for 2 weeks.
2. For patients with relatively resistant streptococcal infection, the following may be used:
   - A high dose of intravenous Penicillin G at 24 million U/d for 4 weeks, or
   - Intravenous Cefazolin at 6 g/d in three equally divided doses or intravenous Ceftriaxone at 2 gm/d for 4 weeks.
   - Intravenous Gentamicin at 3 mg/kg/d may be added to the above regimes for the first 2 weeks of therapy.
   - Intravenous vancomycin at 30 mg/kg/d for 4 weeks may be given for patients who are allergic to penicillin.
3. Infective endocarditis caused by enterococci is managed as follows:
   - Intravenous Ampicillin Sodium at 12 g/d for 4–6 weeks, or
   - Intravenous Penicillin G at 18–30 million U/d for 4–6 weeks.
   - Intravenous Gentamicin at 3 mg/kg/d is added to the above regimes for 4–6 weeks.
   - Intravenous vancomycin at 30 mg/kg/d in patients who are allergic to penicillin. This is combined with intravenous Gentamicin at 3 mg/kg/d for 4–6 weeks.

4. Endocarditis caused by sensitive *S. aureus* (MSSA) with no prosthetic material should be treated as follows:
   - Nafcillin or oxacillin at 12 g/d IV for 6 weeks with optional Gentamycin at 3 mg/kg/d for 3–5 days.
   - If allergy occurs, Cefazolin at 6 g/d IV for 6 weeks plus Gentamycin at 3 mg/kg/d for 3–5 days.
   - For patients who are allergic or resistant to penicillin, vancomycin at 30 mg/kg/d IV is used without the addition of Gentamycin keeping in mind that vancomycin is associated with a high failure rate (up to 35%) in the treatment of MSSA and MRSA bloodstream infection. Alternative agents such as linezolid or daptomycin may be used.
5. PVE is commonly caused by *S. epidermidis* (early infection) and Streptococcal/*S. aureus* (late infection). Fungal infection is also seen in this group. This may be managed as follows:
   - Nafcillin or oxacillin at 12 g/d IV for at least 6 weeks, or
   - Cefazolin at 6 g/d IV for at least 6 weeks.
   - Oral Rifampin at 900 mg/d for at least 6 weeks and Gentamicin at 3 mg/kg/d for 2 weeks should be added to each of the above options.
   - Vancomycin at 30 mg/kg/d IV for oxacillin resistant PVE or cases caused by MRSA.

   Treatment with linezolid instead of vancomycin has been advocated in IE caused by drug resistant organisms, or by MRSA and MSSA and in patients who are seriously ill. In 2007, Munoz reported the use of linezolid in nine patients with endocarditis and 33 similar cases from the literature. Most cases were IE with multi-resistant strains. Linezolid was mainly administered in cases of refractory disease, and also in cases with intolerance, as a sequential therapy and to those with resistant pathogens. Treatment was for a mean of 37 days. This resulted in a successful outcome in 79% of cases and side effects were described in ten cases. The mean follow-up period was 8.5 months. At the time of this publication, linezolid was not a standard therapy for IE. However, the AHA considers it to be a reasonable alternative for cases of endocarditis caused by methicillin resistant *S. aureus* (MRSA) or multi-resistant enterococci. Further data from randomized controlled clinical trials is needed to determine the role of linezolid in the management of endocarditis.[21]
6. IE due to *Haemophilus, Actinobacillus, Cardiobacterium hominis, Eikenella corrodens, Kingella (Kingella kingae)* (HACEK) microorganisms should be treated as follows:
   - Ceftriaxone at 2 g/d IV for 4 weeks, or
   - Ampicillin at 12 g/d IV combined with IV gentamicin at 3 mg/kg/d for 4 weeks, or
   - Ciprofloxacin at 800 mg/d IV for 4 weeks.

Fungal endocarditis is rare and commonly occurs after valve surgery and in individuals who abuse intravenous drugs. The most common organisms are Candida and Aspergillus species. Current antifungal agents have not shown much success in eliminating fungal IE. The only cure for proven fungal IE is surgical excision of the infected valves combined with amphotericin B therapy.

As a rule, when administering vancomycin, a peak level of 30–45 mcg/mL should be attained 1 hour after completion of the intravenous infusion.

SURGICAL TREATMENT

The most crucial aspect of the treatment of IE remains the eradication of infection by administration of the appropriate antibiotics. Despite this, 40–45% of patients will require surgical management.[22] The first successful surgical treatment of IE was performed by Dr Glenn Young in 1963, on a 45 year old bricklayer who presented with sepsis and congestive heart failure due to an incompetent aortic valve and who was unresponsive to antibiotic therapy.[23] The two primary surgical principles of surgical management of IE remain the control of infection and restoration of cardiac morphology.[24]

Surgical management of native or prosthetic valve endocarditis remains the same. This includes the removal of infected tissues and restoration of normal cardiac morphology including valve repair or replacement, repair of affected large vessels, debridement of annular or root abscesses and closures of any fistulas. The best surgical exposure is via a median sternotomy with cardiopulmonary bypass which facilitates the best access for surgical debridement and definitive surgical treatment.

A proportion of IE patients who are deemed low risk can undergo surgery at a later stage as long as they are hemodynamically stable. In some cases, surgery may need to be expedited as emergent (within 24 hours) or urgent (within a few days), irrespective of the duration of antibiotic regime. In other cases, surgery can be postponed to allow one or two weeks of antibiotic treatment under careful clinical and echocardiographic observation before elective surgery is performed. Early consultation with a cardiac surgeon among cardiologist, microbiologist and intensive care physicians is recommended to determine the best therapeutic approach for each patient.

Surgery during the active phase of IE is associated with significant risk. Early surgical intervention during this phase of the disease is indicated in hemodynamic instability (patients in heart failure and cardiogenic shock), uncontrolled infection, para-valvular abscess or cardiac fistula and occurrence of embolic events.[25,26] Prosthetic valve endocarditis, particularly mechanical valves, is also best treated by early surgical intervention.[27,28]

The in-hospital mortality of IE can be up to 26%. Patients with heart failure, peri-annular complication (irreversible structural damage caused by severe infection such as abscess or fistula formation) and/or Staphylococcal infections are at a higher mortality and morbidity risk and will frequently require surgical intervention in the active phase of treatment. Moreover, when all three of these factors are present, the risk of death increases to 79%.[28]

Special consideration must be paid to the timing of surgery in the light of central nervous system complications, which are largely due to cerebral embolization. Pre-operative neurological evaluation and any significant findings should be assessed by thorough physical examination, brain CT (computer tomography) and/or MRI (magnetic resonance imaging). Patients who have had an ischemic stroke should have surgery delayed by two weeks, if possible, due to the risk of a non-hemorrhagic infarct transforming into a hemorrhagic one as a complication of anticoagulation required during cardiopulmonary bypass.[29]

For the same reason, patients who have had a hemorrhagic stroke should be postponed for four weeks, if possible, to reduce the risk of further intracranial bleeding associated with cardiac surgery.

In 2012, Kang published the results of a randomized study in which patients with large left-sided vegetations either underwent early surgery within 48 hours of diagnosis, or received conservative treatment with surgery performed only if complications requiring urgent intervention developed during medical treatment, or if symptoms persisted after the completion of antibiotic therapy.[30] Results showed that compared to conservative treatment, early surgery in patients with IE and large vegetations significantly reduced the composite end point of death from any cause and embolic events by effectively decreasing the risk of systemic embolization.

A larger randomized control trial started in 2008 to test the optimal timing of surgery for IE is still ongoing.[31] It will be interesting to see whether early surgery remains superior for preventing complications in this larger, more diverse group of patients.

Management with a multi-disciplinary team of cardiologists, microbiologists, cardiothoracic surgeons and other specialties, including stroke physicians and intensive care specialists, as well as allied health staff, is imperative for management of IE.

Operative Techniques

Patient with IE undergoing surgery often require long and complex procedures. The surgical team should eliminate any risk of intra-operative surgical contamination; this includes careful surgical draping, double gloving and careful instrument handling while operating on infected tissues. All contaminated material must be discarded before any replacement or reconstruction begins.

Intra-operatively, even if infection is limited to the cusps of a native or prosthetic valve, complete removal of the valve and replacement with a bio prosthetic or mechanical valve is the practice of most units. The old belief was that bio prosthetic (tissue) valves were less prone to infection and were therefore used more often in IE cases, even in the young patient; however more recent studies have shown no evidence that bio prosthetic valves are superior to mechanical valves in patients with active IE.[32]

Pulmonary auto grafts (Ross Procedure) and homografts have also been used by some units with experience in these techniques, particularly in younger patients.[33]

IE involving the aortic annulus, aortic root abscesses or cardiac fistulas are more difficult to manage. These patients frequently require replacement of the entire aortic root and reconstruction of the surrounding structures involved in the abscess. The most important aspect of the operation is the radical debridement and resection of all infected tissue. The residual defect may need to be patched before a prosthetic valve is implanted. Patches can be fashioned from autologous pericardium, glutaraldehyde-fixed bovine pericardium or Dacron fabric to reconstruct the root.

In native mitral valve IE, exposure is best with a full median sternotomy and bicaval cannulation. The valve is approached via a left atriotomy and evaluated for presence of para-valvular abscesses, intracardiac fistulae, or inter-valvular fibrous body/ventricular involvement. Radical resection of all necrotic tissue is performed with a margin of normal tissue (Figure 10.4). If the tricuspid valve is also involved, a trans-septal approach is done through a right atriotomy and both valves can be dealt with within the same procedure.

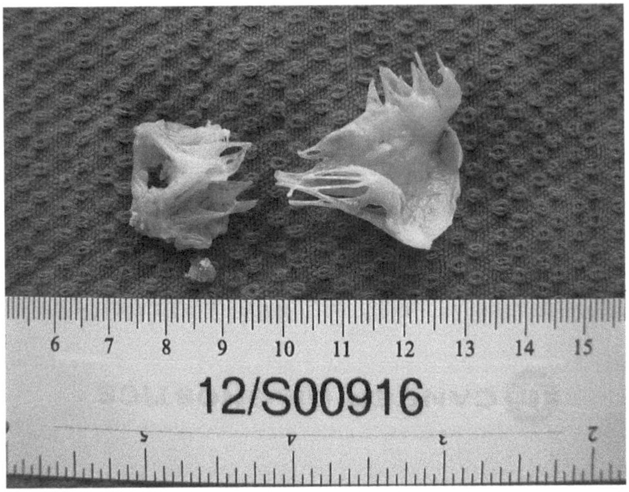

FIGURE 10.4 Gross appearance of mitral valve leaflets showing partially healed endocarditis.

Unlike native aortic valve IE, where valve replacement is the more common procedure, a repair of a native infected mitral valve may be attempted if feasible. Mitral valve repair can be performed safely provided there is sufficient remaining tissue to allow reconstruction without tension. In the event of extensive destruction of the sub-valvular apparatus, prosthetic valve replacement, mechanical or bio prosthetic, is performed.

Prosthetic valve endocarditis has a higher operative mortality than native valve IE. Even with improved antibiotic regimens, outcomes of medical therapy alone without re-operation are poor, particularly for patients with annular involvement or early endocarditis after valve surgery. The approach is via a re-sternotomy with resection of the infected prosthesis and surrounding infected tissues being crucial before the new prosthesis is implanted.

CONCLUSIONS AND FUTURE PROSPECTS

Infective endocarditis remains a major challenge for clinicians where in-hospital mortality may reach 36%. A wide variety of age groups is affected, and if not treated urgently the condition can be fatal or be associated with high morbidity rates. Commonly affecting valves that are structurally abnormal, normal valves are definitely not excluded and are often seen in clinical practice.

A high level of suspicion is essential in reaching a diagnosis, as clinical presentation is frequently vague and non-specific. The mainstay of diagnosis is a positive blood culture supported by echocardiographic evidence of endocardial involvement. Other signs and symptoms of IE are summed up in the Duke Criteria, and these have been paramount in reaching a definite diagnosis over the decades. Once diagnosed, immediate treatment is the rule, since delays are associated with a higher incidence of complications. Patients are usually managed medically with appropriate antibiotic cover and hemodynamic support, and this is successful in the majority of cases. However, even with effective therapy, 40%

of patients with infective endocarditis relapse within two months post treatment. Moreover, approximately 25–30% of patients eventually require surgery, either as early intervention or up to two years after completing therapy. The costs associated with prolonged hospitalization including the need for intensive care, surgery and prolonged antibiotic treatment (especially managing the rapidly increasing occurrence of antibiotic resistant strains such as MRSA and VRSA) are extremely high; placing a severe burden on healthcare systems around the world.

This highlights the importance and need for more research into understanding the molecular interactions that result in the development of infective thrombus on the endocardium and specifically on heart valves. This may lead to prevention of thrombus formation through development of novel therapies, and therefore help to overcome the problem of reaching effective antibiotic concentrations on the heart valve.

Further studies are needed on top of existing ones looking at the emerging role of platelets in infection and immunity, specifically at blocking interactions between IgG and specific platelet receptors. This may prevent platelet receptor clustering and inhibit thrombus formation on the endocardium.

The role of a multi-disciplinary approach to infective endocarditis is paramount, and the involvement of several teams is usually required, including medical, surgical and infectious disease specialists.

If IE is identified at an early stage and treated aggressively, the outcome is usually a favorable one.

References

1. Prendergast B. Diagnostic criteria and problems in infective endocarditis. *Heart* 2004;**90**:611–3.
2. Burke A. *Practical cardiovascular pathology*. 1st ed. Philadelphia PA: Lippincott Williams and Wilkins; 2011. pp. 349–62.
3. Rosai J. *Rosai and Ackermann's surgical pathology*. 10th ed. Edinburgh: Elsevier; 2011. pp. 2274–7.
4. Connor D. *Pathology of infectious diseases*. 1st ed. Stamford Connecticut: Appleton and Lange; 1997. pp. 810–11.
5. Mills S. *Sternberg's diagnostic surgical pathology*. 5th ed. Baltimore MD: Lippincott Williams and Wilkins; 2010. pp. 1219–21.
6. Silver M. *Cardiovascular pathology*. 1st ed. New York: Churchill Livingstone; 1983. pp. 517–49.
7. Silverberg S, DeLellis R, Frable W. *Principles and practice of surgical pathology and cytopathology*. 3rd ed. New York NY: Churchill Livingstone; 1997. pp. 1369–70
8. Mylonakis, E., Calderwood SB. Infective endocarditis in adults. *N Engl J Med* 2001;**345**:1318–30.
9. Houpikan P, Raoult D. Diagnostic methods, current best practices and guidelines for identification of difficult-to-culture pathogens in infective endocarditis. *Infect Dis Clin North Am* 2002;**16**:377–92.
10. Habib G, Hoen B, Tornos P, Prendergast B, Vilacosta I, Moreillon P, et al. Guidelines on the prevention, diagnosis, and treatment of infective endocarditis: the Task Force on the Prevention, Diagnosis, and Treatment of Infective Endocarditis of the European Society of Cardiology (ESC). *Eur Heart J* 2009;**30**:2369–413.
11. Thuny F, Di Salvo G, Belliard O, Avierinos JF, Pergola V, Rosenberg V, et al. Risk of embolism and death in infective endocarditis: prognostic value of echocardiography: a prospective multicenter study. *Circulation* 2005;**112**:69–75.
12. Miller DC. Predictors of outcome in patients with prosthetic valve endocarditis (PVE) and potential advantages of homograft aortic root replacement for prosthetic ascending aortic valve-graft infections. *J Card Surg* 1990;**5**:53.
13. Cabell CH, Jollis JG, Peterson GE, Corey GR, Anderson DJ, Sexton DJ, et al. Changing patient characteristics and the effect on mortality in endocarditis. *Arch Intern Med* 2002;**162**:90–4.

14. Hoen B, Alla F, Selton-Suty C, Beguinot I, Bouvet A, Briancon S. Changing profile of infective endocarditis: results of a 1 year survey in France. *JAMA* 2002;**288**:75–81.
15. Tleyjeh IM, Abdel-Latif A, Rahbi H, Scott CG, Bailey KR, Steckelberg JM, et al. A systematic review of population-based studies of infective endocarditis. *Chest* 2007;**132**:1025–35.
16. Baddour LM, Wilson WR, Bayer AS, Fowler Jr VG, Bolger AF, Levison ME, et al. Infective endocarditis: diagnosis, antimicrobial therapy, and management of complications: a statement for healthcare professionals from the Committee on Rheumatic Fever, Endocarditis, and Kawasaki Disease, Council on Cardiovascular Disease in the Young, and the Councils on Clinical Cardiology, Stroke, and Cardiovascular Surgery and Anesthesia, American Heart Association: endorsed by the Infectious Diseases Society of America. *Circulation* 2005;**111**:394–434.
17. Durack DT, Lukes AS, Bright DK. New criteria for diagnosis of infective endocarditis: utilization of specific echocardiographic finding. Duke Endocarditis Service. *Am J Medicine* 1994;**96**:200–7.
18. Li JS, Sexton DJ, Mick N, Nettles R, Fowler Jr VG, Ryan T, et al. Proposed modifications to the Duke Criteria for the diagnosis of infective endocarditis. *Clin Infect Dis* 2000;**30**:633.
19. Evangelista A, Gonzalez-Alujas MT. Echocardiography in infective endocarditis. *Heart* 2004;**90**:614–7.
20. Greaves K, Mou D, Patel A, Celermajer DS. Clinical criteria and the appropriate use of transthoracic echocardiography for the exclusion of infective endocarditis. *Heart* 2003;**89**:273–5.
21. Munoz P, Rodríguez-Creixems M, Moreno M. Linezolid therapy for infective endocarditis. *Clin Microbiol Infect* 2007;**13**:211–5.
22. Tornos P, Iung B, Permanyer-Miralda G, Baron G, Delahaye F, Ch Gohlke-Bärwolf EG, et al. Infective endocarditis in Europe: lessons from the Euro heart survey. *Heart* 2005;**91**:571–5.
23. Wallace AG, Young GW, Osterhout S. Treatment of acute bacterial endocarditis by valve excision and replacement. *Circulation* 1965;**31**:450–3.
24. David TE, Gavra G, Feindel CM, Regesta T, Armstrong S, Maganti MD. Surgical treatment of active infective endocarditis: a continued challenge. *J Thorac Cardiovasc Surg* 2007;**133**:144–9.
25. Bashore Jr TM, Cabell C, Fowler Jr V. Update on infective endocarditis. *Curr Probl Cardiol* 2006;**31**:274–352.
26. Delahaye F, Celard M, Roth O, de Gevigney G. Indications and optimal timing for surgery in infective endocarditis. *Heart* 2004;**90**:618–20.
27. Thuny F, Beurtheret S, Mancini J, Gariboldi V, Casalta JP, Riberi A, et al. The timing of surgery influences mortality and morbidity in adults with severe complicated infective endocarditis: a propensity analysis. *Eur Heart J* 2011;**32**:2027–33.
28. San Roman JA, Lopez J, Vilacosta I, Luaces M, Sarria C, Revilla A. Prognostic stratification of patients with left-sided endocarditis determined at admission. *Am J Med* 2007;**120**:e1–7.
29. Ting W, Silverman N, Levitsky S. Valve replacement in patients with endocarditis and cerebral septic emboli. *Ann Thorac Surg* 1991;**51**:18.
30. Kang DH, Kim YJ, Kim SH, Sun BJ, Kim DH, Yun SC, et al. Early surgery versus conventional treatment for infective endocarditis. *N Engl J Med* 2012;**366**:2466–73.
31. San Roman JA, Lopez J, Revilla A, Vilacosta I, Tornos P, Almirante B, et al. Rationale, design, and methods for the early surgery in infective endocarditis study (ENDOVAL 1): a multicenter, prospective, randomized trial comparing the state-of-the-art therapeutic strategy versus early surgery strategy in infective endocarditis. *Am Heart J* 2008;**156**:431–6.
32. Moon MR, Miller DC, Moore KA, Oyer PE, Mitchell RS, Robbins RC. Treatment of endocarditis with valve replacement: the question of tissue versus mechanical prosthesis. *Ann Thorac Surg* 2001;**71**:1164.
33. Oswalt JD, Dewan SJ, Mueller MC, Nelson S. Highlights of a ten-year experience with the Ross procedure. *Ann Thorac Surg* 2001;**71**:S332.

SECTION IV

SKIN AND SOFT TISSUE INFECTIONS

CHAPTER

11

Natural Remedies for the Treatment of Wounds and Wound Infection

Kateryna Kon and Mahendra Rai[†,**]*

*Department of Microbiology, Virology and Immunology, Kharkiv National Medical University, Kharkiv, Ukraine, [†]Department of Biotechnology, Sant Gadge Baba Amravati University, Amravati, Maharashtra, India, **Department of Chemical Biology, Institute of Chemistry, University of Campinas, Campinas, SP, Brazil

INTRODUCTION

The treatment of wounds is one of the oldest surgical problems.[1] Methods used for wound healing have been modified during the centuries, but the basis of modern wound treatment is not difficult to recognize among recommendations made by traditional doctors. The description of wound treatments used during the Trojan War (about 1000 BC), such as washing wounds with warm water and application of a powdered root with anti-hemorrhage and analgesic properties,[1] does not differ significantly from modern methods used to stop bleeding, reducing pain and prophylaxis of wound infection.

Development of wound healing methods is based on understanding the wound healing process, which includes four highly integrated and overlapping phases: hemostasis, inflammation, proliferation and tissue remodeling or resolution (Figure 11.1).[2]

The *Hemostasis* phase includes vascular constriction and aggregation of platelets, degranulation, and formation of a fibrin clot. The clot and surrounding wound tissue release pro-inflammatory cytokines and growth factors (transforming growth factor (TGF)-β, platelet-derived growth factor (PDGF), fibroblast growth factor (FGF), and epidermal growth factor (EGF)).

The *Inflammatory* phase includes infiltration of wound area by neutrophils and monocytes with subsequent differentiation to macrophages and by lymphocytes. Neutrophils play a role in clearing the wound area of invading microorganisms and cellular debris. Macrophages have multiple functions, including releasing pro-inflammatory cytokines

Microbiology for Surgical Infections
DOI: http://dx.doi.org/10.1016/B978-0-12-411629-0.00011-8

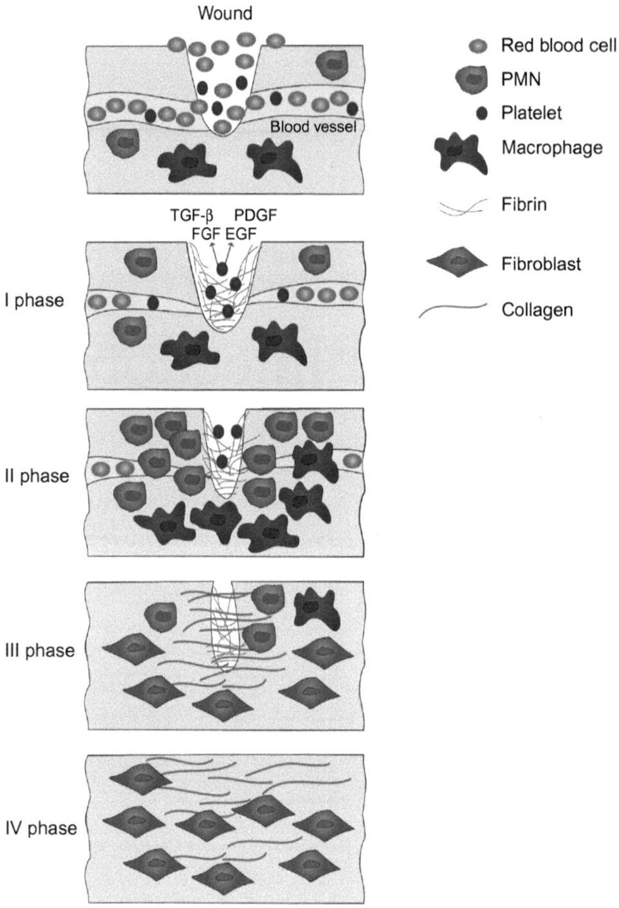

FIGURE 11.1 **Phases of wound healing.** I – Hemostasis phase: constriction of blood vessels, aggregation of platelets, their degranulation with release of pro-inflammatory cytokines and growth factors, formation of fibrin clot; II – Inflammatory phase: infiltration of wound area by neutrophils, lymphocytes, and monocytes with subsequent differentiation to macrophages; III – Proliferative phase: re-epithelialization, angiogenesis, synthesis of collagen and formation of extracellular matrix; IV – Remodeling phase: remodeling of collagen, maturation and regression of blood vessels; PMN – polymorphonuclear cell; TGF-β – transforming growth factor-β, PDGF – platelet-derived growth factor, FGF – fibroblast growth factor, EGF – epidermal growth factor. *Source: Modified from: Beanes SR, Dang C, Soo C, Ting K. Skin repair and scar formation: the central role of TGF-[beta].* Exp Rev Mol Med *2003;5:1–22.*

and activating additional leukocytes. They also resolve inflammation by inducing and clearing apoptotic cells, and finally they stimulate keratinocytes, fibroblasts, and angiogenesis thus promoting tissue regeneration. The role of T-lymphocytes in the wound healing process is still not completely understood.

The *Proliferative* phase consists of re-epithelialization, angiogenesis, synthesis of collagen and formation of extracellular matrix. The most prevalent cells in the reparative dermis are fibroblasts and endothelial cells; they take part in capillary growth, formation of collagen and granulation tissue. The major components of the extracellular matrix within the wound bed are collagen, glycosaminoglycans and proteoglycans produced by fibroblasts.

The *Remodeling* phase consists of collagen remodeling and vascular maturation with regression. Newly formed capillaries regress, returning the vascular density of wound tissue to normal. Together with wound contraction promoted by fibroblasts, these processes lead to the wound architecture approaching the structure of normal tissue.

Wound healing is affected by different systemic and local factors.[2] Systemic factors include age and gender, presence of comorbidities, immune status, stress factors, etc.;

while local factors include the presence of infection and foreign bodies in the wound, level of oxygenation and state of venous sufficiency.

One of the crucial factors which influences wound healing is the presence of infection.[3] The state of infection is classified as *contamination* (presence of non-replicating microorganisms), *colonization* (replicating of microorganisms but without tissue damage), *local infection/critical colonization* (replication of microorganisms with local response of the tissue commencing) and *invasive infection* (microorganisms replicating in the wound and causing subsequent host injury).[2]

Delay in wound healing, particularly in chronic wounds, is often associated with microbial biofilms. Biofilms consist of microorganisms attached to a wound surface and surrounded by extracellular polysaccharide matrix, and recent findings from both *in vitro* and *in vivo* studies demonstrate that this matrix protects the biofilm from some of the inflammatory processes which are key to wound healing[4]; furthermore, impaired wound healing is associated with protease activity and immunological suppression caused by biofilms.[5]

Wound care includes optimizing local and systemic conditions of the patient in conjunction with creating an ideal wound healing environment.[6] There are many different products which have been developed in order to create an optimal, microbe-free, protected, moist area for wound healing, such as protective dressings including gauze, impregnated gauze; antibacterial dressings in the form of ointments, silver- and iodine-based products; autolytic debridement products, including films, hydrocolloids and hydrogels; chemical debridement products with papain/urea or collagenase; absorbent dressings with foams, hydrogels, hydrofibers and alginates; negative pressure wound devices; bioengineered skin substitutes; use of growth factors and biological wound products; and also hyperbaric oxygen.

The continuing development of new and commercial wound healing products indicates the absence of a perfect product which could satisfy all requirements for an optimal wound healing process. Because of this, many recent studies have considered natural wound healing methods, which have been used in traditional medicine for hundreds or even thousands of years, but have not been examined scientifically. They do, however, provide great potential for modern medicine due to their complex healing abilities.

The present chapter aims to review studies of natural wound healing remedies such as plant essential oils, maggots and amphibian skins, and discusses avenues for future studies in this area.

PLANT ESSENTIAL OILS IN THE TREATMENT OF WOUNDS AND WOUND INFECTION

Plant Essential Oils as Alternatives to Antibiotics

Plants have been used in the treatment of different diseases since ancient times, and numerous descriptions of their application in ancient Egyptian, Greek, Chinese and Ayurvedic treatment methods have survived.[7] Plant essential oils (EOs) have attracted particular attention because their volatile nature means that they can be topically applied, or inhaled, creating a complex treatment effect that includes an influence on the emotional state of a patient. There was less interest in EOs after the discovery of antibiotics in the 20th century, but growing antibiotic resistance soon lead to a search for alternative methods, which restored the interest in EOs.

The mechanism of action of an essential oil includes disintegrating the bacterial cytoplasmic membrane and interacting with membrane proteins, such as ATPases and others, disturbing the outer membrane of gram-negative bacteria and releasing of lipopolysaccharides, destabilizing the proton motive force and causing leakage of ions, inhibition of enzyme synthesis and coagulation of the cell content.[8] Therefore, in contrast to conventional antibiotics, EOs have a complex mechanism of action which affects different parts of the bacterial cell, and this makes them promising agents against microorganisms that have developed resistance to antibiotics.[9] Furthermore, EOs can act simultaneously against different kinds of microorganisms – bacteria,[9] fungi,[10] viruses[11] and protozoans,[12] and this feature is useful in infections caused by several microbes.

The *in vitro* activity of EOs has been extensively studied, as evidenced by the large number of research publications available – a PubMed search on antimicrobial essential oils in 2012 found 347 articles demonstrating promising activity of various EOs, which potentially may find application in the prevention and treatment of different infections. Many EOs are reported to be effective against multidrug resistant strains, such as methicillin-resistant *Staphylococcus aureus*, vancomycin-resistant enterococci, multidrug resistant strains of *Pseudomonas aeruginosa*, *Klebsiella pneumoniae*, etc., which indicates their potential use in this era of growing antibiotic resistance.[13] The number of publications which prove *in vivo* activity is much smaller, and therefore this area requires further detailed study.

Studies on the *In Vivo* Efficacy of Plant Essential Oils in the Treatment of Wounds

The revitalization of interest in the practical use of EOs became noticeable in the late 1990s. This growth was demonstrated by the appearance of review articles on the use of essential oils in wound treatment.[14–16] Descriptions of observed clinical effects have been actively published since the 2000s (Table 11.1). For example, Orafidiya et al.[17] showed a wound healing effect of *Ocimum gratissimum* EO in experimental full-thickness excisional and incisional wounds in rabbits. Results in this study were assessed only by evaluating quantitative parameters of wound healing, but in subsequent studies more precise clinical and laboratory tests were implemented.

An interesting study reported a comparison of the wound healing effect of *Lavandula x allardii* EO, *L. x allardii* honey and a standard therapeutic honey (Medihoney) using a rat excisional wound model.[18] The results were assessed by analyzing wound contraction and capillary volume, and the best effect was found with *L. x allardii* honey which promoted more advanced scar maturation.

Wound healing properties were demonstrated by *Rosmarinus officinalis* EO in an alloxan-induced diabetic BALB/c mice model.[19] The authors compared the results obtained using the EO and an aqueous extract by performing a visual examination of the wound and monitoring blood glucose level, body weight, regenerated granulation tissue

TABLE 11.1 Selected Studies on the use of Essential Oils for Wounds Treatment

Essential Oil	Model	Tested Criteria	Results	Reference
Ocimum gratissimum	Topical application of EO in the full-thickness excisional and incisional wound models in rabbits	Determination of quantitative parameters of wound healing	Studied EO demonstrated good wound healing properties	17
Lavandula x allardii EO, *L. x allardii* honey and a standard therapeutic honey (Medihoney)	Topical application of studied substances in the excisional wound models in rats	Analysis of wound contraction and capillary volume	The best effect was in *L. x allardii* honey	18
Rosmarinus officinalis	Application of EO into experimentally induced diabetic wounds in mice	Visual examination, regenerated granulation tissue weight, the percentage of wound contraction, histopathology	EOs showed high effectiveness in the wound healing	19
Melaleuca alternifolia	Water-miscible EO solution (3.3%) was used in a wound cleansing procedure in 11 patients with wounds caused by MRSA	Measuring size of wound by computer planimetry, bacteriological examination of wound content	Majority of wounds reduced in size but were not cleaned of MRSA	20
Abies spp., *Cedrus libani* and *Picea orientalis*	Ointments from EOs were evaluated by linear incision and circular excision animal wound models	Histopathological analysis, anti-inflammatory activity of EOs	The best wound healing activity was in EOs of *C. libani* and *A. cilicica* subsp. *cilicica*	21
Olive oil extract of flowering aerial parts of *Hypericum perforatum*, olive oil, *Origanum* spp., *Salvia triloba* EOs	Ointment from listed components was evaluated in linear incision and circular excision animal wound models	Histopathology, collagenase activity, activity of components against *Staphycoccus aureus* and *Candida albicans*	Formulation exhibited significant wound healing activity	22
Juniperus and *Cupressus* spp.	The same	Histopathological analysis, hydroxyproline content, anti-inflammatory activity of EOs	The best wound healing activity was in EOs from berries of *J. oxycedrus* ssp. *oxycedrus* and *J. phoenicea*	23
From the cones and needles of *Pinus* spp.	The same	Histopathological analysis, hydroxyproline content, anti-inflammatory and anti-hyaluronidase activity of EOs	The best wound healing activity was in EOs from the cones of *P. pinea* and *P. halepensis*	24

(Continued)

TABLE 11.1 (Continued)

Essential Oil	Model	Tested Criteria	Results	Reference
From heartwood samples of *Juniperus virginiana, Juniperus occidentalis*, and *Juniperus ashei*	The same	Histopathological analysis, hydroxyproline content, anti-inflammatory activity of EOs	*J. occidentalis* EO was the most biologically active	25

EO – essential oil; MRSA – methicillin-resistant *Staphylococcus aureus*.

weight, the percentage of wound contraction and finally histopathology. EO demonstrated a better effect than the aqueous extract.

Edmondson et al. enrolled 11 patients to study the effectiveness of tea tree (*Melaleuca alternifolia*) oil application in decolonizing wounds of methicillin resistant *S. aureus* (MRSA).[20] Water-miscible tea tree oil (3.3%) solution was applied during each dressing change, and although the majority of wounds (8 out of 11) reduced in size after treatment, the MRSA was not eradicated.

Tumen et al.[21] evaluated the wound healing activity of *Abies cilicica* subsp. *cilicica, Abies nordmanniana* subsp. *bornmulleriana, Abies nordmanniana* subsp. *equi-trojani, Abies nordmanniana* subsp. *nordmanniana, Cedrus libani* and *Picea orientalis* using linear incision and circular excision wound models in rats and mice. The authors used three groups of animals: one treated with an ointment composed of the studied EOs, one treated with a reference ointment Madecassol®, and a negative control group. The linear incision wound model helps to assess the tensile strength of a treated wound, while the circular excision model is used to evaluate wound contraction and time of wound closure. Wound healing was assessed by histopathological examination of epidermal and dermal remodeling, and anti-inflammatory activity was studied by the Whittle method: by determining the inhibition of acetic acid-induced increase in capillary permeability. The highest wound healing potential was demonstrated by the *Cedrus libani* and *Abies cilicica* subsp. *cilicica* EOs.

By using similar techniques, good wound healing properties were demonstrated in a formulation made from olive oil and several EOs (olive oil extract of flowering aerial parts of *Hypericum perforatum*, olive oil, an equivalent mixture of *Origanum majorana* and *Origanum minutiflorum* EOs, and *Salvia triloba* EO).[22]

Tumen et al.[23] and the same team of authors further evaluated the activity of *Cupressus sempervirens* var. *horizontalis, Cupressus sempervirens* var. *pyramidalis, Juniperus communis* subsp. *nana, Juniperus excelsa, Juniperus foetidissima, Juniperus oxycedrus* subsp. *oxycedrus*, and *Juniperus phoenicea* EOs in wound healing. The best wound healing activity of the EOs and control groups was demonstrated by that of the berries of *J. oxycedrus* subsp. *oxycedrus* and *J. phoenicea*.

This research team also studied the wound healing effect of *Pinus brutia, Pinus halepensis, Pinus nigra, Pinus pinea* and *Pinus sylvestris* EOs using the same methods, together with a determination of the anti-hyaluronidase activity of the EOs.[24] The results indicated good wound healing activity for EOs obtained from the cones of *P. pinea* and *P. halepensis*. The activities of *Juniperus virginiana, Juniperus occidentalis*, and *Juniperus ashei* were similarly compared, and the most biologically active appeared to be the EO of *J. occidentalis*, while the best anti-inflammatory properties were found in *J. virginiana* EO.[25]

One of the limitations of studies on currently available EOs is their non-standard composition, which varies with geographical area of the source, season, and method of preparation. This disadvantage can be overcome by isolating pure EO components and studying their effect; but on the other hand, the multi-component composition of the EO plays an important role in preventing microbial resistance and in its complex healing potential. Another limitation is the absence of multicenter studies, which prevents the recommendation of a particular EO for general practice. Nevertheless, EOs could be alternative methods that are sometimes more effective than conventional antibiotics.

EOs can be used not only in their crude form. Results from *in vitro* and *in vivo* studies indicate other possible applications, such as wound dressing materials containing the most active EO components. One such study was performed by Dias et al.,[26] who developed a wound dressing material impregnated with quercetin and thymol. Quercetin is a flavonoid with anti-inflammatory and antioxidant activity found in fruits, vegetables, herbs, nuts, and seeds,[27] while thymol is a monoterpene phenol with antimicrobial properties that is found in many EOs.[28]

The Wound Healing Mechanism of Essential Oils

The antimicrobial activity of EOs plays an important role in the healing of infected wounds (Figure 11.2). Their healing properties are also due to their anti-inflammatory activity; for example, such activity was demonstrated by α-pinene,[29] which is present at high concentrations in *Juniperus* EOs, particularly that of *J. phoenicea*.

The healing potential of *J. oxycedrus* subsp. *oxycedrus* and *J. phoenicea* EOs[23] may be explained by the high concentration of flavonoids and phenolic acid, which have significant antioxidant activity,[30] and which also promotes wound healing. Tumen et al.[23] reported that levels of hydroxyproline (which indicates enhanced collagen synthesis in regenerated tissues) were stimulated by EOs. Wound healing is also facilitated by anti-hyaluronidase activity of some EOs.[24]

AMPHIBIAN SKIN SECRETIONS IN THE TREATMENT OF WOUNDS

Zootherapeutic practices are very popular in folk medicine. Zootherapy is the use of insects, mammals, fishes, birds, reptiles, mollusks, crustaceans, echinoderms, arachnids, amphibians, annelids or millipedes as a medical treatment.[31] Since ancient times the body parts and metabolic products of these animals have been used for medicinal purposes: healing effects were found for skin, horn, teeth, corporal secretions, blood, excrements and even animal housings, such as nests and cocoons.[32–34]

Recent ethnopharmacological studies have been planned to document in detail the animal species used against different diseases in different geographical areas, such as in Mexico,[35,36] Brazil,[31] Latin America in general,[37] India,[38,39] Albania, Italy, Spain and Nepal,[40] Nigeria,[41] and Israel.[42] Each study listed identified huge numbers of animal

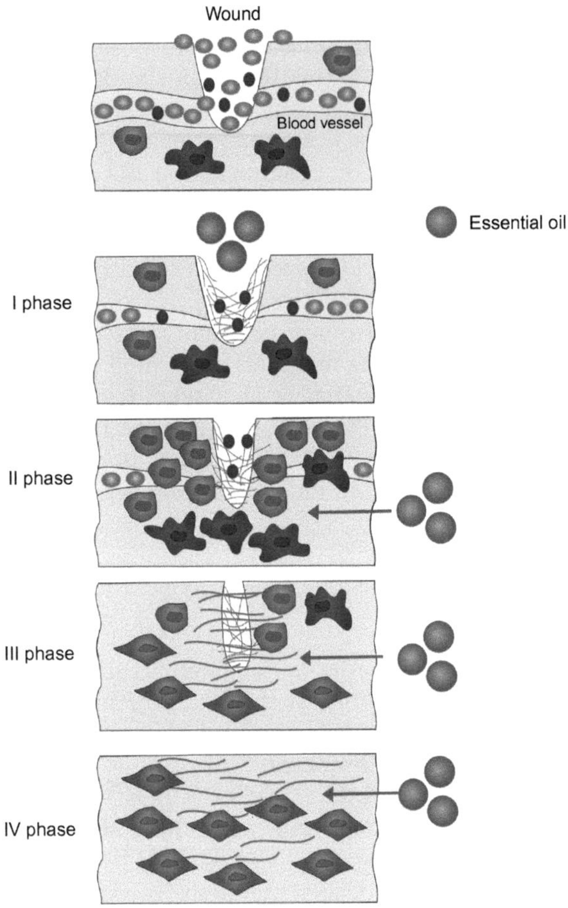

FIGURE 11.2 **Mechanism of wound healing activity of plant essential oils depending on phases of wound healing.** Phase I and next phases are facilitated by antimicrobial activity of EOs. Phase II: anti-inflammatory and antioxidant activity; Phase III and IV: collagen-stimulating activity.

species used in folk medicine, for example, Alves and Alves[37] reviewed researches conducted on the use of Latin America's fauna and reported the use of 584 animal species distributed in 13 taxonomic categories for the treatment of different diseases, with the most frequently used being mammals (130 species), birds (122 species), fishes (110 species) and reptiles (95 species).

The published studies include an impressive diversity of zootherapeutic methods used in traditional medicine; among them are such unusual methods as the application of a spider's web to seal skin lacerations and facilitate healing; a deer horn tip is used to puncture the skin in order to release the poison from a snake bite; the use of dog saliva for the treatment of furuncles; the application of fried snake meat mixed with olive oil to treat wounds[40]; the use of goat's urine for the treatment of insect bites; the penis of Sambhar for the treatment of hydrocele; and pig's fat to treat hemorrhoids.[38] Such remedies have been used for centuries, and recent studies have revealed the scientific basis for some of them. These findings can be further implemented into modern medicine as a source of new drugs.

With respect to wound healing potential, special interest is elicited by amphibian skin, as it was used in traditional medicine as a wound dressing and in the form of an extract. Frog skin was used as a xenograft to provide wound coverage as long ago as 1500 BC,[43] and nowadays it is used to cure wounds in India.[44] Recently, research has provided scientific substantiation for its healing properties.

Almost 20 years ago, the wound healing properties of *Rana tigrina* frog skin were experimentally proven using a rat model.[45] The authors compared the healing of wounds dressed by dorsal frog skin to those with a cotton gauze dressing. The experimental group healed faster than the control group, as shown by the biochemical estimation of wound granulation tissues.

In experiments on mice, Mashreghi et al.[46] studied the wound healing properties of skin secretions extracted from the frog *Rana ridibunda*. Secretions were used in two forms – as a raw extract and as an ultrafiltrated extract, which was prepared using a membrane with cutoff 10 kDa as under 10 kDa. These extracts were applied in the form of an ointment onto the wound surface. The results were assessed histopathologically and by studying the microbial loads of the wounds, and they indicated that the ultrafiltrated extract had a potent accelerating wound healing effect.

Frog skin as a wound cover plays the role of a biological membrane, providing mechanical protection and preventing infection. The presence of biochemical compounds in frog skin also plays a significant role in wound healing. The granular glands of amphibian skin and the non-lymphoid cells on the mucosal surfaces of the respiratory and gastrointestinal tract produce large numbers of compounds whose diversity is surprisingly high even within a single species.[47] These substances include biogenic amines, bufogenines and bufotoxins (steroids), alkaloids and peptides (including smaller oligopeptides, polypeptides and proteins) and yet uncharacterized zetekitoxins.[47]

In several studies, peptides isolated from frogs showed potential antimicrobial activity: ocellatins were active against *Escherichia coli*,[48] temporins A, B, and G, the fragment from positions 1 to 18 of esculentin 1b [Esc(1–18)] and bombinin H2 demonstrated activity against multidrug resistant clinical isolates of *S. aureus*, *Enterococcus faecium*, *P. aeruginosa*, *Stenotrophomonas maltophilia*, and *Acinetobacter baumannii*,[49] Esc(1–18) was active against *E. coli*[50] and *S. maltophilia*[51]; magainin II acted against *Leishmania donovani*[52] and *Burkholderia cepacia* complex strains,[53] bombins showed activity against different gram-positive and gram-negative bacteria and fungi.[54]

Apart from antimicrobial peptides, frog skin contains lipids which also play an important role in wound healing. They possess antimicrobial, pain reducing and proliferation stimulating properties.[45,55,56] Raghavan et al.[56] published interesting findings in which wound healing was stimulated in experiments using lipids from frog skin, either by local application or intraperitoneal injection. Proper remodeling of the wound was shown by maintenance of the balance between synthesis and degradation of the extracellular matrix proteins. This was evidenced by a noticeable increase in the hydroxyproline, hexosamine and uronic acid content during the first few days of wound healing, and subsequent decrease in the next few days. Lipid extracts influenced inflammation and showed immune stimulatory properties, playing an important role in the inflammatory and proliferative stages of wound healing.[56] This knowledge could be used in the improvement of the techniques used for engineering bio-artificial grafts and implants, especially for wounds and burns.

MAGGOTS IN THE TREATMENT OF WOUNDS

Wound Healing and Antibacterial Properties of Maggots

Wound infestation with maggots also has a long history of use over the ages.[57] Maggot therapy employs freshly emerged, sterile larvae of *Lucilia sericata*, the common green-bottle fly, by artificially inducing myiasis in a controlled clinical situation. The first documented descriptions of this method date back to the 16th century; later descriptions of successful maggot applications were found in the writings of Napoleon's surgeon, and, finally, the founder of the modern maggot therapy is believed to be William Baer, Professor of Orthopedic Surgery at The Johns Hopkins School of Medicine in Maryland, USA, who was a pioneer in the use of sterile maggots for wound therapy, and made observations of the maggot therapy of traumatic wounds during World War I on the battlefields in France.[58]

Maggots have three core beneficial effects on wounds: debridement, disinfection and enhancement of wound healing.[58] Maggot debridement therapy is based on the fact that maggots secrete a relatively large group of digestive enzymes to remove devitalized necrotic tissues. These enzymes include carboxypeptidases A and B,[59] leucine aminopeptidase,[59] collagenase,[60,61] serine proteases,[59,62–64] cysteine proteases,[64] metalloproteinase, aspartyl proteinase,[63] and deoxyribonuclease (DNAse).[65] A recent study found α-D-glucosidase, α-D-mannosidase and N-acetylglucosaminidase in maggot excretions/secretions which participate in wound debridement by removing sugars from slough/eschar proteins.[66]

The proteases of maggot excretions/secretions enhance wound healing by stimulating the production of hepatocyte growth factor (HGF) by fibroblasts, and this factor plays a crucial role in re-epithelialization, neovascularization, and the formation of granulation tissue. Furthermore, maggot excretions/secretions promoted production of HGF more effectively with less cytotoxicity than commercially available proteases, such as chymotrypsin and bromelain.[64]

The secreted enzymes do not just take part in wound debridement, they also show antimicrobial activity. Antibacterial properties of maggot excretions/secretions were demonstrated against different gram-positive and gram-negative bacteria including methicillin resistant *S. aureus*,[67–69] *Micrococcus luteus*,[70] *E. coli*,[69,70] methicillin sensitive *S. aureus*,[70] *P. aeruginosa*,[68,69] *Klebsiella pneumoniae*[68,69], *Bacillus cereus*, *Bacillus subtilis*, *Serratia marcescens*, *Enterococcus faecalis*, *Proteus mirabilis* and *Staphylococcus epidermidis*.[69]

In contrast to previously described studies, Cazander et al.[71] did not find antibacterial properties in live maggots added to bacterial suspensions of gram-positive (*S. aureus*, coagulase-negative staphylococcus, *Steprococcus pyogenes*, *E. faecalis*) or gram-negative (*Klebsiella oxytoca*, *P. aeruginosa*) bacteria. Furthermore, all bacteria except *P. aeruginosa* showed an increase in growth compared to control cultures without maggots. The authors hypothesized that the clinical antibacterial effect of maggots could possibly be an indirect antibacterial action occurring by immune-mediated mechanisms,[71] which requires further evaluation.

Purified maggot DNAase showed the ability to degrade genomic DNA of bacteria *per se*, DNA from the slough/eschar of a venous leg ulcer, and, what is especially important, extracellular bacterial DNA from biofilms made of *P. aeruginosa*.[65] Therefore, maggots have potential in combating biofilm-associated infections which currently pose a significant clinical concern.

The Preparation and Sterilization of Maggots

The preparation of maggots for application to a wound involves the following steps: female *Lucilia* flies oviposit onto porcine liver; the eggs are then separated and sterilized; the resultant larvae are sterile upon appearance from the egg, and a careful microbiological examination is carried out to ensure their sterility; the larvae are kept under aseptic conditions until application onto the wound.[58]

The key factor in the preparation of maggots is their sterility, as the outer surfaces of fly eggs are heavily contaminated with different microorganisms, and because of this numerous studies have been devoted to solving the sterility problem. The modern founder of maggot use, Baer,[72] applied a solution containing mercuric chloride (1:1000), 25% alcohol and 0.5% hydrochloric acid in order to sterilize maggots before placing them into the wound. Quite good results were achieved by using 5% formalin and 1% sodium hydroxide, but this did not kill *Clostridium* bacteria,[73] and since that study total sterilization was considered to be impossible. Recent renewed interest in this topic has led to advances in the sterilization methods used for the larvae. Sterilization by ultraviolet light and by the successive use of chlorhexidine and 70% alcohol[57] have been proposed, as has the use of sodium hypochlorite and formaldehyde.[74]

After sterilization, maggots are applied to the lesion. If the lesion is too dry, then gauze soaked in hypertonic saline is placed on it to maintain the necessary moisture content for the larvae. The larvae are placed on the lesion at a density of 5–10 per cm^2. The lesion with larvae is covered using an appropriate dressing in order to absorb the draining exudates but without obstructing the air flow. The dressing and larvae are changed daily until the lesion heals.[75]

The Clinical Use of Maggots for Treating Wounds

Nowadays, maggot therapy is used for the treatment of chronic, long-standing, and infected wounds where conventional therapy has failed.[58,76]

Flies which cause myiasis are grouped into two categories: obligate and facultative parasites. Obligate parasites ingest living tissue to complete their lifecycles, and hence their larvae may produce severe damage in healthy tissue; they are not suitable for use in maggot therapy. In contrast, facultative parasites are able to parasitize living hosts in favorable conditions but usually develop on carrion, and therefore this type of maggot is used for wound healing, with the most common being *L. sericata*.

Maggots of *L. sericata* have been seen to be effective in many clinical studies all over the world (Table 11.2). However, the maggots of another fly, *Lucilia cuprina*, which is an obligate parasite, have also been used successfully and safely in several recent studies, such as the prospective case-control study of Paul et al.[77] and the study of Tantawi et al.,[78] in which maggots were effective in the treatment of diabetic foot ulcers.

TABLE 11.2 Selected Studies on the Use of Maggots for Wounds Treatment

Fly Species	Type of Study	Patients Study Groups	Country	Reference
Lucilia sericata	Prospective controlled	8 patients with pressure ulcers in spinal cord injury	USA	a
L. sericata	Prospective controlled	50 patients with pressure ulcers	USA	b
L. sericata	Retrospective	18 patients with 20 nonhealing ulcers; six diabetic foot wounds were treated with conventional therapy, six with maggot therapy, and eight with conventional therapy first, then maggot therapy	USA	c
L. sericata	Pragmatic, three-arm, randomized controlled trial with an economic evaluation	267 patients with venous or mixed venous/arterial leg ulcers	UK	79
Lucilia cuprina	Prospective case-control study	29 patients with diabetic foot wounds	Malaysia	77
L. sericata	Retrospective	25 patients with diabetic foot ulcers and 18 patients with pressure ulcers after spinal cord injury	China	75
L. cuprina	Case report	2 patients with diabetic foot wounds	Egypt	78
L. sericata	Prospective study	23 patients with diabetic wounds	USA	76
L. sericata	Randomized, multicenter, controlled, prospective phase 3 trial with blinded assessment of outcome measures by a single observer	119 patients with non-healing wounds	France	80
L. sericata	Prospective case-control study	435 patients with leg or diabetic foot ulcers	Israel	d
NA	Retrospective study	41 horses with different wounds	France	e
L. sericata	Case report	Patient with periorbital wounds	Deutschland	f
NA	Case report	Patient with crush injury to the hand	Australia	g

NA – data not available.
[a]*Sherman RA, Wyle F, Vulpe M. Maggot therapy for treating pressure ulcers in spinal cord injury patients.* J Spinal Cord Med *1995;***18***:71–4.*
[b]*Sherman RA. Maggot versus conservative debridement therapy for the treatment of pressure ulcers.* Wound Repair Regen *2002;***10***:208–14.*
[c]*Sherman RA. Maggot therapy for treating diabetic foot ulcers unresponsive to conventional therapy.* Diabetes Care *2003;***26***:446–51.*
[d]*Gilead L, Mumcuoglu KY, Ingber A. The use of maggot debridement therapy in the treatment of chronic wounds in hospitalized and ambulatory patients.* J Wound Care *2012;21:78, 80, 82–5.*
[e]*Lepage OM, Doumbia A, Perron-Lepage MF, Gangl M. The use of maggot debridement therapy in 41 equids.* Equine Vet J *2012;***44 Suppl 43***: 120–5.*
[f]*Pitz S, Renieri G, Gericke A. Poorly healing periorbital wounds. Therapeutic use of maggots.* Ophthalmologe *2012;* **109***: 450–3.*
[g]*Chan QE, Hussain MA, Milovic V. Eating out of the hand, maggots – friend or foe?* J Plast Reconstr Aesthet Surg *2012;* **65***: 1116–8.*

The use of maggots has been evaluated in randomized multicenter studies as well as small study groups. A UK multicenter study comparing maggot debridement of non-healing wounds with treatment using hydrogel did not find any differences in rates of healing, but demonstrated increase in wound pain in the use of maggots. However, the study showed that maggot debridement significantly reduced the time to debridement of sloughy and/or necrotic, chronic venous and mixed venous/arterial leg ulcers, in comparison with hydrogel.[79] A study performed in France compared the effectiveness of maggot debridement and conventional treatment in patients with non-healing wounds, and found that maggot debridement showed a benefit over conventional treatments in the first week of application only; there was no apparent benefit later in the treatment, and hence another type of dressing should be used after two or three applications of maggot debridement.[80]

The results of published studies on maggot wound therapy do not allow a recommendation of this method as a substitute for conventional debridement methods, but it should be taken into account as an appropriate alternative and should be further studied in order to optimize the treatment procedure to reduce pain and discomfort during the application of maggots.

CONCLUSIONS AND FUTURE PROSPECTS

Phyto- and zootherapeutic practices have been used since ancient times, and the current increase in rates of antibiotic resistance, which has led to failure in the treatment of infectious processes, has renewed our interest in traditional medicine. Methods known for many centuries and kept in the memory of traditional healers are currently being restored and evaluated from a scientific point of view. This chapter demonstrates that many modern scientific studies have shown the great potential of the wound healing and antimicrobial properties of plant and animal species.

Plant essential oils have demonstrated antimicrobial and wound healing properties in both *in vitro* and *in vivo* studies. In different studies wounds were reproduced in experimental animal models and successfully treated with essential oils, which was proved microbiologically, histologically and biochemically. Amphibian skin showed potential as a natural wound covering, possessing antimicrobial and wound healing properties. Maggots showed good wound debridement ability and antimicrobial properties in many published studies and have recently been evaluated in randomized multicenter studies, in some of which their beneficial properties compared to conventional treatments were also demonstrated.

Natural remedies should be evaluated as a source of biologically active substances which will provide new alternatives to antibiotics and wound healing medicines. Although some constituents have already been identified and are being evaluated, future studies should be directed to better understanding the role of every chemical constituent of natural substances and their potential application in the pharmaceutical industry. Efforts should also be directed to the study of the combined effect of plant essential oils, substances found in amphibian skin, and maggots, alongside conventional antibiotics, in both *in vitro* and *in vivo* experiments.

References

1. Cope Z. The treatment of wounds through the ages. *Med Hist* 1958;**2**:163–74.
2. Guo S, DiPietro LA. Factors affecting wound healing. *J Dent Res* 2010;**89**:219–29.
3. Edwards R, Harding KG. Bacteria and wound healing. *Curr Opin Infect Dis* 2004;**17**:91–6.
4. Thomson CH. Biofilms: do they affect wound healing? *Int Wound J* 2011;**8**:63–7.
5. Percival SL, Hill KE, Williams DW, Hooper SJ, Thomas DW, Costerton JW. A review of the scientific evidence for biofilms in wounds. *Wound Repair Regen* 2012;**20**:647–57.
6. Murphy PS, Evans GRD. Advances in wound healing: a review of current wound healing products. *Plastic Surg Int* 2012;**2012**:190436.
7. Schelz Z, Hohmann J, Molnar J. Recent advances in research of antimicrobial effects of essential oils and plant derived compounds on bacteria. In: Chattopadhyay D, ed. *Ethnomedicine: a source of complementary therapeutics.* Trivandrum, Kerala, India: Research Signpost; 2010. pp. 179–201.
8. Silva NCC, Fernandes Júnior A. Biological properties of medicinal plants: a review of their antimicrobial activity. *J Venom Anim Toxins incl Trop Dis* 2010;**16**:402–13.
9. Kon KV, Rai MK. Plant essential oils and their constituents in coping with multidrug-resistant bacteria. *Expert Rev Anti-Infect Ther* 2012;**7**:775–90.
10. Policegoudra RS, Goswami S, Aradhya SM, Chatterjee S, Datta S, Sivaswamy R, et al. Bioactive constituents of *Homalomena aromatica* essential oil and its antifungal activity against dermatophytes and yeasts. *J Mycol Med* 2012;**22**:83–7.
11. Elaissi A, Rouis Z, Salem NA, Mabrouk S, ben Salem Y, Salah KB, et al. Chemical composition of eight eucalyptus species essential oils and the evaluation of their antibacterial, antifungal and antiviral activities. *BMC Complement Altern Med* 2012;**12**:81.
12. Rondon FC, Bevilaqua CM, Accioly MP, de Morais SM, de Andrade-Júnior HF, de Carvalho CA, et al. *In vitro* efficacy of *Coriandrum sativum, Lippia sidoides* and *Copaifera reticulata* against *Leishmania chagasi. Rev Bras Parasitol Vet* 2012;**21**:185–91.
13. Kon KV, Rai MK. Combining essential oils with antibiotics and other antimicrobial agents as a way of coping with multidrug-resistant bacteria. In: Rai MK, Kon KV, eds. *Fighting multidrug resistance with herbal extracts, essential oils and their components.* USA: Elsevier; 2013. pp. 149–64.
14. Baker J. Essential oils: a complementary therapy in wound management. *J Wound Care* 1998;**7**:355–7.
15. Carson CF, Riley TV. Safety, efficacy and provenance of tea tree (*Melaleuca alternifolia*) oil. *Contact Dermatitis* 2001;**45**:65–7.
16. Halcón L, Milkus K. *Staphylococcus aureus* and wounds: a review of tea tree oil as a promising antimicrobial. *Am J Infect Control* 2004;**32**:402–8.
17. Orafidiya LO, Agbani EO, Abereoje OA, Awe T, Abudu A, Fakoya FA. An investigation into the wound-healing properties of essential oil of *Ocimum gratissimum* linn. *J Wound Care* 2003;**12**:331–4.
18. Lusby PE, Coombes AL, Wilkinson JM. A comparison of wound healing following treatment with *Lavandula x allardii* honey or essential oil. *Phytother Res* 2006;**20**:755–7.
19. Abu-Al-Basal MA. Healing potential of *Rosmarinus officinalis* L. on full-thickness excision cutaneous wounds in alloxan-induced-diabetic BALB/c mice. *J Ethnopharmacol* 2010;**131**:443–50.
20. Edmondson M, Newall N, Carville K, Smith J, Riley TV, Carson CF. Uncontrolled, open-label, pilot study of tea tree (*Melaleuca alternifolia*) oil solution in the decolonisation of methicillin-resistant *Staphylococcus aureus* positive wounds and its influence on wound healing. *Int Wound J* 2011;**8**:375–84.
21. Tumen I, Akkol EK, Süntar I, Keleş H. Wound repair and anti-inflammatory potential of essential oils from cones of Pinaceae: preclinical experimental research in animal models. *J Ethnopharmacol* 2011;**137**:1215–20.
22. Süntar I, Akkol EK, Keleş H, Oktem A, Başer KH, Yeşilada E. A novel wound healing ointment: a formulation of *Hypericum perforatum* oil and sage and oregano essential oils based on traditional Turkish knowledge. *J Ethnopharmacol* 2011;**134**:89–96.
23. Tumen I, Süntar I, Keleş H, Küpeli Akkol E. A therapeutic approach for wound healing by using essential oils of cupressus and juniperus species growing in Turkey. *Evid Based Complement Alternat Med* 2012;**2012**:728281.
24. Süntar I, Tumen I, Ustün O, Keleş H, Akkol EK. Appraisal on the wound healing and anti-inflammatory activities of the essential oils obtained from the cones and needles of *Pinus* species by *in vivo* and *in vitro* experimental models. *J Ethnopharmacol* 2012;**139**:533–40.

25. Tumen I, Süntar I, Eller FJ, Keleş H, Akkol EK. Topical wound-healing effects and phytochemical composition of heartwood essential oils of *Juniperus virginiana* L., *Juniperus occidentalis* Hook., and *Juniperus ashei* J. Buchholz. *J Med Food* 2013;**16**:48–55.
26. Dias AM, Braga ME, Seabra IJ, Ferreira P, Gil MH, de Sousa HC. Development of natural-based wound dressings impregnated with bioactive compounds and using supercritical carbon dioxide. *Int J Pharm* 2011;**408**:9–19.
27. Weng Z, Zhang B, Asadi S, Sismanopoulos N, Butcher A, Fu X, et al. Quercetin is more effective than cromolyn in blocking human mast cell cytokine release and inhibits contact dermatitis and photosensitivity in humans. *PLoS One* 2012;**7**:e33805.
28. Radulović NS, Blagojević PD, Stojanović-Radić ZZ, Stojanović NM. Antimicrobial plant metabolites: structural diversity and mechanism of action. *Curr Med Chem* 2013;**20**:932–52.
29. Orhan I, Küpeli E, Aslan M, Kartal M, Yesilada E. Bioassay-guided evaluation of anti-inflammatory and antinociceptive activities of pistachio, *Pistacia vera* L. *J Ethnopharmacol* 2006;**105**:235–40.
30. Oztürk M, Tümen İ, Uğur A, Aydoğmuş-Öztürk F, Topçu G. Evaluation of fruit extracts of six Turkish *Juniperus* species for their antioxidant, anticholinesterase and antimicrobial activities. *J Sci Food Agric* 2011;**91**:867–76.
31. Costa-Neto EM. Implications and applications of folk zootherapy in the state of Bahia, Northeastern Brazil. *Sust Dev* 2004;**12**:161–74.
32. Costa-Neto EM, Oliveira MVM. Cockroach is good for asthma: zootherapeutic practices in Northeastern Brazil. *Hum Ecol Rev* 2000;**7**:41–51.
33. Alves RR, Rosa IL. Zootherapeutic practices among fishing communities in North and Northeast Brazil: a comparison. *J Ethnopharmacol* 2007;**111**:82–103.
34. Govender T, Dawood A, Esterhuyse AJ, Katerere DR. Antimicrobial properties of the skin secretions of frog. *S Afr J Sci* 2012;**108**. Available from: http://dx.doi.org/10.4102/sajs.v108i5/6.795.
35. Alonso-Castro AJ, Carranza-Álvarez C, Maldonado-Miranda JJ, Del Rosario Jacobo-Salcedo M, Quezada-Rivera DA, Lorenzo-Márquez H, et al. Zootherapeutic practices in Aquismón, San Luis Potosí, México. *J Ethnopharmacol* 2011;**138**:233–7.
36. Jacobo-Salcedo Mdel R, Alonso-Castro AJ, Zarate-Martinez A. Folk medicinal use of fauna in Mapimi, Durango, México. *J Ethnopharmacol* 2011;**133**:902–6.
37. Alves RR, Alves HN. The faunal drugstore: animal-based remedies used in traditional medicines in Latin America. *J Ethnobiol Ethnomed* 2011;**7**:9.
38. Solavan A, Paulmurugan R, Wilsanand V, Ranjith Sing AJA. Traditional therapeutic uses of animals among tribal population of Tamil Nadu. *Indian J Traditional Knowledge* 2004;**3**:198–205.
39. Mahawar MM, Jaroli DP. Traditional zootherapeutic studies in India: a review. *J Ethnobiol Ethnomed* 2008;**4**:17.
40. Quave CL, Lohani U, Verde A, Fajardo J, Rivera D, Obón C, et al. A comparative assessment of zootherapeutic remedies from selected areas in Albania, Italy, Spain and Nepal. *J Ethnobiol* 2010;**30**:92–125.
41. Soewu DA. Zootherapy and biodiversity conservation in Nigeria. In: Alves RRN, Rosa IL, eds. *Animals in traditional folk medicine*. Springer; 2013. pp. 347–65.
42. Lev E. Healing with animals in the Levant from the 10th to the 18th century. *J Ethnobiol Ethnomed* 2006;**2**:11.
43. Andersson LC, Nettelblad HC, Kratz G. From basic wound healing to modern skin engineering. In: Hakim N, ed. *Artificial organs, new techniques in surgery series 4*. Springer-Verlag London Limited; 2009. pp. 93–105.
44. Kakati LN, Ao B, Doulo V. Indigenous knowledge of zootherapeutic use of vertebrate origin by the Ao tribe of Nagaland. *J Hum Ecol* 2006;**19**:163–7.
45. Sai KP, Babu M. Growth potential of *Rana tigerina* skin lipids in cell cultures. *In Vitro Cell Dev Biol Anim* 1998;**34**:561–7.
46. Mashreghi M, Rezazade Bazaz M, Mahdavi Shahri N, Asoodeh A, Mashreghi M, Behnam Rassouli M, et al. Topical effects of frog "*Rana ridibunda*" skin secretions on wound healing and reduction of wound microbial load. *J Ethnopharmacol* 2013;**145**:793–7.
47. Costa-Neto EM. Animal-based medicines: biological prospection and the sustainable use of zootherapeutic resources. *An Acad Brac Cienc* 2005;**77**:33–43.
48. Nascimento AC, Zanotta LC, Kyaw CM, Schwartz EN, Schwartz CA, Sebben A, et al. Ocellatins: new antimicrobial peptides from the skin secretion of the South American frog *Leptodactylus ocellatus* (Anura: Leptodactylidae). *Protein J* 2004;**23**:501–8.

49. Mangoni ML, Maisetta G, Di Luca M, Gaddi LM, Esin S, Florio W, et al. Comparative analysis of the bactericidal activities of amphibian peptide analogues against multidrug-resistant nosocomial bacterial strains. *Antimicrob Agents Chemother* 2008;**52**:85–91.
50. Marcellini L, Borro M, Gentile G, Rinaldi AC, Stella L, Aimola P, et al. Esculentin-1b(1–18) – a membrane-active antimicrobial peptide that synergizes with antibiotics and modifies the expression level of a limited number of proteins in *Escherichia coli. FEBS J* 2009;**276**:5647–64.
51. Maisetta G, Mangoni ML, Esin S, Pichierri G, Capria AL, Brancatisano FL, et al. *In vitro* bactericidal activity of the N-terminal fragment of the frog peptide esculentin-1b (Esc 1–18) in combination with conventional antibiotics against *Stenotrophomonas maltophilia. Peptides* 2009;**30**:1622–6.
52. Guerrero E, Saugar JM, Matsuzaki K, Rivas L. Role of positional hydrophobicity in the leishmanicidal activity of magainin 2. *Antimicrob Agents Chemother* 2004;**48**:2980–6.
53. Thwaite JE, Humphrey S, Fox MA, Savage VL, Laws TR, Ulaeto DO, et al. The cationic peptide magainin II is antimicrobial for *Burkholderia cepacia*-complex strains. *J Med Microbiol* 2009;**58**:923–9.
54. Simmaco M, Kreil G, Barra D. Bombinins, antimicrobial peptides from Bombina species. *Biochim Biophys Acta* 2009;**1788**:1551–5.
55. Kumar KV, Sai KP, Babu M. Application of frog (*Rana tigerina* Daudin) skin collagen as a novel substrate in cell culture. *J Biomed Mater Res* 2002;**61**:197–202.
56. Raghavan K, Babu M, Rajaram R, Sai KP. Efficacy of frog skin lipids in wound healing. *Lipids Health Disease* 2010;**9**:74.
57. Mohd Masri S, Nazni WA, Lee HL, T Rogayah TA, Subramaniam S. Sterilisation of *Lucilia cuprina* Wiedemann maggots used in therapy of intractable wounds. *Trop Biomed* 2005;**22**:185–9.
58. Nigam Y, Bexfield A, Thomas S, Ratcliffe NA. Maggot therapy: the science and implication for CAM part I-history and bacterial resistance. *Evid Based Complement Alternat Med* 2006;**3**:223–7.
59. Vistnes LM, Lee R, Ksander GA. Proteolytic activity of blowfly larvae secretions in experimental burns. *Surgery* 1981;**90**:835–41.
60. Hobson RP. Studies on the nutrition of blow-fly larvae. I. Structure and function of the alimentary tract. *J Exp Biol* 1931;**8**:110–23.
61. Ziffren SE, Heist HE, May SC, Womack NA. The secretion of collagenase by maggots and its implication. *Ann Surg* 1953;**138**:932–4.
62. Casu RE, Pearson RD, Jarmey JM, Cadogan LC, Riding GA, Tellam RL. Excretory/secretory chymotrypsin from *Lucilia cuprina*: purification, enzymatic specificity and amino acid sequence deduced from mRNA. *Insect Mol Biol* 1994;**3**:201–11.
63. Chambers L, Woodrow S, Brown AP, Harris PD, Phillips D, Hall M, et al. Degradation of extracellular matrix components by defined proteinases from the green-bottle larva *Lucilia sericata* used for the clinical debridement of non-healing wounds. *Br J Dermatol* 2003;**148**:14–23.
64. Honda K, Okamoto K, Mochida Y, Ishioka K, Oka M, Maesato K, et al. A novel mechanism in maggot debridement therapy: protease in excretion/secretion promotes hepatocyte growth factor production. *Am J Physiol Cell Physiol* 2011;**301**:C1423–30.
65. Brown A, Horobin A, Blount DG, Hill PJ, English J, Rich A, et al. Blow fly *Lucilia sericata* nuclease digests DNA associated with wound slough/eschar and with *Pseudomonas aeruginosa* biofilm. *Med Vet Entomol* 2012;**26**:432–9.
66. Telford G, Brown AP, Rich A, English JS, Pritchard DI. Wound debridement potential of glycosidases of the wound-healing maggot, *Lucilia sericata. Med Vet Entomol* 2012;**26**:291–9.
67. Bexfield A, Nigam Y, Thomas S, Ratcliffe NA. Detection and partial characterisation of two antibacterial factors from the excretions/secretions of the medicinal maggot *Lucilia sericata* and their activity against methicillin-resistant *Staphylococcus aureus* (MRSA). *Microbes Infect* 2004;**6**:1297–304.
68. Huberman L, Gollop N, Mumcuoglu KY, et al. Antibacterial properties of whole body extracts and haemolymph of *Lucilia sericata* maggots. *J Wound Care* 2007;**16**:123–7.
69. Bexfield A, Bond AE, Roberts EC, Dudley E, Nigam Y, Thomas S, et al. The antibacterial activity against MRSA strains and other bacteria of a <500 Da fraction from maggot excretions/secretions of *Lucilia sericata* (Diptera: Calliphoridae). *Microbes Infect* 2008;**10**:325–33.
70. Daeschlein G, Mumcuoglu KY, Assadian O, Hoffmeister B, Kramer A. *In vitro* antibacterial activity of *Lucilia sericata* maggot secretions. *Skin Pharmacol Physiol* 2007;**20**:112–5.

71. Cazander G, van Veen KE, Bernards AT, Jukema GN. Do maggots have an influence on bacterial growth? A study on the susceptibility of strains of six different bacterial species to maggots of *Lucilia sericata* and their excretions/secretions. *J Tissue Viability* 2009;**18**:80–7.
72. Baer WS. The treatment of chronic osteomyelitis with the maggot (larva of the blow fly). *J Bone Joint Surg* 1931;**13**:438–75.
73. Simmons SW. Sterilization of blowfly eggs in the culture of surgical maggots for use in the treatment of pyogenic infections. *Am J Surg* 1935;**25**:140–7.
74. Yeong YS, Nazni WA, Santana RL, Mohd Noor I, Lee HL, Mohd Sofian A. Scanning electron microscopic evaluation of the successful sterilization of *Lucilia cuprina* (Wiedemann) utilized in maggot debridement therapy (mdt). *Trop Biomed* 2011;**28**:325–32.
75. Wang SY, Wang JN, Lv DC, Diao YP, Zhang Z. Clinical research on the bio-debridement effect of maggot therapy for treatment of chronically infected lesions. *Orthop Surg* 2010;**2**:201–6.
76. Marineau ML, Herrington MT, Swenor KM, Eron LJ. Maggot debridement therapy in the treatment of complex diabetic wounds. *Hawaii Med J* 2011;**70**:121–4.
77. Paul AG, Ahmad NW, Lee HL, Ariff AM, Saranum M, Naicker AS, et al. Maggot debridement therapy with *Lucilia cuprina*: a comparison with conventional debridement in diabetic foot ulcers. *Int Wound J* 2009;**6**:39–46.
78. Tantawi TI, Williams KA, Villet MH. An accidental but safe and effective use of *Lucilia cuprina* (Diptera: Calliphoridae) in maggot debridement therapy in Alexandria, Egypt. *J Med Entomol* 2010;**47**:491–4.
79. Dumville JC, Worthy G, Soares MO, Bland JM, Cullum N, Dowson C, et al. VenUS II: a randomized controlled trial of larval therapy in the management of leg ulcers. Health Technol Assess 2009;**13**:1–182.
80. Opletalová K, Blaizot X, Mourgeon B, Chêne Y, Creveuil C, Combemale P, et al. Maggot therapy for wound debridement: a randomized multicenter trial. *Arch Dermatol* 2012;**148**:432–8.

CHAPTER

12

Bioactive Molecules of Herbal Extracts with Anti-Infective and Wound Healing Properties

Ali Parsaeimehr[*,†], *Yi-Feng Chen*[†] *and Elmira Sargsyan*[*]

[*]G. S. Davtyan Institute of Hydroponics Problems, National Academy of Sciences, Yerevan, Republic of Armenia, [†]Institute of Biotechnology, Jiangsu Academy of Agricultural Sciences, Nanjing, Jiangsu, China

INTRODUCTION

Throughout history, mankind has screened medicinal plants for their biological activity. We have tried to use the anti-infective, antioxidant, anti-coagulative, anticancer and wound healing activities of medicinal plants, while trying to avoid their poisonous properties; including neurotoxicity, inhibition of cellular respiration, cytotoxicity, and alkylating or intercalating DNA toxins.[1–3] Archaeological discoveries have shown that mankind has made profitable use of herbs since the Paleolithic period, but evidence of the effectiveness of herbal medicine has only become obvious through science and technology. Reports from the World Health Organization (WHO) show that three quarters of the world's population use diverse herbal remedies to treat diseases such as rheumatism, Parkinson's disease, asthma, diabetes, cancer, burns, and infections.[4–6]

In fact, the reason that plants have medicinal properties is because they are rich in phytochemicals. The word "phytochemical" is derived from the Greek "phyto" meaning plant. They can be divided into three main classes:

1. Terpenoids, such as carotenoids, monoterpenes and diterpenes;
2. Phenolic compounds, such as benzoquinones, phenolic aldehydes and xanthonoids;
3. Alkaloids, such as stachydrine, trichodesmine and ancistrocladine.[7]

Microbiology for Surgical Infections
DOI: http://dx.doi.org/10.1016/B978-0-12-411629-0.00012-X

The bioactivity of phytochemicals is rooted in the diversity of their molecular structures, which gives them specific properties. Plants synthesize them for protection, against organisms that would eat them, and against environmental stress; this helps the plant to survive under harsh biotic and abiotic conditions.[8,9] Evidence indicates that phytochemicals are highly effective in fighting against microbial disease, and they have shown promise in the treatment of stubborn infections. In this regard, anti-infective agents such as berberine, piperine, chrysin, galangin, hypericin, catechol, capsaicin etc. have been isolated from plants and tested against bacteria such as *Salmonella typhi, Escherichia coli, Shigella dysenteriae, Helicobacter pylori, Pseudomonas aeruginosa*, and *Staphylococcus aureus*.[10–12]

Much has happened since the first observation of an antimicrobial agent in 1929 by Alexander Fleming, and now different classes of antibiotics such as aminoglycosides, ansamycins, carbacephems, cephalosporins, glycopeptides etc., having different modes of actions, have been discovered and tested.[13–15] Medical researchers have fought bacterial infections with new classes of antibiotics, but unfortunately bacteria have continuously developed their resistance mechanisms, by mutation or the acquisition of new genes through genetic exchange mechanisms such as transduction, transformation, or conjugation; a process largely driven by the abuse of antibiotics.[16] The rapid emergence of "multidrug resistance" as a significant threat to public health is based on this phenomenon, and now a number of resistant pathogens such as *S. aureus, P. aeruginosa, Clostridium difficile, Acinetobacter baumannii, Mycobacterium tuberculosis* have emerged, and some classes and types of antibiotics such as erythromycin, penicillin, methicillin, isoniazid, rifampin and tetracycline are becoming practically useless.[17–19] Evidence indicates that some herbal extracts have acted synergistically with antibiotics such as chloramphenicol, norfloxacin and tetracycline, and hence can be used concurrently with them.[20–22]

A wound is defined as a physical, chemical, immunological, or microbial condition of a tissue associated with loss of function. Wound healing (cicatrization) is an intricate anabolic process through which an organ, tissue or skin starts to repair itself after damage. The process involves cell to cell and cell to matrix signaling, and consists of four phases: hemostatis, inflammation, proliferation and remodeling.[23,24]

The inflammatory phase is characterized by vasoconstriction and platelet aggregation to induce blood clotting, followed by vasodilatation and phagocytosis to produce inflammation at the wound site.[24] Evidence shows that necrotic tissue, interference with the blood supply, lymphatic blockage, diabetes mellitus and bacterial infection delay the wound healing process.[25,26] It is known that the inflammatory phase starts immediately after the injury, and it pulls the trigger of hemostatic mechanisms which prevent blood loss from the wound site. Bioflavonoids such as quercetin, which is derived from several medicinal plants, or COX-1 and COX-2 inhibitors from plants such as *Evodia rutaecarpa, Croton gratissimus* and *Eucomis* species can considerably support this phase.[27–29]

During the proliferative phase of wound healing, fibroblasts produce a variety of substances necessary for wound repair, including glycosaminoglycans and collagen. Shenoy and colleagues[30] reported that alcoholic and aqueous extracts of *Hyptis suaveolens* significantly increased the collagen turnover in the proliferative phase compared to controls. New collagen is formed in the third phase (remodeling), and the tensile strength of the tissue will be increased through the cross linking of collagen by vitamin C-dependent hydroxylation. Upadhyay and colleagues[31] claimed that leaf extracts of *Hippophae rhamnoides* increased endogenous antioxidants, and that these extracts play an important role in angiogenesis, deposition of extracellular matrix and the remodeling phase of wound healing. Clear evidence illustrates that phytochemicals derived from plants are affordable, safe, and have significant potential for treating infections and wounds.[32–34]

MEDICINAL PLANTS WITH WOUND HEALING AND ANTI-INFECTIVE PROPERTIES

Aloe Vera L.

This thickened and fleshy plant (succulent) originates from northern Africa, and belongs to the family Xanthorrhoeaceae. It has been used therapeutically for over 5000 years by Egyptians, Romans and Native Americans. Evidence shows that *A. vera* contains phytochemicals such as pyrocatechol, saponin, acemannan, β-mannan, anthraquinone, C-glycosides, diethylhexylphthalate, bradykininase, oleic acid, phytol, and magnesium lactate along with water-soluble polysaccharides like glucomannan.[35–38] Reports indicate that *Aloe* has been used successfully to treat burns and skin irritations, canker sores (aphthous stomatitis), gastric and duodenal ulcers caused by *Helicobacter pylori*.[39,40] Aloeride is a new, high molecular weight, immunostimulatory polysaccharide (4 to 7 million Da) derived from *A. vera* which can induce the expression of the mRNAs encoding tumor necrosis factor-alpha (TNF-α) and interleukin-1-beta (IL-1β).[38] *In vitro* studies have shown that *A. vera* gel has an inhibitory influence on the production of interleukin 8 and colorectal prostaglandin E_2, and that it also contains antioxidant compounds.[41] Moreover, the antimicrobial activity of *A. vera* against a range of pathogens, including *S. typhi, S. aureus, P. aeruginosa, Klebsiella pneumoniae, M. tuberculosis, E. coli* and *Aspergillus flavus*, is well documented.[42–44]

Mannose-6-phosphate is the main sugar in *A. vera*, and Davis and colleagues[45] reported that this compound improved wound healing process at a dosage of 300 mg/kg. Evidence indicates that the use of *A. vera* extracts for four days at a concentration range of 100–300 mg/kg reduced the action of the wound healing inhibitors, such as sterols and certain amino acids.[46] The influence of *A. vera* on the process of wound healing is comparable to that of mast cell stabilizer drugs such as lodoxamide, and lipid peroxidation inhibitors such as lazaroids, with 82% to 85% tissue survival compared to the control.[47]

Equisetum Arvense L. (Bottlebrush, Cat's Tail)

This perennial plant with underground rhizomes belongs to the family of Equisetaceae and is found in the northern hemisphere. Chemical screening of *E. arvense* indicates that it is rich in calcium, potassium and silicon. It has been used medicinally by ancient and modern cultures to cure wounds and tissue injuries. Currently, compounds such as petrosins, onitin, onitin-9-*O*-glucoside, apigenin, luteolin, kaempferol-3-*O*-glucoside, quercetin-3-*O*-glucoside, luteolin-7-*O*-D-glucoside, apigenol 7-*O*-glucoside, rosmarinic acid, chlorogenic acid, caffeic acid, rutin and apigenol have been isolated from extracts of *E. arvense*.[48–50] Another study reported that kaempferol, quercetin glycosides and caffeic acid derivatives were the main compounds identified in hydro-alcoholic extracts of

Equisetum species.[51] Extracts from this plant have considerable ability to stimulate collagen synthesis and glycosaminoglycan metabolism, and accelerate the wound healing process.[52] The presence of silicon in the tissue of *E. arvense* makes this plant a valuable natural source of silicon, a necessary element for the synthesis and integrity of connective tissue.[53]

Milovanović and colleagues[54] reported that hexahydrofarnesyl acetone (18.34%), *cis*-geranyl acetone (13.74%), thymol (12.09%) and *trans*-phytol (10.06%) were the major constituents in the essential oil of *E. arvense*. Moreover, a 1:10 dilution of the essential oil of *E. arvense* L. showed strong antimicrobial activity against *S. aureus*, *E. coli*, *K. pneumoniae*, *P. aeruginosa*, *S. enteritidis*, *Aspergillus niger* and *Candida albicans*. Flavonoids and phenolic acids give extracts from this plant high antiradical activity. Milovanović and colleagues[51] reported that the total phenol content is 92 to 349 μmol equivalents of chlorogenic acid per g of dried plant material.

Camellia Sinensis (L.) Kuntze (Green Tea)

This evergreen shrub belongs to the Theaceae family. Literature indicates that the Chinese have used this plant since 2700 BC as an antipyretic, sedative, antihypertensive, and anti-inflammatory medicine. *C. sinensis* contains chemicals such as epigallocatechin-3-gallate, acetamide, 5-*N*-ethyl-glutamine, cetophenone, actinidiolide dihydroactinidiolide, afzelechin, amyrin, aniline, apigenin, benzothiazole, arbutin, assamicain, camellianin, caffeine, catechin, carvacrol, catechol, limonene, menthol, vitamin C, tannins and thymol.[55–57] Epigallocatechin-3-gallate (EGCG), the major catechin found in green tea, has been shown to have anti-carcinogenic and antimicrobial properties against a number of bacterial pathogens, including an inhibitory influence on *Stenotrophomonas maltophilia*, which is recognized as an important nosocomial pathogen. Also, catechin derived from *C. sinensis* has the potential to reduce hepatocyte growth factor receptor activity in human colon cancer cells.[58,59]

Theophylline is a colorless crystalline alkaloid, and an isomer of theobromine, which has been extracted from *C. sinensis* leaves and used to treat asthma, emphysema, brachycardia, acute bronchitis and bronchial spasm.[60] It is reported that extracts of *C. sinensis* accelerate keratinocyte cell differentiation and improve the wound healing process. In particular, EGCG can inhibit skin matrix metalloproteinases (MMP-2 and MMP-9) to adjust extracellular matrix alterations.[61] Evidence indicates that extracts of *C. sinensis* can inhibit methicillin resistant *S. aureus* and penicillin resistance in beta-lactamase producing *S. aureus*. This synergy occurs through prevention of penicillin binding protein 2-synthesis and inhibiting the secretion of beta-lactamase.[62] Lee and colleagues[63] reported that the acidic polysaccharide CS-F2 derived from green tea has a selective anti-adhesive influence against certain pathogenic bacteria, while exerting no effects against beneficial and commensal bacteria.

Calendula Officinalis L. (Marigold)

This aromatic plant belongs to the Asteraceae family, and is native to southern Europe, western Asia and North America. *C. officinalis* is known to have antispasmodic, aperient, cholagogic and diaphoretic activities. Compounds such as carotenoids, triterpenic alcohols, zeaxanthin, beta-carotene, lutein, saponin, calendic acid, etheric oils, vitamin C, sitosterols, stigmasterols, umbelliferone and esculetin have been isolated from *C. officinalis* extracts.[64,65] Faradiol 3-*O*-laurate, palmitate and myristate are three major triterpenoid esters with anti-inflammatory activities that were discovered in *C. officinalis* flowers.[66] Zitterl-Eglseer and colleagues[64] reported the isolation of faradiol-3-myristic acid ester, faradiol-3-palmitic acid ester and ψ-taraxasterol, which have anti-edematous activity, from *C. officinalis* using column chromatography and high-performance liquid chromatography (HPLC). It has been reported that both organic and aqueous extracts of *C. officinalis* flowers were nontoxic to human lymphocytic Molt-4 cells, but only the organic extracts exhibited anti-HIV activity in an *in vitro* MTT/tetrazolium-based assay.[67]

Fronza and colleagues[68] claimed that extracts of *C. officinalis* stimulated the proliferation and migration of fibroblasts at low concentrations (10 μg/ml) and it enhanced cell numbers. Experiments by Chandran and Kuttan[69] show that *C. officinalis* flower extracts contain valuable compounds which accelerate the wound healing process, and parameters of the wound healing such as hexosamine and collagen-hydroxyproline were significantly increased in treated samples. The antioxidant defense mechanism was activated and lipid peroxidation and tissue damage marker enzymes (alkaline phosphatase, alanine and aspartate transaminases) were lower in treated samples compared with controls. Bashir and colleagues[70] claimed that *C. officinalis* flower extract contains both spasmolytic and spasmogenic agents, which have cholinergic and calcium channel blocking activities. These extracts have been shown to possess inhibitory influences which are comparable to common antibiotics on *S. aureus*, coagulase positive *Bacillus subtilis, Bacillus cereus, Bacillus stearothermophillus, E. coli, V. cholerae, S. typhi, P. aeruginosa* and *K. pneumoniae*.[71]

Carica Papaya L. (Papaya, Pawpaw)

This dicotyledonous, polygamous tropical tree belongs to the Caricaceae family, and it originates from South America. The fruit of this plant is rich in vitamins B and C, provitamin A, carotenoids and a range of phytochemicals such as chymopapain, papain, danielone, protocatechuic acid, caffeic acid, chlorogenic acid, kaempferol, quercetin, *p*-coumaric acid, oleic acid, linolenic acids, 5,7-dimethoxycoumarin, carpaine, nicotine, and proteolytic enzymes.[72–74] Extracts of *C. papaya* appear to be effective in the desloughing phase and in preventing infections in burns. Aqueous extracts of papaya have been shown to improve the healing process compared to controls.[75]

A new, papain-based wound cleanser containing a proteolytic enzyme showed acceleration of the wound healing process by influencing wound reduction rates and collagen induction. It also showed antibacterial properties, which indicate that it could be appropriate for wound management.[76] Suresh[77] compared and tested extracts of *C. papaya, Cynodon dactylon, Euphorbia hirta, Melia azedarach* and *Psidium guajava* on a range of gram negative and positive bacteria. Their results indicate that of the chosen medicinal plants, extracts of *C. papaya* had the best activities. In another study, extracts from the seeds of papaya fruit inhibited the growth of *B. cereus, E. coli, Enterococcus faecalis, S. aureus, Proteus vulgaris,* and *Shigella flexneri*, respectively; which shows the antimicrobial potential of this plant.[78]

Matricaria Chamomilla L. (Chamomile)

This plant belongs to the Asteraceae family, and has been used since ancient times in Egypt, Greece and Rome. The dried flower is known to contain medicinal compounds including apigenin, luteolin, alpha-bisabolol, herniarin, umbelliferone, chlorogenic acid, caffeic acid, chamazulene, apigenin, luteolin, quercetin and scopoletin-7-glucoside.[79,80]

It has been confirmed that bioactive compounds in chamomile have anti-anxietal activity and the capacity to bind with receptors of the neurotransmitter gamma-aminobutyric acid (GABA). The extracts influence histamine release, so it is suggested that they be used against stress, insomnia and allergies.[81] (Z)- and (E)-2-β-d-glucopyranosyloxy-4-methoxycinnamic acid (GMCA), the glucoside precursor of herniarin and apigenin glucoside were found as native compounds in chamomile.[82] Chamazulene derived from *M. chamomilla* has antioxidant activities and prevents leukotriene synthesis in neutrophilic granulocytes. In this regard, apigenin influences adhesion protein expression and also inhibits IL-1 alpha-induced prostaglandin synthesis.[83] An aqueous extract of chamomile diminishes the tick-borne encephalitis virus (TBEV) titer in kidney cells, and raises resistance in virus-infected samples, as hydro-alcoholic extracts are reported to be effective against poliovirus development at the cellular and viral RNA synthesis stage.[84]

Chamomile extracts were compared with corticosteroids, and the samples treated with chamomile completed wound healing in a shorter period (nine days) compared to the other treatments. It is notable that in the chamomile group, the lesions were completely induced after five days.[85] It has been reported that essential oil and methanol extracts of chamomile were completely effective against *Staphylococcus* and *Candida*, and derived α-bisabolol and chamazulene recorded the strongest inhibitory influence on both gram positive and gram negative bacteria, while spiro ethers were weaker in antimicrobial action.[86]

Echinacea Purpurea (L.) Moench (Black Sampson)

This ornamental plant belongs to the Asteraceae family, and is native to North America. It contains valuable compounds, including echinacoside, echinacin, humulene, caryophyllene, borneol, germacrene D, palmitic acid, luteolin, kaempferol, quercetin, quercetagetin, apigenin, isorhamnetin and pyrrolizidine alkaloids.[87,88] Lii and colleagues[89] identified five new compounds in *E. purpurea* extracts: 1 beta, 6 alpha-dihydroxy-4(14)-eudesmene, (2E, 4E, 8Z, 10E)-N-isobutyl-2,4,8,10-dodecatetraenamide, (2E, 4E, 8Z, 10Z)-N-isobutyl-2,4,8,10-dodecatetraenamide, cerotic acid and hyxacosyl alcohol. In addition, two hydroxylated (8-hydroxy-pentadeca-(9*E*)-ene-11,13-diyn-2-one and 8-hydroxy-pentadeca-(9*E*,13*Z*)-dien-11-yn-2-one) and two dicarbonylic (pentadeca-(9*E*)-ene-11,13-diyne-2,8-dione and pentadeca-(9*E*,13*Z*)-dien-11-yne-2,8-dione) compounds were isolated from *n*-hexane extracts of *Echinacea pallida*. Results showed that the hydroxylated compounds have low cytotoxicity against human pancreatic adenocarcinoma cells, while the other isolated compounds showed moderate activity.[90] It is hypothesized that purified polysaccharides, alkamides and caffeic acid derivatives from *Echinacea* species possess immunostimulatory activity. In this regard, Rininger and colleagues[91] reported that *Echinacea* stimulates murine macrophage cytokine secretion and enhances the viability of proliferation of human peripheral blood mononuclear cells (PBMCs). *Echinacea* extracts have the ability to influence cell morphology via their effect on fibroblast synthesis and collagen formation; the extracts also had a significant impact on the healing process in arsenic-induced dermal necrosis and can protect and minimize arsenic effects.[92] *Echinacea* inactivates influenza and respiratory syncytial viruses (RSV) and certain respiratory bacteria, including *Haemophilus influenzae, Legionella pneumophila, S. aureus, Mycobacterium smegmatis,* as well as reducing inflammatory influences on epithelial cells.[93,94]

Allium Sativum L. (Garlic)

This plant belongs to the Amaryllidaceae family and is native to central Asia. The literature indicates that this plant was well known in many nations, and has a very long history of medicinal usage as an anti-asthmatic, anti-cholesteremic, antiseptic, anti-spasmodic, febrifuge, stimulant and vasodilator. Recently, compounds such as allicin, alliin, ajoene, diallyl disulfide, diallyl trisulfide and A, B, C vitamins have been isolated from garlic.[95,96]

It is reported that *A. sativum* shows antiviral activity against viruses including herpes simplex virus, parainfluenza virus, vaccinia virus, vesicular stomatitis virus and human rhinovirus. The presence of thiosulfinates appeared to be correlated with the highest levels of antivirus activity.[97,98] Additionally, diallyl sulfide, a thioether and a component derived from garlic, has the potential to inhibit dimethyihydrazine-induced colon cancer.[96] It has been reported that the levels of free radicals increase in the blood during renal ischemia reperfusion injury, as shown by increases in creatinine and decreases in glutathione. Garlic extract promotes glutathione levels and decreases levels of the free radicals myeloperoxidase and renal malondialdehyde.[99] Allicin derived from *A. sativum* was shown to have an inhibitory influence on *Mycobacterium tuberculosis.* Allicin is known as an antibacterial agent, and reduces RNA synthesis in bacteria. Moreover garlic extracts showed an inhibitory influence on *S. aureus, A. niger, Penicillium cyclopium, Fusarium oxysporum* and *Salmonella* serotype Enteritidis.[100]

Ginkgo Biloba L. (Maidenhair Tree)

This member of the Ginkgoaceae family originated almost 300 million years ago, and is native to China. The Chinese used *G. biloba* to cure asthma, bronchitis and cardiovascular diseases. Currently Ginkgo is used clinically to treat a range of circulatory disorders and dementia. Chemical screening of this plant revealed that it contains a range of valuable compounds, such as ginkgolides (A, B, C, J and M), bilobalide,3,3′-dimethoxy-4,4′-dihydroxy-stilbene, aglycones, quercetin, kaempferol, isorhamnetin, bilobol and cardanol.[101,102]

Extracts of Ginkgo are known to play a defensive role in the development of atherosclerosis, and to inhibit a range of free radicals including nitric oxide, superoxide, hydroxyl, oxoferryl and peroxyl radicals. Ginkgolide B (derived from Ginkgo) decreases platelet aggregation by reducing the platelet-activating factor produced by neutrophils, monocytes, macrophages and endothelial cells during shock and ischemic injury.[103] Bairy and Rao[104] claimed that *G. biloba* extract improves the breaking strength and hydroxyl-proline content of granulation tissue without altering its mass. The wound healing process also improved in treated samples. It has been reported that extract of Ginkgo was effective in reducing levels of vascular endothelial growth factor and interleukin 8, and also played a significant role in the wound healing process. *G. biloba* enhanced the activity of superoxide dismutase and glutathione peroxidase enzymes and reduced lipid peroxidation activity in cranium irradiated samples.[105]

Wang and Ng[106] isolated an antifungal protein with a molecular weight of 13 kDa from seeds of *G. biloba*, which inhibited the growth of *Botrytis cinerea, Mycosphaerella arachidicola, F. oxysporum, Rhizoctonia solani,* and *Coprinus comatus*. The authors also claimed that Ginkbilobin, another compound derived from Ginkgo, showed moderate antibacterial activity against *S. aureus, P. aeruginosa* and *E. coli*. Moreover, two compounds (2-hydroxy-6-(8-pentadecenyl) salicylic acid and 2-hydroxy-6-(10-heptadecenyl) salicylic acid) isolated from the chloroform fraction of *G. biloba* showed significant antibacterial activity against vancomycin-resistant *Enterococcus* bacteria.[107]

Azadirachta Indica A. Juss (Neem)

This medicinal plant belongs to the family of Meliaceae and originates from India and Myanmar. The evidence indicates that *A. indica* was used in ancient times to cure cholera, diarrhea, wounds, rheumatism, diabetes, eczema, ringworm, and scabies. Valuable compounds such as margosic acid, limonoids, azadirachtin, azadiradione, nimbin, salannin, stigmasterol, nimbiol, sugiol, α-terpinene terpinen-4-ol, 4-cymene, epoxyazadiradione and vitamin E have been isolated from *A. indica* and identified. Neem extracts are antihyperglycemic, in that they block the inhibitory influences of serotonin on insulin secretion mediated by glucose.[108,109] It is also documented that volatile oils and extracts of Neem promote the wound healing process; in this regard, the intramuscular application of Neem oil diminished the surface area of wounds and improved their tensile strength; moreover complete epithelization was observed within 15 days.[110]

Chattopadhyay[111] claimed that *A. indica* leaf extract possesses hepatoprotective activity against paracetamol-induced hepatic damage, and that it can reduce the content of glutathione, Na^+ K^+-ATPase activity and content of thiobarbutiric acid in paracetamol hepatotoxicity. The antibacterial and antifungal activity of Neem against *K. pneumoniae, Staphylococcus epidermitis, P. aeroginosa, S. aureus, E. coli, S. typhi, Propionibacterium acnes, F. oxysporum, Drechslera oryzae* and *Alternaria tenuis* has also been documented.[112] Rajasekaran[113] also reported that Neem extract showed remarkable antimicrobial activity against skin pathogens, including *Streptococcus pyogenes* and *S. aureus*.

Chromolaena Odorata (L.) King & H.E. Robins (Devil Weed)

This tropical, perennial, scrambling shrub plant originates from North America and tropical parts of Asia, and belongs to the Asteraceae family. Historical evidence indicates that extracts of this plant were used to treat malarial fever, headache, toothache, burns, ischuria and skin infections. *C. odorata* contains valuable compounds such as rutin, isoquercitrin, triterpene α-amyrin, 5,7-dihydroxy-6,4′-dimethoxyflavone, 1,2-dehydropyrrolizidine alkaloids, caryophyllene, kaempferol, kaempferol-3-*O*-rutinoside, geijerene, β-cubebene, germacrene, α and β-pinene.[112–115] It is known that terpenoid compounds are the main components of the volatile oil of *C. odorata,* including trans-caryophyllene, delta-cadinene, alpha-copaene, caryophyllene oxide, germacrene-D, and delta-humulene.[116]

Antioxidants agents derived from *C. odorata* have been shown to help in curing wounds via their antioxidant activity. Oxidants are known to cause cell damage and to inhibit the wound healing process. It is reported that an ethanol extract of *C. odorata* possesses strong antioxidant activities and so protects fibroblasts and keratinocytes. In addition, phenolic acids (protocatechuic, p-hydroxybenzoic, p-coumaric, ferulic and vanillic acids) and complex mixtures of lipophilic flavonoid aglycones (flavanones, flavonols, flavones and chalcones) were considered to be the most influential and important antioxidants for protecting cultured skin cells against oxidative damage.[114,115]

Additionally, the influence of *C. odorata* on fibroblasts and endothelial cells involved in the wound healing process is well accepted. *C. odorata* extract at concentrations of 10 μg/mL and 100 μg/mL enhanced the growth of fibroblasts and endothelial cells.[114] Scientific studies using *C. odorata* extract have revealed antimicrobial and anticoagulation activities and an increase in tissue remodeling in the wound healing process.[117]

Irobi[118] stated that an ethanol extract of *C. odorata* produced assessable zones of inhibition (6.5–16 mm) against *Bacillus thuringiensis, Bacillus stearothermophilus, S. aureus, E. coli, Pseudomonas sp., S. faecalis* and *Klebsiella sp.* Notably, Atindehou and colleagues[119] reported two new antimicrobial compounds isolated from *C. odorata* that were identified as 3′,4′,5,6,7-pentamethoxyflavone (sinensetin) and 4′,5,6,7-tetramethoxyflavone (scutel larein tetramethyl ether).

Furthermore, three flavonoid compounds isolated from *C. odorata* and identified as eriodictyol-7,4′-dimethyl ether, naringenin-4′-methyl ether and 2′4-dihydroxy-4′5′6′-trimethoxychalcone have demonstrated significant efflux inhibitory influence on methicillin resistant *S. aureus.*

Alternanthera Sessilis (L.) R.Br. ex DC. (Sessile Joyweed)

This annual or perennial herbaceous plant belongs to the Amaranthaceae family, and is native to the tropical regions of Central and South America. Extracts of this plant have been used to treat hypertension, diabetes, hepatitis, bronchitis, chronic liver congestion, gonorrhea and asthma. The plant contains valuable compounds including β-sitosterol, β-carotene, 12-hydroxy-9-cis-octadecenoic acid, palmitic, myristic, ricinoleic, oleic and linoleic acids.[120,121] The immunomodulatory activities of the *Alternanthera* genus were evaluated based on antibody production in male albino Swiss mice. The results indicated that the *Alternanthera* extract significantly improved IgM (by 64%) and IgG2a (by 50%). Aqueous *Alternanthera* extract also increased the life span and diminished the number of viable tumor cells (59%) in mice with Ehrlich ascites carcinoma.[122] A chloroform extract of the leaves of *A. sessilis* administered orally at a dose of 200 μg/mL to Albino rats showed significant results on parameters of the wound healing process, including contraction of wound area, time for complete epithelization, wound breaking strength, granuloma dry weight, and granuloma breaking strength.[123] Extracts of *A. sessilis* showed inhibitory influences on the growth of *E. coli, P. vulgaris, S. pyogenes, B. subtilis, P. aeruginosa, S. typhi, V. cholerae,* and *K. pneumonia.*[123]

Mussaenda Frondosa L. (Wild Mussenda)

This tropical plant belongs to the Rubiaceae family, and was used to cure jaundice, ulcers, leprosy, asthma, hyperacidity, fever, cough, and as a diuretic. At present some compounds including isovanillin, naphthalene, decahydro-2-methoxy, caryophyllene, β sitosterol-glucoside, quercetin, rutin, hyperin, oleic, ferulic and sinapic acid have been identified and isolated from extracts of *M. frondosa*.[124] Also, a new iridoid, shanzhiol, has been isolated from the *M. roxburghii* and shown to have mild antibacterial activity against *S. aureus* and *E. coli*. Its hypolipidemic effect is also well documented, and is thought to be due to the inhibition of hepatic cholesterogenesis and antioxidant activity.[125] Extracts of *M. frondosa* were shown to be completely effective in different wound healing models such as excision, incision and dead space wound; all healing parameters improved through its use. Moreover, alcoholic and aqueous extracts of *M. frondosa* decrease levels of aspartate transaminase, alkaline phosphatase, alanine transaminase, bilirubin, cholesterol, triglyceride, and improve levels of protein, high-density lipoprotein and antioxidants including superoxide dismutase, glutathione peroxidase, and catalase enzymes.[126] A comparison of the antimicrobial agent silver sulfadiazine (which has been used as a cream on burns) and extracts of *M. frondosa* showed that an alcoholic extract of *M. frondosa* and silver sulfadiazine clearly enhanced epithelialization and collagen bundle synthesis.[127]

Lantana Camara L. (Spanish Flag)

This ornamental shrub with aromatic leaves belongs to the family Verbanaceae, and is native to tropical America. It has been used to treat a wide range of disorders from tumors to malaria and respiratory system infections. Valuable compounds including, α-copaene, β-elemene, tetradecane, E-caryophyllene, phytol, davanone, globulol, cubeol, bicyclogermacrene, germacrene-D, isocaryophyllene, lantadene-A and valecene have isolated from this plant.[128,129] A novel triterpene, 22β-acetoxylantic acid, has been isolated from *L. camara* and shown to have antimicrobial activity against *S. aureus* and *S. typhi*. This compound and 22β-dimethylacryloyloxy lantanolic acid, isolated from *L. camara*, also possess antimutagenic activity.[130] Screening the chemical composition of *L. camara* revealed that the major components of its essential oil are germacrene-D (19.8%) and E-caryophyllene (19.7%). In another experiment, bicyclogermacrene (19.42%), isocaryophyllene (16.70%), valecene (12.94%) and germacrene-D (12.34%) were reported as the main constituents of *L. camara* essential oil and its synergistic action with aminoglycosides against *E. coli* was also confirmed.[131]

It has been documented that a methanolic extract of *L. camara* leaves could cure gastric ulcers and halted the progress of duodenal ulcers in treatments, and also that it has antipyretic and carminative activities.[132] A study of the antihyperglycemic and wound healing activity of *L. camara* showed that its aqueous extract significantly decreased blood glucose concentrations in alloxan-induced hyperglycemic treatments. The leaf juice was found to be more active than extracts in improving the wound healing process.[133,134] Moreover, *L. camara* was able to act as a curative agent in tissue repair processes, its extract considerably enhanced the rate of wound contraction (98%), and *L. camara* was able to promote the synthesis of collagen and reduce the mean wound healing time.[135]

CONCLUSIONS AND FUTURE PROSPECTS

The floral richness of our world has given us the opportunity to isolate and use new bio-molecules with specific modes of action, but undoubtedly progress will not be possible without cooperation among scientists who specialize in different areas of sciences. In addition to medicinal plants, other sources of bio-molecules from algae, mushrooms, bacteria and yeasts, must be screened seriously in order to find novel medicines. The safety of these diverse compounds must be confirmed and an optimum range of usage must be determined. New technologies in the area of biotechnology, chemistry, microbiology and genetics must be introduced; this area certainly needs talent, responsibilities, effort and funding.

References

1. Sumner J. *The natural history of medicinal plants. Timber Press*; 2000.
2. Sofowora A. *Medicinal plants and traditional medicine in Africa*. John Wiley & Sons, Ltd; 1982.
3. Bhattacharjee SK. *Handbook of medicinal plants*. Aavishkar Publishers; 2000.
4. Kufer J, Heinrich M, Förther H, Pöll E. Historical and modern medicinal plant uses – the example of the Ch'orti 'Maya and Ladinos in Eastern Guatemala. *J Pharm Pharmacol* 2005;**57**:1127–52.
5. Thring TSA, Weitz FM. Medicinal plant use in the Bredasdorp/Elim region of the Southern Overberg in the Western Cape Province of South Africa. *J. Ethnopharmacol* 2006;**103**:261–75.
6. Leonti M, Cabras S, Weckerle CS, Solinas MN, Casu L. The causal dependence of present plant knowledge on herbals – contemporary medicinal plant use in Campania (Italy) compared to Matthioli (1568). *J Ethnopharmacol* 2010;**130**:379–91.
7. Scalbert A, Andres-Lacueva C, Arita M, Kroon P, Manach C, Urpi-Sarda M, et al. Databases on food phytochemicals and their health-promoting effects. *J Agric Food Chem* 2011;**59**:4331–48.
8. Ahuja I, de Vos RC, Bones AM, Hall RD. Plant molecular stress responses face climate change. *Trends Plant Sci* 2010;**15**:664–74.
9. Van Dam NM, Van der Meijden E. A role for metabolomics in plant ecology. *Annu Rev Plant Bio Plant Metabol* 2011;**58**:87–107.
10. Miyasaki Y, Nichols WS, Morgan MA, Kwan JA, Van Benschoten MM, Kittell PE, et al. Screening of herbal extracts against multi-drug resistant *Acinetobacter baumannii*. *Phytother Res* 2010;**24**:1202–6.
11. Rauha JP, Remes S, Heinonen M, Hopia A, Kähkönen M, Kujala T, et al. Antimicrobial effects of Finnish plant extracts containing flavonoids and other phenolic compounds. *Int J Food Microbiol* 2000;**56**:3–12.
12. Thuille N, Fille M, Nagl M. Bactericidal activity of herbal extracts. *Int J Hyg Environ Health* 2003;**206**:217–21.
13. Berdy J. Recent developments in antibiotic research and classification of antibiotics according to chemical structure. *Adv Appl Microbiol* 1974;**18**:309–406.
14. Abraham EP, Newton GG. Chemistry and classification of antibiotics. *Br Med Bull* 1960;**16**:3–10.
15. Wong WR, Oliver AG, Linington RG. Development of antibiotic activity profile screening for the classification and discovery of natural product antibiotics. *J. Biol. Chem* 2012;**19**:1483–95.
16. Wright GD, Sutherland AD. New strategies for combating multidrug-resistant bacteria. *Trends mol med* 2007;**13**:260–7.
17. Piddock LJ. Multidrug-resistance efflux pumps – not just for resistance. *Nature Rev Microbiol* 2006;**4**:629–36.
18. McGowan JE. Resistance in non-fermenting gram-negative bacteria: multidrug resistance to the maximum. *Am J Med* 2006;**34**:29–37.
19. George AM, Hall RM, Stokes HW. Multidrug resistance in *Klebsiella pneumoniae*: a novel gene, ramA, confers a multidrug resistance phenotype in *Escherichia coli*. *Microb* 1995;**141**:1909–20.
20. Hemaiswarya S, Kruthiventi AK, Doble M. Synergism between natural products and antibiotics against infectious diseases. *Phytomed* 2008;**15**:639–52.

21. Nascimento GG, Locatelli J, Freitas PC, Silva GL. Antibacterial activity of plant extracts and phytochemicals on antibiotic-resistant bacteria. *Braz J Microbiol* 2000;**31**:247–56.
22. Ma XH, Zheng CJ, Han LY, Xie B, Jia J, Cao ZW, et al. Synergistic therapeutic actions of herbal ingredients and their mechanisms from molecular interaction and network perspectives. *Drug Discov Today* 2009;**14**:579–88.
23. Davydov I, Larichev AB, Abramov A, Men'kov KG. Concept of clinico-biological control of the wound process in the treatment of suppurative wounds using vacuum therapy. *Vestn Khir Im I I Grek* 1991;**146**:132–6.
24. Kirsner RS, Eaglstein WH. The wound healing process. *Dermatol Clin* 1993;**11**:629–40.
25. Puratchikody A, Devi CN, Nagalakshmi G. Wound healing activity of *cyperus rotundus* linn. *Indian J Pharm Sci* 2006;**68**:97–101.
26. Sahu K, Verma Y, Sharma M, Rao KD, Gupta PK. Non-invasive assessment of healing of bacteria infected and uninfected wounds using optical coherence tomography. *Skin Res Technol* 2010;**16**:428–37.
27. Moon TC, Murakami M, Kudo I, Son KH, Kim HP, Kang SS, et al. A new class of COX-2 inhibitor, rutaecarpine from *Evodia rutaecarpa*. *Inflamm Res* 1999;**48**:621–5.
28. Taylor JLS, Van Staden J. COX-1 and COX-2 inhibitory activity in extracts prepared from Eucomis species, with further reference to extracts from *E. autumnalis*. *S Afr J Bot* 2002;**68**:80–5.
29. Ndhlala AR, Aderogba MA, Ncube B, Van Staden J. Anti-oxidative and cholinesterase inhibitory effects of leaf extracts and their isolated compounds from two closely related croton species. *Molecules* 2013;**18**:1916–32.
30. Shenoy C, Patil MB, Kumar R. Wound healing activity of *Hyptis suaveolens* (L.) Poit (Lamiaceae). *Inter J Pharm Res* 2009;**1**:737–44.
31. Upadhyay NK, Kumar R, Siddiqui MS, Gupta A. Mechanism of wound-healing activity of *Hippophae rhamnoides* L. leaf extract in experimental burns. *Evid Based Complement Alternat Med* 2009;**2011**:1–9.
32. Grierson DS, Afolayan AJ. Antibacterial activity of some indigenous plants used for the treatment of wounds in the Eastern Cape, South Africa. *J Ethnopharmacol* 1999;**66**:103–6.
33. Aiyegoro OA, Afolayan AJ, Okoh AI. Synergistic interaction of *Helichrysum pedunculatum* leaf extracts with antibiotics against wound infection associated bacteria. *Bio Res* 2009;**42**:327–38.
34. Raina R, Parwez S, Verma PK, Pankaj NK. Medicinal plants and their role in wound healing. *Online J Vet Res* 2008;**3**:1–7.
35. Reynolds T, Dweck AC. *Aloe vera* leaf gel: a review update. *J Ethnopharmacol* 1999;**68**:3–37.
36. Vogler BK, Ernst E. *Aloe vera*: a systematic review of its clinical effectiveness. *Brit J Gen Pract* 1999;**49**:823–8.
37. Tanaka M, Misawa E, Ito Y, Habara N, Nomaguchi K, Yamada M, et al. Identification of five phytosterols from *Aloe vera* gel as anti-diabetic compounds. *Biol Pharm Bull* 2006;**29**:1418–22.
38. Pugh N, Ross SA, ElSohly MA, Pasco DS. Characterization of Aloeride, a new high-molecular-weight polysaccharide from *Aloe vera* with potent immunostimulatory activity. *J Agric Food Chem* 2001;**49**:1030–4.
39. Grindlay D, Reynolds T. The *Aloe vera* phenomenon: A review of the properties and modern uses of the leaf parenchyma gel. *J Ethnopharmacol* 1986;**16**:117–51.
40. Yusuf S, Agunu A, Diana M. The effect of *Aloe vera* and *A. Berger* (Liliaceae) on gastric acid secretion and acute gastric mucosal injury in rats. *J Ethnopharmacol* 2004;**93**:33–7.
41. Langmead L, Makins RJ, Rampton DS. Anti-inflammatory effects of *Aloe vera* gel in human colorectal mucosa *in vitro*. *Aliment Pharmacol Ther* 2004;**19**:521–7.
42. Alemdar S, Agaoglu S. Investigation of *in vitro* antimicrobial activity of *Aloe vera* juice. *J Anim Vet Adv* 2009;**8**:99–102.
43. Pandey R, Mishra A. Antibacterial activities of crude extract of *Aloe barbadensis* to clinically isolated bacterial pathogens. *Appl Biochem Biotechnol* 2010;**160**:1356–61.
44. Lawrence R, Tripathi P, Jeyakumar E. Isolation, purification and evaluation of antibacterial agents from *Aloe vera*. *Braz J Microbiol* 2009;**40**:906–15.
45. Davis RH, Donato JJ, Hartman GM, Haas RC. Anti-inflammatory and wound healing activity of a growth substance in *Aloe vera*. *J Am Podiatr Med Assoc* 1994;**84**:77–81.
46. Davis RH, DiDonato JJ, Johnson RW, Stewart CB. *Aloe vera*, hydrocortisone, and sterol influence on wound tensile strength and anti-inflammation. *J Am Podiatr Med Assoc* 1994;**84**:614–21.
47. Panda H. *Aloe Vera* Handbook Cultivation, Research Finding, Products, Formulations, Extraction & Processing. *National Institute of Industrial Research;* 2003.

48. Oh H, Kim DH, Cho JH, Kim YC. Hepatoprotective and free radical scavenging activities of phenolic petrosins and flavonoids isolated from *Equisetum arvense*. *J Ethnopharmacol* 2004;**95**:421–4.
49. Asgarpanah J, Roohi E. Phytochemistry and pharmacological properties of *Equisetum arvense* L. *J Med Plants Res* 2012;**6**:3689–93.
50. Sandhu NS, Kaur S, Chopra D. *Equietum arvense*: pharmacology and phytochemistry – a review. *Asian J Pharm Clin Res* 2010;**3**:146–50.
51. Milovanović V, Radulović N, Todorović Z, Stanković M, Stojanović G. Antioxidant, antimicrobial and genotoxicity screening of hydro-alcoholic extracts of five Serbian Equisetum species. *Plant Foods Hum Nutr* 2007;**62**:113–9.
52. Perry CC. Biosilicification – structure, regulation of structure and model studies. In: Jutzi P, Schubert U, eds. *Silicon chemistry – from molecules to extended systems*. Weinheim: Wiley-VCH; 2003. pp. 475–87.
53. Holzhüter G, Narayanan K, Gerber T. Structure of silica in *Equisetum arvense*. *Anal Bioanal Chem* 2003;**376**:512–7.
54. Radulović N, Stojanović G, Palić R. Composition and antimicrobial activity of *Equisetum arvense* L. essential oil. *Phytother Res* 2006;**20**:85–8.
55. Chandra S, Gonzalez de Mejia E. Polyphenolic compounds, antioxidant capacity, and quinone reductase activity of an aqueous extract of *Ardisia compressa* in comparison to mate (*Ilex paraguariensis*) and green (*Camellia sinensis*) teas. *J Agric Food Chem* 2004;**52**:3583–9.
56. Shi CY, Yang H, Wei CL, et al. Deep sequencing of the *Camellia sinensis* transcriptome revealed candidate genes for major metabolic pathways of tea-specific compounds. *BMC Genomics* 2011;**12**:131.
57. Blagojevic P, Radulovic N, Palic R, Stojanovic G. Chemical composition of the essential oils of Serbian wild-growing *Artemisia absinthium* and *Artemisia vulgaris*. *J Agric Food Chem* 2006;**54**:4780–9.
58. Gordon NC, Wareham DW. Antimicrobial activity of the green tea polyphenol (−)-epigallocatechin-3-gallate (EGCG) against clinical isolates of *Stenotrophomonas maltophilia*. *Int J Antimicrob Ag* 2010;**36**:129–31.
59. Ravindranath MH, Saravanan TS, Monteclaro CC. Epicatechins purified from green tea (*Camellia sinensis*) differentially suppress growth of gender-dependent human cancer cell lines. *Evid Based Complement Alternat Med* 2006;**3**:237–47.
60. Anaya AL, Cruz-Ortega R, Waller GR. Metabolism and ecology of purine alkaloids. *Front Biosci* 2006;**11**:2354–70.
61. Donà M, Dell'Aica I, Calabrese F, Benelli R, Morini M, Albini A, et al. Neutrophil restraint by green tea: inhibition of inflammation, associated angiogenesis, and pulmonary fibrosis. *J Immunol* 2003;**170**:4335–41.
62. Yam TS, Hamilton-Miller JM, Shah S. The effect of a component of tea (*Camellia sinensis*) on methicillin resistance, PBP2'synthesis, and beta-lactamase production in *Staphylococcus aureus*. *J Antimicrob Chemother* 1998;**42**:211–6.
63. Lee JH, Shim JS, Lee JS, Kim JK, Yang IS, Chung MS, et al. Inhibition of pathogenic bacterial adhesion by acidic polysaccharide from green tea (*Camellia sinensis*). *J Agric Food Chem* 2006;**54**:8717–23.
64. Zitterl-Eglseer K, Sosa S, Jurenitsch J, Schubert-Zsilavecz M, Della Loggia R, Tubaro A, et al. Anti-oedematous activities of the main triterpendiol esters of marigold (*Calendula officinalis* L.). *J Ethnopharmacol* 1997;**57**:139–44.
65. Komissarenko NF, Chernobai VT, Derkach AI. Flavonoids of inflorescences of *Calendula officinalis*. *Chem Nat Compd* 1988;**24**:675–80.
66. Danielski L, Campos LM, Bresciani LF, Hense H, Yunes RA, Ferreira SR. Marigold (*Calendula officinalis* L.) oleoresin: Solubility in SC-CO_2 and composition profile. *Chem Eng Process: Process Intensification* 2007;**46**:99–106.
67. Kalvatchev Z, Walder R, Garzaro D. Anti-HIV activity of extracts from *Calendula officinalis* flowers. *Biomed Pharmacother* 1997;**51**:176–80.
68. Fronza M, Heinzmann B, Hamburger M, Laufer S, Merfort I. Determination of the wound healing effect of Calendula extracts using the scratch assay with 3T3 fibroblasts. *J Ethnopharmaco* 2009;**126**:463–7.
69. Chandran PK, Kuttan R. Effect of *Calendula officinalis* flower extract on acute phase proteins, antioxidant defense mechanism and granuloma formation during thermal burns. *J Clin Biochem Nutr* 2008;**43**:58–64.
70. Bashir S, Janbaz KH, Jabeen Q, Gilani AH. Studies on spasmogenic and spasmolytic activities of *Calendula officinalis* flowers. *Phytother Res* 2006;**20**:906–10.

71. Sakharkar PR, Kasiram K, Patil AT. Antimicrobial activity of *Calendula officinalis*. *Hamdard Medicus* 2000;**43**:5–7.
72. Pino JA, Almora K, Marbot R. Volatile components of papaya (*Carica papaya* L., Maradol variety) fruit. *Flavor Frag J* 2003;**18**:492–6.
73. Sánchez-Moreno C, Plaza L, de Ancos B. Quantitative bioactive compounds assessment and their relative contribution to the antioxidant capacity of commercial orange juices. *J Sci Food Agr* 2003;**83**:430–9.
74. Canini A, Alesiani D, D'Arcangelo G, Tagliatesta P. Gas chromatography – mass spectrometry analysis of phenolic compounds from *Carica papaya* L. leaf. *J Food Comp Anal* 2007;**20**:584–90.
75. Anuar NS, Zahari SS, Taib IA, Rahman MT. Effect of green and ripe *Carica papaya* epicarp extracts on wound healing and during pregnancy. *Food Chem Toxicol* 2008;**46**:2384–9.
76. Singh D, Singh R. Papain incorporated chitin dressings for wound debridement sterilized by gamma radiation. *Radiat Phys Chem* 2012;**81**:1781–5.
77. Suresh K, Deepa R, Harisaranraj R, Vaira Achudhan V. Antimicrobial and phytochemical investigation of the leaves of *Carica papaya* L, *Cynodon dactylon* L. Pers, *Euphorbia hirta* L, *Melia azedarach* L. and *Psidium guajava* L. *Ethnobot Leaflets* 2008;**12**:1184–91.
78. Dawkins G, Hewitt H, Wint Y, Obiefuna PC, Wint B. Antibacterial effects of *Carica papaya* fruit on common wound organisms. *W Indian Med J* 2003;**52**:290–2.
79. Fonseca FN, Tavares MF, Horváth C. Capillary electrochromatography of selected phenolic compounds of *Chamomilla recutita*. *J Chrom A* 2007;**1154**:390–9.
80. Rodriguez-Fragoso L, Reyes-Esparza J, Burchiel S, Herrera-Ruiz D, Torres E. Risks and benefits of commonly used herbal medicines in Mexico. *Toxicol Appl Pharmacol* 2008;**227**:125–35.
81. Avallone R, Zanoli P, Puia G, Kleinschnitz M, Schreier P, Baraldi M. Pharmacological profile of apigenin, a flavonoid isolated from *Matricaria chamomilla*. *Biochem Pharmacol* 2000;**59**:1387–94.
82. Kováčik J, Klejdus B, Bačkor M, Repčák M. Phenylalanine ammonia-lyase activity and phenolic compounds accumulation in nitrogen-deficient *Matricaria chamomilla* leaf rosettes. *Plant Sci* 2007;**172**:393–9.
83. Jing Y. Clinical references on chamomile. *Anticancer Res* 1995;**15**:1147–52.
84. Nemecz G. Herbal pharmacy: chamomile. *US Pharmacist* 2000;**23**:115–23.
85. Martins MD, Marques MM, Bussadori SK, Martins MAT, Pavesi VCS, Mesquita-Ferrari RA, et al. Comparative analysis between *Chamomilla recutita* and corticosteroids on wound healing. An *in vitro* and *in vivo* study. *Phytother Res* 2009;**23**:274–8.
86. Owlia P, Rasooli I, Saderi H. Antistreptococcal and antioxidant activity of essential oil from *Matricaria chamomilla* L. *Res J Biol Sci* 2007;**2**:150–60.
87. Molgaard P, Johnsen S, Christensen P, Cornett C. HPLC method validated for the simultaneous analysis of cichoric acid and alkamides in *Echinacea purpurea* plants and products. *J Agric Food Chem* 2003;**51**:6922–33.
88. Vimalanathan S, Kang L, Amiguet VT. *Echinacea purpurea* aerial parts contain multiple antiviral compounds. *Pharm Biol* 2005;**43**:740–5.
89. Lii JR, Gao XF, Ai TM. Lipid compounds from *Echinacea purpurea*. *China J Chin Mat Med* 2002;**27**:40–2.
90. Pellati F, Calò S, Benvenuti S, Adinolfi B, Nieri P, Melegari M. Isolation and structure elucidation of cytotoxic polyacetylenes and polyenes from *Echinacea pallida*. *Phytochemistry* 2006;**67**:1359–64.
91. Rininger JA, Kickner S, Chigurupati P, McLean A, Franck Z. Immunopharmacological activity of *Echinacea* preparations following simulated digestion on murine macrophages and human peripheral blood mononuclear cells. *J Leukoc Biol* 2000;**68**:503–10.
92. Rezaie A, Najafzadeh H, Mohammadian B, Heidari M. Effects of *Echinacea Purpurea* on wound healing after arsenic induced skin necrosis. *Zahedan J Res Med Sci* 2013;**15**:29–33.
93. Hudson JB. The multiple actions of the phytomedicine *Echinacea* in the treatment of colds and flu. *J Med Plant Res* 2010;**4**:2746–52.
94. Sharma M, Vohra S, Arnason JT, Hudson JB. *Echinacea* extracts contain significant and selective activities against human pathogenic bacteria. *Pharm Biol* 2008;**46**:111–6.
95. Weber ND, Andersen DO, North JA, Murray BK, Lawson LD, Hughes BG. *In vitro* virucidal effects of *Allium sativum* (garlic) extract and compounds. *Planta Med* 2007;**58**:417–23.
96. Wargovich MJ. Diallyl sulfide, a flavor component of garlic (*Allium sativum*), inhibits dimethyihydrazine-induced colon cancer. *J Carcinog* 1987;**8**:487–9.
97. Ankri S, Mirelman D. Antimicrobial properties of allicin from garlic. *Microb Infect* 1999;**1**:125–9.

98. Abubakar EMM. Efficacy of crude extracts of garlic *Allium sativum* (Linn.) against nosocomial *Escherichia coli, Staphylococcus aureus, Streptococcus pneumoniea* and *Pseudomonas aeruginosa*. *J Med Plants Res* 2009;**3**:179–85.
99. Dausch JG, Nixon DW. Garlic: a review of its relationship to malignant disease. *Prev Med* 1990;**19**:346–61.
100. Delaha EC, Garagusi VF. Inhibition of mycobacteria by garlic extract (*Allium sativum*). *Antimicrob Agents* 1985;**27**:485–6.
101. Hasler A, Sticher O, Meier B. Identification and determination of the flavonoids from *Ginkgo biloba* by high-performance liquid chromatography. *J Chromatogr* 1992;**605**:41–8.
102. Itil T, Martorano D. Natural substances in psychiatry (*Ginkgo biloba* in dementia). *Psychopharmacol Bull* 1995;**31**:147–58.
103. Potron G, Petitfrere E, Nguyen P. Platelet activating factor and ischemia. *Clin Hemorheol Micro* 1993;**13**:155–76.
104. Bairy KL, Rao CM. Wound healing profiles of *Ginkgo biloba*. *J Nat Remed* 2001;**1**:25–7.
105. Trompezinski S, Bonneville M, Pernet I, Denis A, Schmitt D, Viac J. *Gingko biloba* extract reduces VEGF and CXCL-8/IL-8 levels in keratinocytes with cumulative effect with epigallocatechin-3-gallate. *Arch Dermatol Res* 2010;**302**:183–9.
106. Wang H, Ng TB. Ginkbilobin a Novel antifungal protein from *Ginkgo biloba* seeds with sequence similarity to embryo-abundant protein. *Biochem Biophys Res Commun* 2000;**279**:407–11.
107. Valasidis A, Kyrgidis A, Tzellos G, Triaridis SK, Antoniades HK. Novel antimicrobial agents for the management of maxillofacial and neck infections. *Recent Pat Antiinfect Drug Discov* 2012;**7**:95–103.
108. Balandrin MF, Lee SM, Klocke JA. Biologically active volatile organosulfur compounds from seeds of the neem tree, *Azadirachta indica* (Meliaceae). *J Agric Food Chem* 1988;**36**:1048–54.
109. Biswas K, Chattopadhyay I, Banerjee RK, Bandyopadhyay U. Biological activities and medicinal properties of neem (*Azadirachta indica*). *Curr Sci* 2002;**82**:1336–45.
110. Pandey IP, Ahmed SF, Chhimwal S, Pandey S. Chemical composition and wound healing activity of volatile oil of leaves of *Azadirachta indica* A. juss. *Adv Pur App Chem* 2012;**1**:62–6.
111. Chattopadhyay RR. Possible mechanism of hepatoprotective activity of *Azadirachta indica* leaf extract: Part II. *J Ethnopharmacol* 2003;**89**:217–9.
112. El-Mahmood AM, Ogbonna OB, Raji M. The antibacterial activity of Azadarichta indica (Neem) seeds extracts against bacterial pathogens associated with eye and ear infections. *J Med Plant Res* 2010;**4**:1414–21.
113. Rajasekaran C. Investigations on antibacterial activity of leaf extracts of *Azadirachta indica* A. Juss (Meliaceae): A traditional medicinal plant of India. *Ethnobotanic Leaflets* 2008;**1**:161.
114. Phan T, Wang L, See P, Grayer RJ, Chan SY, Lee ST. Phenolic compounds of *Chromolaena odorata* protect cultured skin cells from oxidative damage: implication for cutaneous wound healing. *Biol Pharm Bull* 2001;**24**:1373–9.
115. Thang PT, Teik LS, Yung CS. Anti-oxidant effects of the extracts from the leaves of *Chromolaena odorata* on human dermal fibroblasts and epidermal keratinocytes against hydrogen peroxide and hypoxanthine–xanthine oxidase induced damage. *Burns* 2001;**27**:319–27.
116. Ling B, Zhang M, Kong C, Pang X, Liang G. Chemical composition of volatile oil from *Chromolaena odorata* and its effect on plant, fungi and insect growth. *J Appl Eco* 2003;**14**:744.
117. Lamond D, Farnell S. Pressure ulcers. *Adv Skin Wound Care* 1998;**11**:320.
118. Irobi ON. Antibiotic properties of ethanol extract of *Chromolaena odorata* (Asteriaceae). *Pharm Biol* 1997;**35**:111–5.
119. Ménonvè A, Latifou L, Bernard G, Jean MS, Minjie Z, Alain VD, et al. Isolation and identification of two antibacterial agents from *Chromolaena odorata* L. active against four diarrheal strains. *Adv Microb* 2013;**3**:115–21.
120. Ashok Kumar CK, Revathi K, Mohanalakshmi S. A review of edible herbs as haematinics. *Int J Pharm* 2012;**2**:44–53.
121. Hosamani KM, Ganjihal SS, Chavadi DV. *Alternanthera triandra* seed oil: A moderate source of ricinoleic acid and its possible industrial utilisation. *Ind Crop Prod* 2004;**19**:133–6.
122. Guerra RNM, Pereira HA, Silveira LMS, Olea RSG. Immunomodulatory properties of *Alternanthera tenella* Colla aqueous extracts in mice. *Brz J Med Bio Res* 2003;**36**:1215–9.
123. Jalalpure SS, Agrawal N, Patil MB, Chimkode R, Tripathi A. Antimicrobial and wound healing activities of leaves of *Alternanthera sessilis* Linn. *Int J Green Pharm* 2008;**2**:141.

124. Basavaraja BM, Vagdevi HM, Srikrishna LP, Hanumanthapa BC, Joshi SD, Vaidya VP. Antimicrobial and analgesic activities of various extracts of *Mussaenda frondosa*. L. Bark. *J Glob Pharm Tech* 2011;**3**:14–7.
125. Vidyalakshmi KS, Nagarajan S, Vasanthi HR, Rajamanickam V. Hepatoprotective and antioxidant activity of two iridoids from *Mussaenda* 'dona aurora'. *Z Naturforsch* 2009;**64**:329–34.
126. Vadivel E, Gopalakrishnan S. GC-MS analysis of some bioactive constituents of *Mussaenda frondosa* Linn. *Inter J Pharm Bio Sci* 2011;**2**:313–20.
127. Patil SA, Joshi VG, Sambrekar SN. Wound healing effects of *Mussaenda frondosa* extracts on second degree superficial burned rat. *Res J Pharm Pharmacody* 2012;**4**:163–8.
128. Passos JL, Barbosa LCA, Demuner AJ, Alvarenga ES, Silva CMD, Barreto RW. Chemical characterization of volatile compounds of *Lantana camara* L. and *L. radula* Sw. and their antifungal activity. *Molecules* 2012;**17**:11447–55.
129. Patel S. A weed with multiple utility: *Lantana camara*. *Rev Environ Sci Biotechnol* 2011;**10**:341–51.
130. Barre JT, Bowden BF, Coll JC, de Jesus J, de la Fuente VE, Janairo GC, Ragasa Y. A bioactive triterpene from *Lantana camara*. *Phytochemistry* 1997;**45**:321–4.
131. Sousa EO, Silva NF, Rodrigues FF, De Jesus J, De La Fuente VE, Janairo GC, et al. Chemical composition and resistance-modifying effect of the essential oil of *Lantana camara* Linn. *Pharmacogn Mag* 2010;**6**:79–82.
132. Kazmi I, Rahman M, Afzal M. Anti-diabetic potential of ursolic acid stearoyl glucoside: A new triterpenic gycosidic ester from *Lantana camara*. *Fitoterapia* 2012;**83**:142–6.
133. Dash GK, Suresh P, Ganapaty S. Studies on hypoglycemic and wound healing activities of *Lantana camara* Linn. *J Nat Remedies* 2001;**1**:105–10.
134. Karakas FP, Karakas A, Boran Ç, Türker AU, Yalçin FN, Bilensoy E. The evaluation of topical administration of *Bellis perennis* fraction on circular excision wound healing in Wistar albino rats. *Pharmaceuti Biol* 2012;**50**:1031–7.
135. Nayak BS, Raju SS, Eversley M, Ramsubhag A. Evaluation of wound healing activity of *Lantana camara* L. a preclinical study. *Phytother Res* 2009;**23**:241–5.

CHAPTER

13

The Role of Honey and Propolis in the Treatment of Infected Wounds

*Ananias Pascoal**, *Xesus Feás*†, *Teresa Dias**, *Luís G. Dias** *and Leticia M. Estevinho**

*CIMO-Mountain Research Center, Department of Biology and Biotechnology, Agricultural College of Bragança, Polytechnic Institute of Bragança, Bragança, Portugal, †Department of Organic Chemistry, Faculty of Sciences, University of Santiago de Compostela, Lugo, Spain

INTRODUCTION

Honey

Honey is defined as the natural substance produced by *Apis mellifera* bees from plant nectar, from secretions of the living parts of plants, or from the excretions of plant sucking insects feeding on the living parts of plants.[1,2] Honeybees collect, transform, and combine this with specific substances of their own, and then store it and leave it in the honeycomb to ripen and mature. This natural product is generally composed of a complex mixture of carbohydrates and other less common substances, such as organic acids, amino acids, proteins, minerals, vitamins, lipids, aroma compounds, flavonoids, pigments, waxes, pollen grains, several enzymes and other phytochemicals.[3–5] However, the specific composition depends on many factors, such as the nectar composition of the source plant, the species of bee, the climate, environmental and seasonal conditions, agricultural practices and treatment of honey during extraction and storage.[6–9]

As a natural, unprocessed and easily digested food, honey plays an important role in the human diet, and is also used in pharmaceutical and cosmetic industries.[10]

Modern science has studied the medical significance of honeys in healing wounds and burns,[11] oncology care,[12] as well as its antioxidant and antimicrobial properties.[3,13,14]

DOI: http://dx.doi.org/10.1016/B978-0-12-411629-0.00013-1

Propolis

Propolis (bee glue) is a beehive product prepared by bees of the *Apis mellifera* species, using resinous substances collected from various plants.[15,16] These substances are mixed with the β-glycosidase enzyme of their saliva, partially digested, and added to bee wax to form the final product.

This product, commonly referred as a "natural antibiotic", has an important role in protecting, reinforcing and repairing the hives. In addition, any invaders are wrapped in propolis and die by asphyxia, then the bodies are preserved, which prevents the putrefaction of the honeycomb.[17–19] It is also a barrier to molds and yeasts, bacteria, viruses, and rain. Propolis is mainly composed of resins (50%), waxes (30%), essential oils (10%), pollen (5%), and other organic compounds (5%), among which are phenolic compounds, esters, flavonoids, terpenes, aromatic aldeydes and alcohols.[20–23] However, it is worth mentioning that this composition differs significantly with botanical and geographical origin and depends on climatic conditions, terrain, water availability and other environmental factors.[18,24–28]

It has been reported that this product has several pharmacological properties, such as antibiotic, anti-neoplastic, anti-inflammatory,[26] anti-diarrheatic, antimicrobial,[29,30] antiviral,[31] antioxidant,[32] hepatoprotective, cardiostatic,[33] antitumoral,[23,34] anti-ulcerus and immunostimulatory activity.[35,36] Other studies highlight its efficiency as an anti-HIV-1,[37] anti-neurodegenerative[38] and anti-tuberculosis agent.[39]

Due to these medicinal properties, propolis is generating increasing interest from the pharmaceutical, food and cosmetic industries, and it is currently being introduced in products for human consumption, such as drinks and food additives.[32,40]

PHENOLIC AND FLAVONOID COMPOUNDS

Honey

Phenolic compounds are one of the most important groups of plant secondary metabolites, comprising at least 8000 different known structures. Their chemical structures range from simple phenolic molecules to high molecular weight polymers of approximately 30000 Da.[41] These compounds are reported to exhibit anticarcinogenic, anti-inflammatory, anti-atherogenic, anti-thrombotic, immune modulating, analgesic, antineoplatic behavior, antiviral activity, antiulcer properties, hormone, antibiotic and cardiovascular activities, among others, and exert these functions as antioxidant activities.[42,43] In addition, flavonoids inhibit lipid peroxidation, platelet aggregation, capillary permeability and the activity of enzyme systems including cyclo-oxygenase and lipoxygenase.[10,44] The honey phenolics and the flavonoids present in honey are divided into three classes with similar structure; flavonols, flavones and flavanones. These are important due to their contribution to honey color, taste and flavor and also due to their beneficial effects on health. In honey, propolis and royal jelly, most of the phenolic compounds are present in the form of flavonoids, whose concentration depends on various factors, including the plant species used by the bees, the health of the plant, and seasonal and environmental factors.[6,7,45,46] So, the properties of honey are expected to vary with the composition of the active compounds it contains.[47] According to references 48 and 49, the principal flavonoids present in honey, propolis and royal jelly are flavonoles (quercetin, kaempherol, galangin, fisetin), flavanones (pinocembrin, naringin, hesperidin), and flavones (apigenin, acacetin, chrysin, luteolin).

According to one study, *p*-hydroxibenzoic acid, cinnamic acid, naringenin, pinocembrin and chrysin were the phenolic compounds present in most of the Portuguese honey samples analyzed by high-performance liquid chromatography with a diode array detector.[10] Since then many other studies have been performed to look at the total phenolics and flavonoids of honey from different regions and countries.[50–54] Luteolin, quercitrin and isorhamnetin were the flavonoids detected in Meliponinae honey (MH) from the Ecuadorian stingless bee.[55] Other studies analyzed local, unprocessed, multifloral honey[56] and observed that the main flavonoids were luteolin, quercetin, apigenin, kaempferol, isorhamnetin, acacetin, tamarixetin, chrysin, and galangin. In general, the results of a range of studies provide clear information about the the similarities of phenolic compounds found in honey samples. However, even though the results obtained by references 10, 50, 51, and 54 partly agree with other papers, there is great qualitative and quantitative variability in the flavonoid content of bee honeys from different geographical areas. It can be stated that more lipophilic flavonoids (pinocembrin, chrysin, genkwanin and tectochrysin) are found in propolis and beeswax; therefore they are regarded as not being related to the floral origin of honey.[57] The rest of the flavonoids are considered to originate from the nectar or the pollen and are therefore directly related to the botanical origin of the honey, which is entirely dependent on the nectar foraged by the bees.[58]

Propolis

The total quantity of polyphenols and flavonoids in propolis are important parameters for evaluating its quality and biological potential.[32] The specific phenolic composition of propolis is extremely dependent on the plants found around the hive, as well on the geographic and climatic characteristics of the location.[20] Existing studies on the chemical composition of European propolis have only determined and identified the phenolic compounds and flavonoids.[18,26] These compounds have been extracted using different solvents: water, methanol and ethanol. Higher concentrations of flavonoids were obtained in hydro-alcoholic extracts than in methanol or water solvents.[26] These results agree with the data presented in reference 32, which considers propolis from the northeast of Portugal. However, reference 59 obtained inferior values when analyzing propolis from the south of the same country. This discrepancy may be due to the great distances between the locations of origin, and differences in apicultural practices. In fact, the results obtained by reference 26 suggest that propolis from different places contains different concentrations of polyphenols.

It is generally accepted that propolis from temperate climatic zones, like Europe, North America and the non-tropical regions of Asia, originate mainly from the bud exudates of *Populus* species and their hybrids, and are rich in flavonoids, phenolic acids and their esters,[20,60] while propolis from tropical regions, where no poplars and birches exist, are rich in prenylated benzophenons, diterpenes and flavonoids.[61–63]

Other studies revealed that the caffeic acid phenethyl ester (CAPE) is a biologically active ingredient of propolis with several interesting biological properties, including enhancing apoptosis,[64] reducing metastasis,[65] and increasing radiation sensitivity[66] of cancer cells. Additionally, the propolis from tropical zones, in particular the southeastern region of Brazil, were shown to be rich in prenylated phenylpropanoids,[20] although non-typical flavonoids from "poplar type" propolis, such as kaempferide and isosakuranetin, have also been found.[67,68] Moreover, Cuban propolis has recently caught the attention of scientists because its peculiar enrichment in polyisoprenylated benzophenones makes it chemically distinct from both European and Brazilian bee glue.[69]

Recently, new compounds have been isolated from propolis, such as 3,5-diprenyl-4-hydroxycinnamic acid (artepillin C) from Brazilian green propolis, which may be a key immunomodulatory compound.[70] It was reported that CAPE and artepillin C are the two most immunopotent chemicals found in propolis.[71] Artepillin C, which is a low-molecular-weight phenolic compound, is uniquely found in Brazilian green propolis and is one of its main constituents.

ANTIMICROBIAL ACTIVITY IN THE TREATMENT OF INFECTED WOUNDS AND IN WOUNDS HEALING

Honey

It has been reported that the bacteria in wounds increases oxygen consumption and thus reduces the level of oxygen available to the wound tissues to a point where tissue growth is impaired.[72] The consequences of bacterial infection are non-healing wounds, an increase in wound size, the development of ulcers and abscesses, failure of skin grafts, inflammation, swelling, and pain.[73] Studies have revealed that more than 80 bacteria and yeasts are efficiently inhibited by honey.[74,75] Emerging evidence from clinical studies suggests that honey is at least as effective as conventional treatments for healing wounds, particularly in very refractory cases such as in diabetics, the elderly, and extensively burned patients.[76,77] Antibiotic resistance is a public health problem with severe consequences, such as the increasingly difficult treatment of chronic wounds. In this context, it is essential to find alternative treatments; hence bee products and their derivatives may be very useful in this area. Indeed, many studies have assessed the effectiveness of honey for healing wounds that are not responsive to conventional antibiotics and antiseptics.[78–81]

Several studies revealed that honey has a broad spectrum antibacterial activity at the concentrations used in dressings, with unreported microbial resistance in laboratory wounds.[82] In the last decade, the antimicrobial activity of honey (new, stored, heated, heated and stored and exposed to ultraviolet light) was tested at concentrations in aqueous dressing medium varying between 10 and 100% (w/v) for its activity against common human pathogens.[83,84] This biological activity was tested in acidic, neutral, and alkaline media. In one study,[83] surgical wounds were made on the dorsa of mice and infected with *Staphylococcus aureus* or *Klebsiella* sp. and then treated with local application of honey four times a day. Appropriate antibiotics were also used and compared with the control values. The application of honey four times a day was compared with the effect of conventional antibiotics in rats with bacterial conjunctivitis induced by *Escherichia coli*, *Proteus* sp.,

S. aureus, *Klebsiella* sp. *Haemophilus influenza* , and *Pseudomonas aeruginosa*. It was verified that the growth of all the isolates was completely inhibited by 30–100% honey concentrations, with *E. coli*, *P. aeruginosa*, and *H. influenza* being the most sensitive. The effects of storage on the biological activities of the product were investigated, and the study showed that heating at 80°C for 1 hour followed by 5 years storage decreased the antimicrobial activity of honey, but other conservation methods, such as exposure to ultraviolet light, increased its activity against some microorganisms.[83]

Studies have assessed the *in vitro* antimicrobial activity of different concentrations of honey ether extracts against *Candida albicans*, the most prevalent fungal pathogen in humans, using agar-well diffusion and broth micro-dilution methods.[56] Even though honey flavonoids inhibited the growth of the yeasts, they did not kill them, and did not directly affect their cytoplasmic membrane.

Honey has also been reported to be useful in the healing of infected postoperative wounds,[85] where its effectiveness combines with other useful characteristics; it is non-irritating, non-toxic, self-sterile, bactericidal, nutritive, and easily applied.[86] Therefore, honey has advantages as a wound dressing because it provides a moist healing environment and prevents bacterial growth even when the wounds are heavily infected. It is an effective means of rendering heavily infected wounds sterile quickly and of inhibiting the growth of antibiotic resistant strains of bacteria. Its viscosity acts as a further barrier against wound infection.[87]

In a clinical study involving 59 patients with wounds and ulcers,[88] most of which had failed to respond to conventional treatments, fresh honey was applied daily. The bacteria isolated from these wounds (*E. coli*, *S. aureus*, *Proteus mirabilis*, mixed coliforms, *Klebsiella* sp., and *Enterococcus faecalis*) were all susceptible to honey *in vitro*. Another clinical study comprising 50 patients with wounds compared the use of honey as a treatment with the antiseptic Savlon (15% cetrimide, 1.5% chlorohexidine gluconate).[89] Honey cleared 60% of the treated wounds in 6 days whereas Savlon cleared only 36% of wounds within the same period. Another study used honey on nine infants with large infected surgical wounds that had failed to heal with intravenous antibiotics, cleaning the wound with aqueous 0.05% chlorhexidine solution and applying fusidic acid ointment.[90] Marked clinical improvement was seen in all cases after five days of treatment with honey, and all wounds were closed, clean and free of infection after 21 days of this application. Other studies report that two randomized controlled clinical trials have compared honey with silver sulfadiazine ointment on partial-thickness burns. Both showed that honey gave better control of infection.[91] It was observed that honey allowed successful skin grafting in wounds infected with *Pseudomonas* which had not responded to other treatments.[79] Other authors report the efficiency of the use of honey in patients suffering from wound breakdown after operation for carcinoma of the vulva.[92]

It was observed that manuka honey promoted the healing in venous leg ulcers, giving effective de-sloughing and a lower incidence of infection than the control.[93] In a further study of the use of honey as a wound dressing, honey was evaluated in the treatment of wounds of various etiologies, including diabetic ulcers, and the results indicated that it promoted granulation and epithelization of the wounds, reduced odor and had a dehydrating effect.[87] There are reports of a reduction in edema and pain[94] and improvements in healing outcomes[93] being observed clinically in various types of wound.

A recently published study[82] assessed a series of geographically and chemically-defined New Zealand manuka, kanuka and manuka-kanuka blended honeys containing varying concentrations of methylglyoxal (MGO) and hydrogen peroxide to determine their effectiveness in inhibiting the growth of different species of bacteria (*Bacillus subtilis, S. aureus, E. coli and P. aeruginosa*). The author observed that the three manuka honey samples analyzed had the highest MGO contents, and were the most effective in inhibiting the growth of *B. subtilis, E. coli* and *S. aureus,* with all the observed levels of growth inhibition being similar. Low honey concentrations (1–4%) caused significant lag-phase extension and bacterial growth was completely inhibited once concentrations reached 8–16%. Therefore the author demonstrated the potency of natural honey as an antimicrobial wound dressing, and that a range of effects arise from the variety of active compounds present, which not only allow active honey to have a broad spectrum of activity, apparently free of undesirable side effects, but also reduce the potential for resistant microbial populations to evolve.

Propolis

The research reported in reference 95 indicates that propolis delays growth and progression of skin changes in the early stage of infection with Herpes simplex and does not cause cytotoxicity. It was reported that propolis extracts can be used locally in viral infections.[31] The study in reference 96 verified the synergism between ciprofloxacin and propolis in the treatment of experimental *S. aureus* keratitis. It was also suggested that the simultaneous use of propolis and antibiotics may reduce the acquisition of resistance, and consequently remove the need for more powerful antibiotics.[97] Propolis may also show synergistic effects with antimicrobial drugs, and its association with commercially disposable drugs is a field of interest in the development of new products by the pharmaceutical industry.[98] It was reported that propolis diminished the resistance of the bacterial wall to antibiotics (amoxicillin, ampicillin and cefalexin)[99] and acted synergistically with antibiotics acting on ribosomes (chloramphenicol, tetracycline and neomycin).[100] Another study showed that propolis included in an ointment base influences the healing process and stimulates the proliferation of keratinocytes, in a way that depends on the wound healing stage.[101] A study of propolis ointment demonstrated its superior wound healing effect in comparison with the classical treatment of silver sulfadiazine.[102] The possibility of treating burn lesions with propolis was reported in clinical studies.[103] In a previous study of the use of propolis to treat cutaneous erythema, it was found to induce severe collagenization, but the use of propolis in open wounds results in normal collagenization. This leads to the conclusion that propolis acts differently upon the collagenization process depending on the lesion type.[104]

OTHER APPLICATIONS OF HONEY AND PROPOLIS IN HUMAN HEALTH

Honey

Honey has been used as a traditional medicine for centuries, by different cultures and for the treatment of a range of disorders. Reference 105 opines that honey with a high level of antibacterial activity has the potential to reduce the risk of dental caries. In addition to the carioprotective effect of New Zealand manuka honey, Molan and co-workers have shown from his extensive work on the influence of honey on oral health that honey prevents dental plaque, gingivitis, and periodontics.[106] Other workers in different laboratories have also shown that honey is non-cariogenic or less cariogenic than sucrose, which may be due to the protective role of its constituents, among which are inorganic ions and other colloidal components.

The case of a patient with persistent fistulas, resistant to conventional therapy, that were completely healed after six months of treatment with honey has also been reported.[107] Recent studies[108] compared the clinical and mycological cure rates of a novel mixture consisting of honey and yogurt with local antifungal agents for treating patients with vulvo-vaginal candidiasis during pregnancy, and obtained very promising results.

Several authors have reported the antibacterial activity of honey against bacteria causing life-threatening infections in humans. The effectiveness of honey in killing *Proteus* spp, *S. aureus, E. coli* and *P. aeruginosa* bacterial strains involved in urinary tract infections, diarrhea, septicemia, diabetic foot ulcers, community acquired nosocomial infections and wound infections has been widely reported.[109–111] Other studies also highlight the activities of honey in killing *Mycobacterium* tuberculosis.[112] Another study tested a monofloral honey, and found that it successfully eradicated the bacterial flora in the conjunctival sac after seven days of continuous administration. The study involved patients in the perioperative period of cataract surgery or vitrectomy.[113] The results obtained indicate that honeydew honey may act as a prophylactic agent of endophthalmitis. Another study[114] tested the *in vitro* antiviral effect of manuka and clover honeys at concentrations ranging from 0–6% (w/v) against a clinical varicella zoster virus. Their results showed that both types of honey showed antiviral activity *in vitro* against this virus with an approximate EC50 = 4.5 % (w/v). Other studies reported the use of honey in treating diarrhea,[115] and found that it may effect repair of the intestinal mucosa damaged by the infection;[116] it was used to treat peptic ulcers and gastritis infected by *Helicobacter pylori*,[117] to treat Fournier's gangrene, and found to promote rapid healing,[118] especially in first and second degree burns.[119]

Propolis

The research carried out in reference 120 indicated the effectiveness of a propolis extract in reducing the growth of bacteria that belong to the red complex (*Fusobacterium nucleatum, Actinobacillus actinomycetemcomitans, Porphyromonas gingivalis, Prevotella intermedia, Campylobacter rectus* and *Tannerella forsythia*). The studies in references 121–123 confirmed the antibacterial properties of propolis in relation to pathogens of periodontitis. According to some authors,[121] the antibacterial effects are due to flavonoids, phenol acids, and their esters. The results in reference 124 showed the clinical effectiveness of a toothpaste and gel containing a 3% ethanolic extract of propolis in a group of patients with a greater risk of gingivitis caused by dental plaque. As propolis rinses the mouth, and propolis-based toothpastes stop the growth of pathogens of gingivitis and periodontitis, these materials seem to be promising not only as preventive but also as therapeutic agents.[122,125] The results obtained by reference 126 showed, however, that propolis extracts in concentrations that effectively reduce pathogenic organisms for periodontal diseases are cytotoxic for the gingival fibroblasts.

The first report of the biological activity of Portuguese propolis demonstrated that propolis displays effective antiproliferative activity against human renal cancer cells. Furthermore, a comparison of the *in vitro* responses of normal and malignant cells to propolis extracts revealed that cancerous cells were more sensitive in terms of growth inhibition than normal cells. The study[127] reported the main constituents of an aqueous extract of Brazilian green propolis from Southeast Brazil and the effects of the orally administered aqueous extract on distinct phases of wound healing (cellular recruitment and extracellular matrix deposition) in a murine sponge model. Their results showed a decrease in cellular recruitment after treatment with propolis. It has been reported that phenolic compounds exert a wide range of biological actions, such as antioxidant activity by scavenging free radicals.[101] Propolis has been found to exert indirect cytotoxic effects on malignant cancer cell lines by activating macrophages.[128] Its anti-inflammatory role and wound healing promoting properties have also been reported: studies[129–131] have examined the *in vitro* effect of a Chilean propolis ethanolic extract on human spermatozoa treated with benzo[a]pyrene and exogenous reactive oxygen species. The work demonstrates that the natural drug under investigation is able to protect genomic DNA from damage induced by benzo[a]pyrene, hydrogen peroxide and hydrogen peroxide in combination with adenosine 5V-diphosphate (ADP) and ferrous sulfate ($FeSO_4$), producing a significant reduction in intracellular oxidants. A propolis extract was shown to possess the capacity to protect sperm membranes from oxidative attack, suggesting that it may have a role in protection against male infertility.

CONCLUSIONS AND FUTURE PROSPECTS

According to the literature referred to in this chapter, beehive products are useful, not only in the treatment of infected wounds, but also in other medical conditions.

Even though some studies provide enough information and trials to hypothesize the introduction of beehive products on the clinical practice, it must be ensured that the quality parameters are met, for example, that they are free of microbiological contaminations and pesticides. Indeed, the general applicability of honey requires its safe preparation and is hampered by the insufficient knowledge of its composition and mechanisms of action, as well as lack of standardization. One possibility would be to apply the beehive products topically, initially as a complement to conventional therapy. In addition, the lack of reported resistance supports the use of honey and propolis as a new approach for treating multidrug resistant wounds.

Taking into account the current limitations, future studies must thoroughly characterize the biological activities of these products, by analyzing the underlying mechanisms of action and isolating the most beneficial compounds, mainly those related to preventing microorganisms developing drug resistance.

References

1. Codex Alimentarius. Revised Codex Standard for Honey, Codex STAN 12–1981, Rev. 1 (1987), Rev. 2. (2001).
2. EU. Council directive 2001/110/CE relating to honey. *Off J Eur Communities* 2002;**L10**:47–52.
3. Gomes S, Dias L, Moreira L, Rodrigues P, Estevinho L. Physicochemical, microbiological and antimicrobial properties of commercial honeys from Portugal. *Food Chem Toxicol* 2010;**48**:544–8.
4. Lazarevic JS, Palic RM, Radulovic NS, Ristic NR, Stojanovic GS. Chemical composition and screening of the antimicrobial and antioxidative activity of extracts of *Stachys* species. *J Serb Chem Soc* 2010;**75**:1347–59.
5. Erejuwa OO, Sulaiman SA, Wahab MSA. Honey: A novel antioxidant. *Molecules* 2012;**17**:4400–23.
6. Al-Mamary M, Al-Meeri A, Al-Habori M. Antioxidant activities and total phenolic of different types of honey. *Nutr Res* 2002;**22**:1041–7.
7. Yao L, Datta N, Tomás-Barberán FA, Ferreres F, Martos I, Singanusong R. Flavonoids, phenolic acids and abscisic acid in Australian and New Zealand *Leptospermum* honeys. *Food Chem* 2003;**81**:159–68.
8. Marchini LC, Reis VDA, Moreti ACCC. Composição fisico-química de amostras de pólen coletado por abelhas africanizadas *Apis mellifera* (Hymenoptera: Apidae) em Piracicaba, Estado de São Paulo. *Ciência Rural* 2006;**36**:949–53.
9. Iglesias A, Feás X, Rodrigues S, Seijas JA, Vázquez-Tato MP, Dias LG, et al. Comprehensive study of honey with protected denomination of origin and contribution to the enhancement of legal specifications. *Molecules* 2012;**17**:8561–77.
10. Estevinho L, Pereira AP, Moreira L, Dias LG, Pereira E. Antioxidant and antimicrobial effects of phenolic compounds extracts of northeast Portugal honey. *Food Chem Toxicol* 2008;**46**:3774–9.
11. Molan PC. Potential of honey in the treatment of wounds and burns. *Am J Clin Dermatol* 2001;**2**:13–9.
12. Bardy J, Slevin NJ, Mais KL, Molassiotis A. A systematic review of honey uses and its potential value within oncology care. *J Clin Nurs* 2008;**17**:2604–23.
13. Theunissen F, Grobler S, Gedalia I. The antifungal action of three South African honeys on *Candida albicans*. *Apidologie* 2001;**32**:371–9.
14. Akbulut M, Ozcan MM, Coklar H. Evaluation of antioxidant activity, phenolic, mineral contents and some physicochemical properties of several pine honeys collected from Western Anatolia. *Int J Food Sci Nutr* 2009;**60**:577–89.
15. Souza BM, Palma MS. Peptides from *Hymenoptera* venoms. In: Lima MH, Pimenta AMC, Martin-Eauclaire MF, Zingali R, Rochat H, eds. *Animal toxins: state of the art – perspectives in health and biotechnology*. Editora UFMG; 2009. pp. 345–67.
16. Santos VR. Propolis: alternative medicine for the treatment of oral microbial diseases. In: Sakagamai H, ed. *Alternative medicine*; 2012. pp. 133–69.
17. Salatino A, Teixeira E, Negri G, Message D. Origin and chemical variation of Brazilian propolis. *Evid Based Complement Alternat Med* 2005;**2**:33–8.
18. Falcão SI, Vilas-Boas M, Estevinho LM, Barros C, Domingues MR, Cardoso S. Phenolic characterization of Northeast Portuguese propolis: usual and unusual compounds. *Analyt Bioanalyt Chem* 2010;**396**:887–97.
19. Umthong S, Phuwapraisirisan P, Puthong S, Chanchao C. *In vitro* antiproliferative activity of partially purified *Trigona laeviceps* propolis from Thailand on human cancer cell lines. *BMC Complement Alternat Med* 2011;**11**:30–7.
20. Bankova VS, De Castro SL, Marcucci MC. Propolis: Recent advances in chemistry and plant origin. *Apidologie* 2000;**31**:3–15.
21. Gómez-Caravaca AM, Gómez-Romero M, Arráez-Román D, Segura-Carretero A, Fernández-Gutiárrez A. Advances in the analysis of phenolic compounds in products derived from bees. *J Pharm Biomed Anal* 2006;**41**:1220–34.
22. Moreira LL, Dias T, Dias LG, Rogão M, Da Silva JP, Estevinho LM. Propolis influence on erythrocyte membrane disorder (hereditary spherocytosis): a first approach. *Food Chem Toxicol* 2011;**49**:520–6.
23. Valente MJ, Baltazar AF, Henrique R, Estevinho LM, Carvalho M. Biological activities of Portuguese propolis: protection against free radical induced erythrocyte damage and inhibition of human renal cancer cell growth *in vitro*. *Food Chem Toxicol* 2011;**49**:86–92.
24. Moţ AC, Soponar F, Sârbu C. Multivariate analysis of reflectance spectra from propolis: geographical variation in Romanian samples. *Talanta* 2010;**81**:1010–5.
25. Cardoso S, Ribeiro M, Ferreira IL, Rego AC. Northeast Portuguese propolis protects against staurosporine and hydrogen peroxide-induced neurotoxicity in primary cortical neurons. *Food Chem Toxicol* 2011;**49**:2862–8.

26. Silva JC, Rodrigues S, Feás X, Estevinho LM. Antimicrobial activity, phenolic profile and role in the inflammation of propolis. *Food Chem Toxicol* 2012;**50**:1790–5.
27. Cheng H, Qin ZH, Guo XF, Hu XS, Wu JH. Geographical origin identification of propolis using GC-MS and electronic nose combined with principal component analysis. *Food Res Int* 2013;**51**:813–22.
28. Więckiewicz W, Miernik M, Więckiewicz M, Morawiec T. Review Article: does propolis help to maintain oral health? *Evid Based Complement Alternat Med* 2013;**2013**:1–8.
29. Lu LC, Chen YW, Chou CC. Antibacterial activity of propolis against *Staphylococcus aureus*. *Int J Food Microbiol* 2005;**102**:213–20.
30. Bruschi ML, Dota KFD, Consolaro MEL, Svidzinski TIE. Antifungal activity of brazilian propolis microparticles against yeasts isolated from vulvovaginal candidiasis. *Evid Based Complement Alternat Med* 2011;**2011**:1–8.
31. Schnitzler P, Neuner A, Nolkemper S, Zundel C, Nowack H, Sensch KH, et al. Antiviral activity and mode of action of propolis extracts and selected compounds. *Phytotherapy Research* 2010;**24**:20–8.
32. Moreira L, Dias LG, Pereira JA, Estevinho LM. Antioxidant properties, total phenols and pollen analysis of propolis samples from Portugal. *Food Chem Toxicol* 2008;**46**:3482–5.
33. Libério SA, Pereira AL, Araújo MJ, Dutra RP, Nascimento FR, Monteiro-Neto V, et al. The potential use of propolis as a cariostatic agent and its actions on mutans group streptococci. *J Ethnopharmacol* 2009;**125**:1–9.
34. Szliszka E, Czuba ZP, Domino M, Mazur B, Zydowicz G, Krol W. Ethanolic extract of propolis (EEP) enhances the apoptosis- inducing potential of TRAIL in cancer cells. *Molecules* 2009;**14**:738–54.
35. Orsatti CL, Missima F, Pagliarone AC, Bachiega TF, Búfalo MC, Araújo Jr JP, et al. Propolis immunomodulatory action *in vivo* on Toll-like receptors 2 and 4 expression and on pro-inflammatory cytokines production in mice. *Phytother Res* 2010;**24**:1141–6.
36. Nassar SA, Mohamed AH, Soufy H, Nasr SM, Mahran KM. Immunostimulant effect of Egyptian propolis in rabbits. *Scientific World Journal* 2012;**2012**:1–9.
37. Gekker G, Hu S, Spivak M, Lokensgard JR, Peterson PK. Anti-HIV-1 activity of propolis in CD4+lymphocyte and microglial cell cultures. *J Ethnopharmacol* 2005;**102**:158–63.
38. Chen J, Long Y, Han M, Wang T, Chen Q, Wang R. Water soluble derivative of propolis mitigates scopolamine-induced learning and memory impairment in mice. *Pharmacol Biochem Behav* 2008;**90**:441–6.
39. Yildirim Z, Hacievliyagil S, Kutlu NO, Aydin NE, Kurkcuoglu M, Iraz M, et al. Effect of water extract of Turkish propolis on tuberculosis infection in guinea-pigs. *Pharmacol Res* 2004;**49**:287–92.
40. Daugsch A, Moraes C, Fort P, Park Y. Brazilian red propolis – chemical composition and botanical origin. *Evid Based Complement Alternat Med* 2008;**5**:435–41.
41. Bravo L. Polyphenols: chemistry, dietary sources, metabolism and nutritional significance. *Nutr Rev* 1998;**56**:317–33.
42. Vinson JA, Hao Y, Su X, Zubik L. Phenol antioxidant quantity and quality in foods: vegetables. *J Agrict Food Chem* 1998;**46**:3630–4.
43. Uthurry CA, Hevia D, Gomez-Cordoves C. Role of honey polyphenols in health. *J ApiProduct ApiMed Sci* 2011;**3**:141–59.
44. Viuda-Martos M, Ruiz-Navajas Y, Fernández-López J, Pérez-Álvarez JA. Functional properties of honey, propolis, and royal jelly. *J Food Sci* 2008;**73**:117–24.
45. Ksouri R, Megdiche W, Debez A, Falleh H, Grignon C, Abdelly C. Salinity effects on polyphenol content and antioxodant activities in leaves of the halophyte *Cakile maritima*. *Plant Physiol Biochem* 2007;**45**:244–9.
46. Küçük M, Kolayli S, Karaoglu S, Ulusoy E, Baltaci C, Candan F. Biological activities and chemical composition of three honeys of different types from Anatolia. *Food Chem* 2007;**100**:526–34.
47. Bansal V, Medhi B, Pandhi P. Honey – a remedy rediscovered and its therapeutic utility. *Kathmandu Univ Med J* 2005;**3**:305–9.
48. Cushnie TPT, Lamb AJ. Detection of galangin-induced cytoplasmic membrane damage in *Staphylococcus aureus* by measuring potassium loss. *J Ethnopharmacol* 2005;**101**:243–8.
49. Fiorani M, Accorsi A, Blasa M, Diamantini G, Piatti E. Flavonoids from Italian multifloral honeys reduce the extracellular ferricyanide in human red blood cells. *J Agric Food Chem* 2006;**54**:8328–34.
50. Ferreira ICFR, Aires E, Barreira JCM, Estevinho LM. Antioxidant activity of Portuguese honey samples: different contributions of the entire honey and phenolic extract. *Food Chem* 2009;**114**:1438–43.

51. Gomes T, Feás X, Iglesias A, Estevinho LM. Study of organic honey from the Northeast of Portugal. *Molecules* 2011;**16**:5374–86.
52. Alvarez-Suarez-Suarez JM, Tulipani S, Diaz D, Estevez Y, Romandini S, Giampieri F, et al. Antioxidant and antimicrobial capacity of several monofloral Cuban honeys and their correlation with color, polyphenol content and other chemical compounds. *Food Chem Toxicol* 2010;**48**:2490–9.
53. Chang X, Wang J, Yang S, Chen S, Song Y. Antioxidative, antibrowning and antibacterial activities of sixteen floral honeys. *Food Funct* 2011;**2**:541–6.
54. Estevinho LM, Feás X, Seijas JA, Vazquez-Tato MP. Organic honey from Trás-Os-Montes region (Portugal): chemical, palynological, microbiological and bioactive compounds characterization. *Food Chem Toxicol* 2012;**50**:258–64.
55. Guerrini A, Bruni R, Maietti S, Poli F, Rossi D, Paganetto G, et al. Ecuadorian stingless bee (Meliponinae) honey: a chemical and functional profile of an ancient health product. *Food Chem* 2009;**114**:1413–20.
56. Candiracci M, Citterio B, Diamantini G, Blasa M, Accorsi A, Piatti E. Honey flavonoids, natural antifungal agents against *Candida albicans*. *Int J Food Prop* 2011;**14**:799–808.
57. Soler C, Gil M, García-Viguera C, Tomás-Barberán FA. Flavonoid patterns of French honeys with different floral origin. *Apidologie* 1995;**26**:53–60.
58. Vit P, Soler C, Tomás-Barberán FA. Profiles of phenolic compounds of *Apis mellifera* and *Melipona* spp honeys from Venezuela. *Eur Food Res Technol* 1997;**204**:43–7.
59. Miguel MG, Nunes S, Dandhlen SA, Cavaco AM, Antunes MD. Phenols and antioxidant activity of hydro-alcoholic extracts of propolis from Algarve, South of Portugal. *Food Chem Toxicol* 2010;**48**:3418–23.
60. Bankova V, Popova M, Bogdanov S, Sabatini A. Chemical composition of European propolis: Expected and unexpected results. *Zeitschrift für Naturforschung, Teil C* 2002;**57**:530–3.
61. Banskota AH, Tezuka Y, Adnyana IK, Ishii E, Midorikawa K, Matsushige K, et al. Hepatoprotective and anti-*Helicobacter pylori* activities of constituents from Brazilian propolis. *Phytomedicine* 2001;**8**:16–23.
62. Bankova V. Recent trends and important developments in propolis research. *Evid Based Complement Alternat Med* 2005;**2**:29–32.
63. Ahn MR, Kumazawa S, Usui Y, Nakamura J, Matsuka M, Zhu F, et al. Antioxidant activity and constituents of propolis collected in various areas of China. *Food Chem* 2007;**101**:1383–92.
64. Draganova-Filipova MN, Georgieva MG, Peycheva EN, Miloshev GA, Sarafian VS, Peychev LP. Effects of propolis and CAPE on proliferation and apoptosis of McCoy-Plovdiv cell line. *Folia Medica* 2008;**50**:53–9.
65. Liao HF, Chen YY, Liu JJ, Hsu ML, Shieh HJ, Liao HJ, et al. Inhibitory effect of caffeic acid phenethyl ester on angiogenesis, tumor invasion, and metastasis. *J Agric Food Chem* 2003;**51**:7907–12.
66. Chen YJ, Liao HF, Tsai TH, Wang SY, Shiao MS. Caffeic acid phenethyl ester preferentially sensitizes CT26 colorectal adenocarcinoma to ionizing radiation without affecting bone marrow radioresponse. *Int J Radiat Oncol Biol Phys* 2005;**63**:1252–61.
67. Marcucci MC, Bankova VS. Chemical composition, plant origin and biological activity of Brazilian propolis. *Curr Topics Phytochem* 1999;**2**:115–23.
68. Park YK, Alencar SM, Aguiar CL. Botanical origin and chemical composition of Brazilian propolis. *J Agric Food Chem* 2002;**50**:2502–6.
69. Cuesta-Rubio O, Frontana-Uribe BA, Ramırez-Apan T, Cardenas J. Polyisoprenylated benzophenones in Cuban propolis; biological activity of nemorosone. *Zeitschrift für Naturforschung* 2002;C57:372–8.
70. Cheung KW, Sze DM, Chan WK, Deng RX, Tu W, Chan GC. Brazilian green propolis and its constituent, artepillin C inhibits allogeneic activated human CD4 T cells expansion and activation. *J Ethnopharmacol* 2011;**138**:463–71.
71. Chan GC-F, Cheung K-W, Sze DM-Y. The immunomodulatory and anticancer properties of propolis. *Clin Rev Allergy Immunol* 2013;**44**:262–73.
72. Silver IA. The physiology of wound healing. In: Hunt TK, ed. *Wound healing and wound infection: theory and surgical practice*. New York: Appleton-Century-Crofts; 1980. pp. 11–28.
73. Molan P. Why honey is effective as a medicine. *Bee World* 2001;**82**:22–40.
74. Blair S, Carter D. The potential for honey in the management of wound and infection. *Australian Infection Control* 2005;**10**:24–31.

75. Irish J, Carter D, Shokohi T, Blair S. Honey has an antifungal effect against *Candida* species. *Med Mycol* 2006;**44**:289–91.
76. Kegels F. Clinical evaluation of honey-based products for lower extremity wounds in a home care settings. *Wounds UK* 2011;**7**:46–53.
77. Smaropoulos E, Romeos S, Dimitriadou C. Honey-based therapy for paediatric burns and dermal trauma compared to standard hospital protocol. *Wounds UK* 2011;**7**:33–40.
78. Dunford C, Cooper R, White RJ, Molan P. The use of honey in wound management. *Nursing Standard* 2000;**15**:63–8.
79. Robson V, Ward RG, Molan PC. The use of honey in split-skin grafting. *10th conference of the european wound management association*. Harrogate, UK; 2000.
80. Betts JA, Molan PC. A pilot trial of honey as a wound dressing has shown the importance of the way honey is applied to wounds. *11th conference of the european wound management association*. Dublin, Ireland; 2001.
81. Natarajan S, Williamson D, Grey JA, Harding KG, Cooper RA. Healing of an MRSA-colonised, hydroxyurea-induced leg ulcer with honey. *J Dermatol Treat* 2001;**12**:33–6.
82. Lu J, Carter DA, Turnbull L, Rosendale D, Hedderley D, Stephens J, et al. The Effect of New Zealand Kanuka, manuka and clover honeys on bacterial growth dynamics and cellular morphology varies according to the species. *PLoS ONE* 2013;**8**:1–12.
83. Al-Waili NS. Investigating the antimicrobial activity of natural honey and its effects on the pathogenic bacterial infections of surgical wounds and conjunctiva. *J Med Food* 2004;**7**:210–22.
84. Al-Waili N, Al-Ghamdi A, Ansari MJ, Al-Attal Y, Salom K. Synergistic effects of honey and propolis toward drug multi-resistant *Staphylococcus aureus, Escherichia coli* and *Candida albicans* isolates in single and polymicrobial cultures. *Int J Med Sci* 2012;**9**:793–800.
85. Al-Waili NS, Saloom KY. Effects of topical honey on post-operative wound infections due to gram positive and gram negative bacteria following caesarean sections and hysterectomies. *Eur J Med Res* 1999;**4**:126–30.
86. Bulman MW. Honey as a surgical dressing. *Middlesex Hosp J* 1953;**55**:188–9.
87. Efem SEE. Clinical observations on the wound healing properties of honey. *Br J Surg* 1988;**75**:679–81.
88. Khristov G, Mladenov S. Honey in surgical practice: the antibacterial properties of honey. *Khirurgiya* 1961;**14**:937–45.
89. Hamdy MH, El-Banby MA, Khakifa KI, Gad EM, Saven EM. *The antimicrobial effect of honey in the management of septic wounds. Proc 4th int conf on apiculture in tropical climates*, vol. **56**. El Cairo, International Bee Research Association; 1988.
90. Vardi A, Barzilay Z, Linder N, Cohen HA, Paret G, Barzilai A. Local application of honey for treatment of neonatal postoperative wound infection. *Acta Paediatrica* 1998;**87**:429–32.
91. Subrahmanyam M. A prospective randomized clinical and histological study of superficial burn wound healing with honey and silver sulfadiazine. *Burns* 1998;**24**:157–61.
92. Radwan SS, El-Essawy AA, Sarhan MM. Experimental evidence for the occurrence in honey of specific substances active against microorganisms. *Zentralbl Mikrobiol* 1984;**139**:249–55.
93. Gethin G, Cowman S. Manuka honey vs. hydrogel – a prospective, open label, multicenter, randomized controlled trial to compare de-sloughing efficacy and healing outcomes in venous ulcers. *J Clin Nurs* 2008;**10**:1365–2702.
94. Molan PC. The role of honey in the management of wounds. *J Wound Care* 1999;**8**:423–6.
95. Shimizu T, Takeshita Y, Takamori Y, Kai H, Sawamura R, Yoshida H, et al. Efficacy of Brazilian propolis against Herpes Simplex Virus type 1 infection in mice and their modes of antiherpetic efficacies. *Evid Based Complement Alternat Med* 2011;**2011**:1–9.
96. Oksuz H, Duran N, Tamer C, Cetin M, Silici S. Effect of propolis in the treatment of experimental *Staphylococcus aureus* keratitis in rabbits. *Ophtalmic Res* 2005;**37**:328–34.
97. Onlen Y, Duran N, Atik E, Savas L, Altug E, Yakan SA, et al. Antibacterial activity of propolis against MRSA and synergism with topical mupirocin. *J Altern Complement Med* 2007;**13**:713–8.
98. Sforcin JM, Bankova V. Propolis: Is there a potential for the development of new drugs? *J Ethnopharmacol* 2011;**133**:253–60.
99. Orsi RO, Sforcin JM, Funari SRC, Fernandes Jr A, Bankova V. Synergistic effect of propolis and antibiotics on the Salmonella Typhi. *Br J Microbiol* 2006;**37**:108–12.

100. Orsi RO, Fernandes A, Bankova V, Sforcin JM. The effects of Brazilian and Bulgarian propolis *in vitro* against Salmonella Typhi and synergism with antibiotics acting on the ribosome. *Nat Prod Res* 2012;**26**:430–7.
101. Sehn E, Hernandes L, Franco SL, Gonçalves CCM, Baesso ML. Dynamics of reepithelialisation and penetration rate of a bee propolis formulation during cutaneous wounds healing. *Anal Chim Acta* 2009;**635**:115–20.
102. Gregory SR, Piccolo N, Piccolo MT, Heggers JP. Comparison of propolis skin cream to silver sulfadiazine: a naturopathic alternative to antibiotics in treatment of minor burns. *J Alternat Complement Med* 2002;**8**:77–83.
103. Fearnley J. *Bee propolis: natural healing from the hive using propolis to treat disease*. London: Souvenir Press; 2001.
104. Andriţoiu CV, Prisăcaru AI, Andriescu C, Popa IM. Evaluation of the efficacy of some bee products upon the chemically induced cutaneous erythema in Wistar rats. *Ann Roman Soc Cell Biol* 2012;**17**:258–64.
105. Molan PC. The potential of honey to promote oral wellness. *Gen Dent J* 2001;**49**:584–9.
106. Molan PC, Betts JA. Clinical usage of honey as a wound dressing: an update. *J Wound Care* 2004;**13**:353–6.
107. Vlcekova P, Krutakova B, Takac P, Kozanek M, Salus J, Majtan J. Alternative treatment of gluteofemoral fistulas using honey: a case report. *Int Wound J* 2012;**9**:100–3.
108. Abdelmonem AM, Rasheed SM, Mohamed AS. Bee-honey and yogurt: a novel mixture for treating patients with vulvovaginal candidiasis during pregnancy. *Arch Gynecol Obstet* 2012;**286**:109–14.
109. Agbagwa OE, Frank-Peterside N. Effect of raw commercial honeys from Nigeria on selected pathogenic bacteria. *Afr J Microbiol Res* 2010;**4**:1801–3.
110. Chauhan A, Pandey V, Chacko KM, Khandal RK. Antibacterial activity of raw and processed honey. *Electronic J Biol* 2010;**5**:58–66.
111. Sherlock O, Dolan A, Athman R, Power A, Gethin G, Cowman S, et al. Comparison of the antimicrobial activity of ulmo honey from Chile and manuka honey against methicillin-resistant *Staphylococcus aureus, Escherichia coli* and *Pseudomonas aeruginosa*. *BMC Complementary Altern Med* 2010;**10**:1–5.
112. Hassanein SM, Gebreel HM, Hassan AA. Honey compared with some antibiotics against bacteria isolated from burn wound infections of patients in Ain Shams University hospital. *J Am Sci* 2010;**6**:301–20.
113. Cernak M, Majtanova N, Cernak A, Majtan J. Honey prophylaxis reduces the risk of endophthalmitis during perioperative period of eye surgery. *Phytother Res* 2012;**26**:613–6.
114. Shahzad A, Cohrs RJ. *In vitro* antiviral activity of honey against varicella zoster virus (VZV): a translational medicine study for potential remedy for Singles. *Translational Biomedical* 2012;**3**:1–7.
115. Haffejee IE, Moosa A. Honey in the treatment of infantile gastroenteritis. *Br Med J* 1985;**290**:1866–7.
116. Linnett P. Honey for equine diarrhea. *Control Therapy* 1996;**906**.
117. Dooley CP, Cohen H. The clinical significance of *Campylobacter pylori*. *Ann Intern Med* 1989;**108**:70–9.
118. Hejase MJ, Simonin JE, Bihrle R, Coogan CL. Genital Foumier's gangrene: experience with 38 patients. *Urology* 1996;**47**:734–9.
119. Burlando F. Sull'azione terapeutica del miele nelle usdoni. *Minerva Dennatologica* 1978;**113**:699–706.
120. Koo H, Gomes BPFA, Rosalen PL, Ambrosano GMB, Park YK, Cury JA. *In vitro* antimicrobial activity of propolis and *Arnica montana* against oral pathogens. *Arch Oral Biol* 2000;**45**:141–8.
121. Santos FA, Bastos EMA, Uzeda M, Carvalho MAR, Farias LM, Moreira ESA, et al. Antibacterial activity of Brazilian propolis and fractions against oral anaerobic bacteria. *J Ethnopharmacol* 2002;**80**:1–7.
122. Feres M, Figueiredo LC, Barreto IM, Coelho MH, Araujo MW, Cortelli SC. *In vitro* antimicrobial activity of plant extracts and propolis in saliva samples of healthy and periodontally-involved subjects. *J Int Acad Periodontol* 2005;**7**:90–6.
123. Koru O, Toksoy F, Acikel CH, Tunca YM, Baysallar M, Uskudar Guclu A, et al. *In vitro* antimicrobial activity of propolis samples from different geographical origins against certain oral pathogens. *Anaerobe* 2007;**13**:140–5.
124. Tanasiewicz M, Sucha-Nowak M, Dawiec M, Król W, Skaba D, Twardawa H. Influence of hygienic preparations with a 3% content of ethanol extract of brazilian propolis on the state of the oral cavity. *Adv Clin Exp Med* 2012;**21**:81–92.
125. Pereira EMR, Da Silva JLDC, Silva FF, De Luca MP, Ferreira EF, Lorentz TCM, et al. Clinical evidence of the efficacy of a mouthwash containing propolis for the control of plaque and gingivitis: a phase II study. *Evid Based Complement Alternat Med* 2011;**2011**:1–7.
126. Sonmez S, Kirilmaz L, Yucesoy M, Yücel B, Yilmaz B. The effect of bee propolis on oral pathogens and human gingival fibroblasts. *J Ethnopharmacol* 2005;**102**:371–6.

127. De Moura SAL, Negri G, Salatino A, Lima LDC, Dourado LPA, Mendes JB, et al. Aqueous extract of Brazilian green propolis: primary components, evaluation of inflammation and wound healing by using subcutaneous implanted sponges. *Evid Based Complement Alternat Med* 2011;**2011**:1–8.
128. Orsolic N, Basic I. Immunomodulation by water-soluble derivative of propolis: a factor of antitumor reactivity. *J Ethnopharmacol* 2003;**84**:265–73.
129. Moura SAL, Negri G, Salatino A, Lima LDC, Dourado LPA, Mendes JB, et al. Aqueous extract of Brazilian green propolis: primary components, evaluation of inflammation and wound healing by using subcutaneous implanted sponges. *Evid Based Complement Alternat Med* 2009;**2011**:1–9.
130. Tani H, Hasumi K, Tatefuji T, Hashimoto K, Koshino H, Takahashi S. Inhibitory activity of Brazilian green propolis components and their derivatives on the release of cys-leukotrienes. *Bioorg Med Chem* 2010;**18**:151–7.
131. Russo A, Troncoso N, Sanchez F, Garbarino JA, Vanella A. Propolis protects human spermatozoa from DNA damage caused by benzo[a]pyrene and exogenous reactive oxygen species. *Life Sci* 2006;**78**:1401–6.

CHAPTER

14

Necrotizing Soft Tissue Infections

Donald E. Fry

Department of Surgery, Northwestern University Feinberg School of Medicine, Chicago, Illinois, and the Department of Surgery, University of New Mexico School of Medicine, Albuquerque, New Mexico

INTRODUCTION

Soft tissue infections occur commonly and are frequent causes for patients to seek medical care.[1] These infections may be simple cellulitis for which only minimal treatment may be necessary, or they may be localized pyogenic infections. Most of these cutaneous infections follow cuts, punctures, or abrasions of the skin and a minority may evolve from infections of the hair follicle or glandular elements of the integument. Minor skin infections may assume a more severe nature, as are seen with erysipelas, and local tissue invasion may be seen with lymphadenitis or lymphangitis. In general, common soft tissue infections are confined to the skin, but pyogenic infections may extend into the subcutaneous fat especially if the inciting wound has extended through the entire thickness of the skin layers. The pathogens associated with these common soft tissue infections are *Staphylococcus aureus* or *Streptococcus pyogenes*.

On occasion, the infection arising from the skin and skin structures becomes a severe and life-threatening soft tissue event. These will be collectively called necrotizing soft tissue infections (NSTIs) in this presentation.[2] They have a wide continuum of clinical and pathologic presentation and have collectively been referred to by the misnomer of necrotizing fasciitis. The misnomer evolved from the clinical reality that while sometimes observed to be associated with necrosis of the investing fascia of the underlying muscle beneath the subcutaneous fat of the soft tissues, many of these infections are seen superficial to the fascia, and the fascial layer of tissue over the underlying skeletal muscle remains viable. The hallmark of NSTI is the rapid progression of the invasive infection, which is associated with necrosis of major components of the soft tissues and also with a robust systemic response of the host that constitutes clinical sepsis and even septic shock. The frequency of these infections has been steadily increasing (Figure 14.1). This chapter will highlight the pathogenesis, pathogens, and treatment strategies that are necessary to treat these infrequent but life-threatening infections.

Microbiology for Surgical Infections
DOI: http://dx.doi.org/10.1016/B978-0-12-411629-0.00014-3

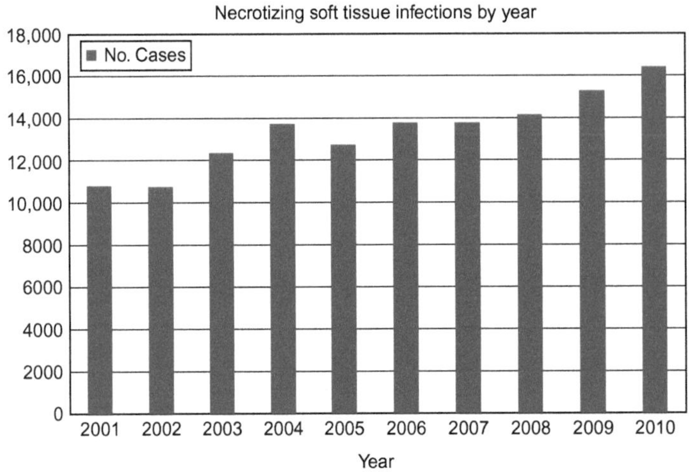

FIGURE 14.1 Illustrates the annual rates of necrotizing fasciitis infections among hospitalized patients in the United States from 2001–2010. *(Agency for Healthcare Research and Quality. www.ahrq.gov/hcupnet) Accessed October 20, 2013.*

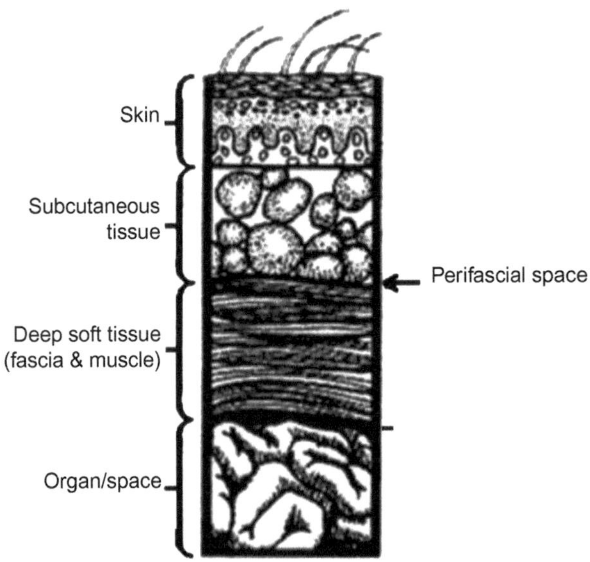

FIGURE 14.2 Illustrates the location of the perifascial space overlying the skeletal muscle. *(Adapted from Mangram, A.J., Horan, T.C., Pearson, M.L., Silver, L.C., and Jarvis, W.R. Guideline for Prevention of Surgical Site Infection, 1999. Centers for Disease Control and Prevention (CDC) Hospital Infection Control Practices Advisory Committee. Am J Infect Control 27, 97–132.)*

PATHOGENESIS OF NSTI

NSTI, like other soft tissue infections, commonly follows a mechanical wound that may be a laceration, cut, scrape, or small superficial burn. The critical determinant for the development of a necrotizing soft tissue infection is contamination of the tissue by a virulent organism, and access of the pathogen to the *perifascial* space. As is illustrated in Figure 14.2, the perifascial space is the potential and avascular tissue plane that resides immediately superficial to, and immediately beneath, the investing fascia of the muscle. This potential space can be validated by the gentle dissecting surgical finger being able to tease the subcutaneous fat from the superficial surface of the fascia without any appreciable blood loss. Similarly, if the fascia is incised at the time of an operation, the investing fascia can be easily separated from the surface of the skeletal muscle beneath with little or no bleeding.

Access of the developing NSTI to the perifascial space may occur because the inciting injury has directly penetrated to that anatomical level, or may be the consequence of vertical extension of a more superficial infection to the depth of this space. Puncture wounds from metal objects may penetrate to the depth of the fascia and result in the associated clostridial myonecrosis. Superficial wounds of the skin that occur over sites with very shallow subcutaneous fat, such as over the olecranon of the elbow or patellar surfaces, permit the evolving infection to rapidly disseminate laterally along this avascular potential space, over and beneath the fascia. In other select circumstances, the suboptimal host responsiveness of the patient or the unusually aggressive virulence of the contaminating pathogen results in the vertical invasion of a more superficial infection to the level of the perifascial space, with subsequent lateral invasion being the consequence.

The combination of acute microbial infection, usually with a particularly aggressive pathogen, and an impaired host results in rapid progression of infections laterally along this avascular anatomic plane. No anatomic partitions prevent the lateral progression. Invasive infection creates a robust inflammatory response that results in thrombosis of the nutrient vessels that supply the overlying skin and subcutaneous tissues. Importantly, the extent of the infection extends well beyond the perimeter of the evident surface bullae, blistering, cutaneous necrosis, or any clinical evidence that is identified over the site of infection. This rapid and lateral extension of the infection yields the characteristic finding of NSTI: *pain and tenderness that is out of proportion to the inciting injury or the clinically apparent infection!*

DIAGNOSIS

The diagnosis of NSTI is a clinical one.[3] Some have advocated pathological assessment of the subcutaneous tissues to establish the presence of necrosis, but frozen sections are notoriously untrustworthy. Permanent sections that may be more accurate in the definition of tissue necrosis require a time interval for completion that will be too long for clinical relevance. These infections progress rapidly, and the clinician must have a high index of suspicion when a clinical interpretation of pain and tenderness is disproportionate to the apparent cutaneous evidence of infection or injury. In thin patients with shallow subcutaneous fat, the induration of the advancing infection can be palpated and an advancing edge can be recognized by the discriminating examiner through careful examination. Soft tissue crepitus can be similarly palpated. Crepitus has commonly been associated with clostridial myonecrosis, but soft tissue crepitus can be seen occasionally with streptococci and with polymicrobial infections, especially when anaerobic pathogens are present. Infections that are superficial to the investing skeletal muscle fascia will commonly have erythema about the site of injury and with selected bacterial strains, such as Group A streptococci, the erythema will advance rapidly. When NSTI is suspected but not certain, it may be desirable to mark the perimeter of the erythema at the time of presentation so that rapid advancement can be readily recognized. Sloughed skin and blistering will be recognized but may be totally absent during the critical early period of the infection when intervention has the greatest prospect of success. When drainage or sloughing of skin is present, it represents an opportunity for a gram stain that can be very helpful in the identification of the responsible pathogen. Patients with obesity may have the infection quite deep from the skin surface and have minimal or no findings of erythema or skin changes until the infection is far advanced.

If the infection is deep to the muscle fascia, then extensive necrosis will often be present, but crepitus may not be palpable even when gas-forming pathogens are present. Inappropriate and severe pain that is enhanced by movement of the extremity, or clinical evidence of an evolving compartment syndrome is consistent with invasive infection beneath the fascia. Extensive infection from myonecrosis and increased compartment pressures may actually have the patient present with reduced pain sensitivity or frank anesthesia in extremity infections. Conventional roentgenograms may demonstrate deep tissue air, and computerized tomography or magnetic resonance imaging may define evidence of soft tissue edema or induration. Most often, the diagnosis of NSTI is made by the astute interpretation of the overall clinical findings and not by imaging studies. Confirmation in questionable cases is validated by surgical exploration of the area.

Laboratory tests are commonly employed in the evaluation of these patients. The patients may have marked leukocytosis with counts greater than 25,000 cells/mm^3, but may also have profound leukopenia when the disease is far advanced or especially if fulminate as is seen with clostridial or streptococcal NSTIs. The use of white blood cell counts, C-reactive protein, hemoglobin concentrations, serum sodium, serum creatinine, and serum glucose to create a scoring system for the diagnosis of NSTI has been proposed.[4] This scoring system has not been prospectively validated and is yet to be confirmed as being better than an experienced clinician in making this diagnosis.[5–7]

A common component of the evolving infection in these patients is the systemic inflammatory response syndrome (SIRS). The patients may be quite febrile with temperatures greater than 40°C. Paradoxically they curiously may have no febrile response at all, and very severe cases may present with hypothermia of less than 36°C. As noted above, the white blood count is also highly variable with dramatic leukocytosis or with leukopenia of less than 2,000 per mm^3. Peripheral immature white cell forms (e.g., "bands") are almost invariably present regardless of the patients' white blood cell count. A common finding in the NSTI patient is a severe metabolic acidosis that will feature very high concentrations of lactate. Septic shock will soon evolve in the untreated NSTI patient with severe lactic acidosis, and the clinical scenario requires immediate surgical intervention if the patient is to survive the infection. Toxic shock syndrome is used to describe the fulminate decompensation of the NSTI patient infected with specific strains of *S. pyogenes* or *S. aureus* that produce a specific exotoxin.[8,9] As soon as evidence of SIRS is present, the patients rapidly become hypoxemic and require ventilator support. Renal failure and evidence of hepatic injury commonly follow. The International Normalized Ratio (INR) will commonly be abnormal but may or may not be associated with a clinical coagulopathy in these patients.

RISK FACTORS

Common risk factors are associated with the occurrence of necrotizing soft tissue infection regardless of the bacterial type. The incidence of NSTIs by age group is illustrated in Figure 14.3. The mortality and morbidity of the disease is greater in patients over 70 years of age. Clinical conditions that are associated with poor tissue oxygenation or diminished tissue perfusion are particularly significant.[10,11] Thus, patients with congestive heart failure, severe cardiac arterial disease, and chronic lung disease are particularly vulnerable. Peripheral vascular arterial disease is a common predisposing risk factor for lower extremity necrotizing infection. Obesity has been cited as yet another predisposing clinical variable.

Clinical immunosuppression associated with advance HIV infection, corticosteroid treatments, or cancer chemotherapy increases the risk. Type 2 diabetics, obesity, chronic renal failure, and patients with hypoalbuminemia and dependent edema of any cause similarly have increased risks. A review of a reported series of patients with these infections identifies that 30–45% of patients actually have no identified risk factors and are likely to develop these infections because of exposure to pathogens of exceptional virulence, or because the presence of tissue necrosis, foreign bodies, and blood clot at the site of injury greatly enhances the advance of local infection.

Anatomic location appears to have different risks for NSTI. Lower extremities appear to be the most common location, with the upper extremities next in frequency. The predisposition of the lower extremities is likely due to the greater frequency of injuries associated with this area, and its poorer blood supply in elderly patients and those with diabetes. Truncal NSTIs are less common than the extremity infections, but may be secondarily involved if invasive infections of the extremities are not promptly managed. Head and neck NSTIs are least common, in all likelihood because of the better perfusion of these tissues. However, dental infections,[12] head and neck injuries and operations, cancers, and

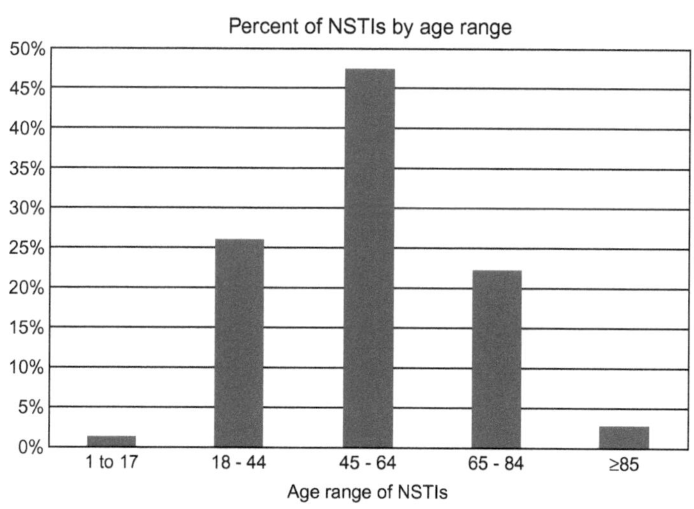

FIGURE 14.3 Illustrates the ages of patients with necrotizing fasciitis infections among hospitalized patients in the United States from 2001–2010. *(Agency for Healthcare Research and Quality. www.ahrq.gov/hcupnet) Accessed October 20, 2013.*

aggressive oropharyngeal infections[13] are all associated with these infections in the head and neck area.

PATHOGENS OF NSTI

Classification of necrotizing soft tissue infections is usually based upon the bacteriology of the infection. Infections are identified as monomicrobial or polymicrobial. The usual pathogens of monomicrobial infections are *Clostridium perfringens*, *S. pyogenes*, and *S. aureus*. Polymicrobial infections are the result of the mixed microflora that have contaminated the soft tissue from the index injury and may include gram-positive, gram-negative, and anaerobic species. In most reports, polymicrobial infections are the most common NSTIs.[14] However, the distribution of monomicrobial and polymicrobial infections may be highly variable among different institutions,[15] and one report identifies an increase in monomicrobial pathogens in the last decade.[16] In recent years, it has been this author's opinion that *C. perfringens* with typical "gas gangrene" has become increasing uncommon, while the evolution of the community-associated methicillin resistant *S. aureus* is a pathogen that is identified with increased frequency. Many believe that streptococcal necrotizing infections are increasing. In addition, the polymicrobial infections that occur among trauma patients may occur in injury or surgical wounds that are secondarily contaminated with hospital-acquired microflora, and accordingly, the occurrences of very resistant gram positive and gram negative bacteria participating in these infections appear to be increasing.

The prevention of NSTIs is very difficult, because many of these infections have already advanced to the invasive phase at clinical presentation, but it is important to recognize a severely contaminated wound as having the potential for invasive infection and to manage it accordingly. Severe soft tissue injuries with obvious environmental contamination require aggressive early removal of dead tissue, debris, and grossly apparent contamination to hopefully avoid the evolution of NSTI. NSTIs will complicate abdominal surgical wounds severely contaminated with colonic content secondary to traumatic or other (perforated diverticulitis) perforations. Delayed primary closure has been advocated as a preventive strategy to minimize the risks of necrotizing infection, although the effectiveness of this method remains controversial.

While specific therapeutic measures will be discussed with each type of necrotizing infection, it is important to emphasize that the common denominator for management is effective debridement of the primary focus of infection.[17] The severe advancing infection with its multitude of microbial toxins and the influence of severe tissue inflammation results in necrotic tissue that requires debridement, and tissue suppuration that needs to be drained. Surgical intervention remains the mainstay of managing these patients regardless of the causative pathogen. Antibiotic therapy specific to the causative pathogen is also important in hopefully limiting the advancement of the infection into adjacent tissues (Table 14.1). Because the systemic inflammatory response may be quite robust in these patients, most will require aggressive systemic support to maintain oxygenation and tissue perfusion. Thus, volume support, ventilator-assistance, and inotropic drugs are commonly

TABLE 14.1 Potential Antibiotic Choices for the Treatment of Monomicrobial and Poly Microbial NSTIs. Dosing is Recommended by the Author. All Dosages are for Adults

Pathogen	Antibiotic Choice	Comment
Clostridium myonecrosis	Penicillin 2 million units every 2 hours; Clindamycin 1200–1500 mg Q6–8 hrs. If penicillin allergic, use vancomycin 30 mg/kg/daily in two doses.	Aggressive dosing strategy because of systemic inflammatory response. Clindamycin reduces toxin production.
Streptococcus pyogenes	Penicillin 2 million units every 2 hours; Clindamycin 1200–1500 mg every 6–8 hrs. If penicillin allergic, use vancomycin 30 mg/kg/daily in two doses.	Aggressive dosing strategy because of systemic inflammatory response. Clindamycin reduces toxin production.
Methicillin-sensitive *Staphylococcus aureus* (MSSA)	Nafcillin 1–2 grams every 4 hours; or, Cefazolin 1–2 grams every 6–8 hours.	Aggressive dosing is recommended. Dosing should be de-escalated once the systemic inflammatory response is resolving.
Methicillin-resistant *Staphylococcus aureus* (MRSA)	Vancomycin 30 mg/kg/day in two doses; or linezolid 600 mgs every 12 hours, or daptomycin 6 mg/kg/daily.	Presumptive coverage of MRSA is employed. De-escalated if MSSA.
Polymicrobial Infection: Use all three if cultures are not available		
Gram Positive coverage	Vancomycin 30 mg/kg/day in two doses; or linezolid 600 mgs every 12 hours, or daptomycin 6 mg/kg/daily.	Presumptive coverage of MRSA is employed. De-escalated if MSSA.
Gram Negative coverage	Piperacillin/Tazobactam 3.375–4.5 grams every 6 hours; or imipenem 1 gram Q 8 hours, or Meropenem 1 gram every 8 hours, or cefepime 1–2 gm every 8 hours.	Treatment will need modification after culture and sensitivity results are available.
Anaerobic coverage	Clindamycin 1200–1500 mg every 6–8 hours, or metronidazole 500 mg every 6–8 hours.	Clindamycin may offer reduction of toxic production from sensitive organisms.

required in these patients, to give the surgical debridement and antimicrobial therapy an opportunity to stem the systemic effects of the infection.

CLOSTRIDIAL NECROTIZING INFECTION

Clostridial myonecrosis fortunately remains an uncommon infection of the soft tissue. It is customarily associated with *C. perfringens*. It is even more rarely seen with the non-perfringens subgroups of clostridial species. The clostridial bacteria normally occur as a spore in nature and are ubiquitous in distribution. The spores are seldom identified as even transient colonists of human skin and are likely introduced because the wounding object was contaminated, hence resulting in infection; this explains the association with foot injuries from rusty nails. *C. perfringens* is an obligate anaerobe, and the oxygen content of the normal environment and of normal tissues will not favor the transition of the spore to the vegetative phase of the bacteria which cause clinical disease. Thus, the wounding process must not only introduce the spore into the tissue, but there also needs to be hematoma or necrotic tissue present from the injury to provide an appropriate local environment to promote the infection in otherwise well vascularized skeletal muscle. Once a small local infection is initiated, the vegetative phase of the bacteria produces a potent array of exotoxins which kill adjacent tissues. The cycle of exotoxins, tissue death, and advancing microbial proliferation into the newly dead muscle and soft tissues results in a rapidly advancing infection with extensive necrosis. The cycle from injury to a fulminate septic shock and imminent death may require only 24 hours. Many patients will have advanced to the point of "no return" at the time of clinical presentation, either because they are delayed in seeking medical attention, or their transfer from one institution to another results in far advanced disease.

While the fulminate clostridial myonecrosis is the commonly recognized form of *C. perfringens* soft tissue infection, there is a less virulent and less aggressive clostridial infection that is superficial to the fascia within the subcutaneous fat tissues. A similar pathophysiology exists, in which hematoma or necrotic tissue must be present to promote the infection, but the fatty tissue provides a poorer nutrient supply and the diffusion of toxins is much slower in the fatty environment. A locally aggressive infection that is associated with a cellulitis-like presentation of the overlying skin and palpable crepitus is commonly present. These patients have far fewer systemic manifestations of the fulminate myonecrosis infection. These superficial infections, like the classic myonecrosis, require adequate local debridement, but are found to be superficial to the muscle fascia and have a far more favorable prognosis with appropriate treatment. On occasion, the infection that is superficial to the investing muscle fascia will have a rapidly advancing character in the perifascial space like that seen with streptococcal NSTIs, but fulminate clostridial infections are typically myonecrotic.

The typical physical findings of clostridial myonecrosis are important to recognize, but are inconsistently present in an infection that itself occurs very infrequently. "Dish water" drainage at the injury site and palpable crepitus are important, but may not be present in deep muscle infection. There may be extreme leukocytosis or leucopenia, thrombocytopenia, coagulopathy, acute kidney injury, liver enzyme abnormalities, and myoglobinemia/myoglobinuria may all be present. The "port wine" appearance of the urine at the time of presentation of a septic patient should create the suspicion of clostridial or streptococcal myonecrosis. Cultures of wound drainage and blood cultures are usually positive but yield results that are too delayed temporally to affect the clinical decision. The gram stain of drainage with gram positive rods is of great value in establishing the diagnosis (Figure 14.4).

Once the diagnosis is made, the patient is ideally managed aggressively with volume resuscitation to blood pressure and urine output endpoints. However, in advanced cases the blood pressure will remain volatile, and urine output may not be re-established until aggressive surgical therapy is undertaken. *C. perfringens* is very sensitive to penicillin G and a strategy of 24 million units given every 24 hours is used after a large initial dose before the operating room. Clindamycin in a dose of 1200–1500 mgs every 6–8 hours is used to inhibit toxin production by the pathogen. Clindamycin alone is a good choice for penicillin-allergic patients. All antibiotics are administered rapidly as the patient is moved to the operating room for definitive treatment.

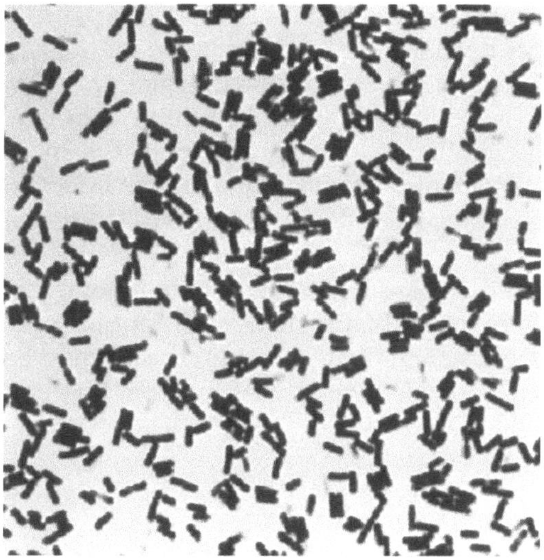

FIGURE 14.4 A photomicrograph of *Clostridium perfringens*, 1000× magnification. *(From the Public Health Image Library, Centers for Disease Control, Content provider: CDC/Don Stalons. http://phil.cdc.gov/phil/details.asp.) Accessed October 20, 2013.*

The operating room debridement begins with an incision over the inciting injury. Limited infection or necrosis is usually encountered until the muscle fascia is open. The dead muscle is usually black and all must be excised. Dead muscle compartments in the distal extremities commonly require amputation if survival is to be achieved. Debridement onto the trunk requires excision of muscle that may appear questionably alive but does not contract. The wounds are packed open with topical antiseptics and are treated like large burn wounds with silver sulfadiazine. In this author's opinion, patients that survive the initial 24 hours following the initial debridement should be returned to the operating room daily until no dead tissue is seen. Antibiotics are continued until clinical sepsis resolves and granulation tissue is identified in the wound. Immunoglobulin preparations have been used but are not of proven value.

Considerable interest has focused upon the use of hyperbaric oxygen therapy for these patients.[18] Prospective data does not validate the use of hyperbaric oxygen despite commercial claims of benefit.[19] Hyperbaric oxygen will not penetrate the dead muscle, and it causes a generalized vasoconstriction that may compromise the vasodilated state of the septic patient that is associated with an increased cardiac output. Patients on ventilators pose serious logistic issues in the use of hyperbaric oxygen. Additional evidence is necessary before this mode of treatment can be recommended for the clostridial NSTI patient.

Clostridium septicum and *Clostridium sordellii* are lesser known clostridial pathogens that are associated with unusual soft tissue infections.[20] These organisms are associated with seeding of muscle and soft tissue injuries from a remote, blood-borne source. These uncommon infections are usually identified in areas where a non-penetrating bruise or contusion has occurred, with the local edematous injury serving as a vulnerable site for a subclinical inoculum of blood-borne clostridial to lodge and cause a life-threatening NSTI. *C. septicum* myonecrosis has been the most common NSTI reported by this mechanism. The portal of entry of the bacterium is thought to be the colon from either colon cancer or diverticulitis with metastatic infection to a distant muscle site. The patients present with increasing pain and tenderness, commonly in a lower extremity. Systemically the patients have all of the signs and symptoms of an NSTI as detailed above, but they do not have a penetrating injury to explain the genesis of the infection. Because of the unusual presentation of an uncommon disease, the diagnosis is commonly mistaken until too late. Aggressive debridement and antibiotic coverage similar to that employed in the more common *C. perfringens* infections are necessary. Patients surviving the initial infection require a gastrointestinal evaluation to define the portal of entry of the clostridial pathogen.

Similarly, metastatic *C. sordellii* myonecrotic NSTIs are associated with myometrial infections following medical or self-induced interruption of pregnancy. *C. sordellii* may also be identified in intravenous drug abusers from contaminated injections and may be associated with NSTI at the injection site or at remote skeletal muscles sites from injection of blood-borne pathogens.[21]

STREPTOCOCCAL NECROTIZING INFECTION

Streptococcal necrotizing soft tissue infection may involve only the skin and subcutaneous tissue, it may cause a necrotizing myositis, and it may involve all components of soft tissues at the same time.[22] These infections are associated with superficial wounds and small burns. A troublesome association of streptococcal NSTIs has been seen in chicken pox, where children may introduce streptococci by scratching the pruritic vesicles.[23] Postoperative infections are seen following inguinal hernia[24] and even at laparoscopic trochar sites.[25] Infections following injury can occur at any anatomic site but are most common in the extremities. Metastatic streptococcal NSTIs with myositis/myonecrosis can be identified without any injury at the site.[26] Contusions of soft tissue, joint effusions, and old scars can be sites to be seeded by presumed subclinical pharyngeal bacteremia with *S. pyogenes*. The author has seen streptococcus seed infection during an episode of lower extremity NSTI that hematogenously spread to an infiltrated IV site of the upper extremity.

The diagnosis is made by the identification of pain and tenderness that is dramatically out of proportion to the inciting injury. Marked cutaneous and rapidly advancing erythema, dramatic tissue edema, lymphangitis and lymphadenitis, and all of the systemic signs and symptoms of the human septic response are recognized. Selected organisms will produce superantigens that are extremely effective in activating the toxic shock syndrome from streptococci which results in the patients with evolving NSTI rapidly progressing to a severe hypotensive state.[27] Drainage from the wound will have gram positive cocci in chains when examined by the gram stain (Figure 14.5). Blood cultures are usually positive but the results are available too late to be clinically relevant in making the diagnosis. Like all NSTIs, streptococcal infections require diagnosis based upon a high index of suspicion in the clinical presentation.

Like clostridial myositis, streptococcal NSTIs require rapid volume support, systemic antibiotics, and surgical debridement. These patients are typically massively vasodilated and require crystalloid administration as rapidly as possible while they are being mobilized to the operating room. Antibiotics of penicillin G and clindamycin are given in the

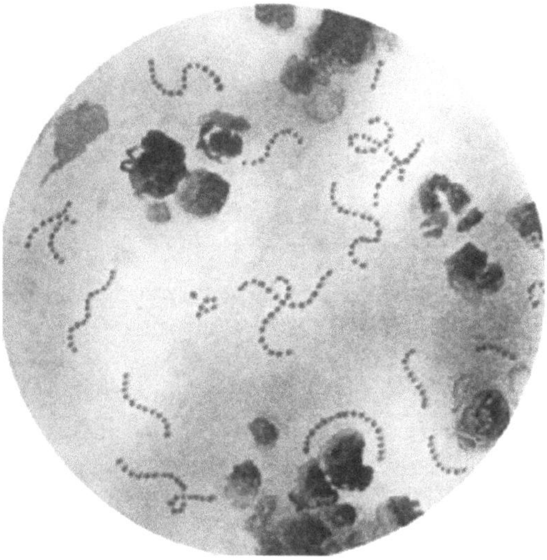

FIGURE 14.5 A photomicrograph of *Streptococcus pyogenes*, 900× magnification. *(From the Public Health Image Library, Centers for Disease Control, Content provider: CDC. http://phil.cdc.gov/phil/details.asp.) Accessed October 20, 2013.*

same aggressive fashion as with clostridial myonecrosis. At operation, a tan inflammatory exudate is identified, extending along the surface of the investing fascia, but the fascia is commonly glistening and very much alive. The subcutaneous fat is dead or dying in the areas beneath all of the skin areas involved with erythema, and all of it needs to be excised. Avoiding the excision of all dead tissue in the interest of future reconstruction of the area is generally a mistake. With myonecrosis beneath the fascia, all dead tissue needs to be debrided. There are variants of streptococcal infection in muscle where invasive infection occurs over the surface and between the muscle groups but does not actually result in overt myonecrosis of the muscle groups. These infections are managed by draining all infected material and vigorous irrigation of the tissues. The viability of the areas of muscle can be difficult to ascertain and the responsiveness of the muscle contraction response to stimulation may indicate viability. As with all debridments of NSTIs, a return to the operating room every 24 hours is necessary until only viable tissue remains.

A very difficult area with streptococcal infections of the extremities is the need for amputation. If the muscle is spared or the debridement only requires excision of superficial amounts of muscle, then preservation of the extremity can be attempted. If an entire muscle group is necrotic, then amputation above the level of the completely necrotic muscle group is necessary. The decision to amputate is difficult, and the surgeon should not be caught in the trap of trying to salvage an extremity at the expense of risking the patient's life.

An area of considerable interest in severe invasive Group A streptococcal infection is the use of intravenous immunoglobulin for treatment.[28] The concept is that active immunoglobulin would bind superantigens and perhaps other exotoxins released by the streptococcal pathogen and avoid the severe inflammatory cascade that is seen in many of these infections. While there is experimental work to encourage this line of research, clinical evidence of efficacy is not present as yet. If immunoglobulin is to work, it certainly will need to be administered at the earliest possible time in the evolution of these infections.

Superantigen that is already bound to the lymphocyte target cells will not likely be neutralized by the belated administration of immunoglobulin.

Like clostridial NSTIs, there has been interest in using hyperbaric oxygen therapy. Evidence is anecdotal with good evidence not available.

STAPHYLOCOCCAL NECROTIZING INFECTION

Staphylococcal infections have traditionally been viewed as being very pyogenic and locally destructive, but they have not been associated with NSTIs. However, recent reports have identified an increase in frequency, and staphylococci may now be the leading cause of monomicrobial NSTIs.[29] This increase in NSTIs due to staphylococci is being attributed to the uniquely virulent characteristics of community-acquired methicillin resistant *Staphylococcus aureus* (CA-MRSA).[30] These infections are associated with the same risk factors that were defined earlier for all patients with NSTIs. The staphylococcal infections appear to evolve from conventional pyogenic soft tissue infections, but presumably access the perifascial space and become laterally invasive. Staphylococcal necrotizing infection has the same signs and symptoms as streptococcal infections, but its progression is clinically slower than streptococcal or clostridial infection. Blood borne seeding of non-penetrating injury sites with subsequent NSTI infection has been reported, and presumably follows similar pathologic processes to those seen in streptococcal infection.[31] Staphylococcal infections are also identified with metastatic spread to remote sites including heart valves.

The diagnosis of staphylococcal NSTI has the same characteristic findings as the infections seen with other bacteria. Identification of the specific organism is usually easily done by culture of pus, exudates, or tissue, but like other infections treatment decisions need to be made before the culture results are available. Gram stains will identify gram-positive cocci in clusters that have a distinctly different appearance to streptococci (Figure 14.6). Initial antibiotic coverage should include coverage for CA-MRSA.

Surgical debridement is the treatment. Like other NSTIs, the debridement includes removal of all necrotic tissue, drainage of pockets of suppuration, and should be extended until only bleeding and viable tissues remain. Reoperation under general anesthesia is usually required, although these infections are less aggressive after effective initial debridement and bedside debridement may be considered in selected cases. Silver sulfadiazine or mafenide acetate management of the open wound is employed.

Systemic antibiotics are employed (Table 14.1). All staphylococcal infections should be assumed to be methicillin resistant, and intravenous vancomycin, linezolid, or daptomycin are appropriate choices. Culture results need to dictate whether alternative antibiotics are necessary. If the infection is with methicillin-sensitive bacteria, then the antibiotics can be changed to an expanded-spectrum penicillin or a cephalosporin antibiotic. If oral antibiotic therapy is continued in the patient with CA-MRSA, then oral trimethoprim-sulfamethoxazole or clindamycin may be used as indicated by sensitivities from the culture results. Antibody therapy for staphylococcal infections continues to be pursued, but the constantly changing antigenic character of these pathogens has made a successful immunotherapy elusive.[32]

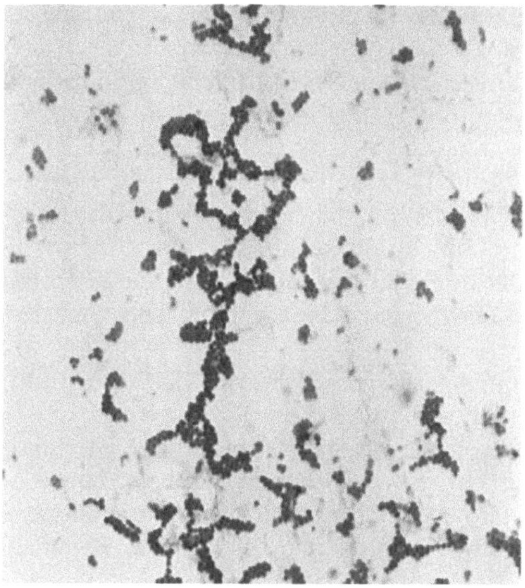

FIGURE 14.6 A photomicrograph of *Staphylococcus aureus*, 250× magnification. *(From the Public Health Image Library, Centers for Disease Control, Content provider: CDC/Richard Facklam, Ph.D. http://phil.cdc.gov/phil/details.asp.) Accessed October 20, 2013.*

POLYMICROBIAL NECROTIZING INFECTIONS

Polymicrobial NSTIs include a heterogeneous group of infections that involve the trunk, extremities, and other areas. These NSTIs will be seen in traumatic and surgical wounds that have been exposed to major colonic or environmental contamination. Farming accidents are notorious for polymicrobial infections. Diabetic foot ulcers are commonly thought to originate with staphylococcal infections, but after repeated hospitalizations and multiple courses of antibiotics, they become polymicrobial infections. Pressure ulcers of the sacrum and pelvis are exposed to colonic bacteria and will be the source of necrotizing infection. Fournier's gangrene of the scrotum is seen in diabetic patients and other vulnerable hosts that complicate perirectal abscesses, hemorrhoidectomy, and even inguinal hernia procedures.[33]

These infections begin as localized soft tissue infections but become NSTIs when the infection extends into the perifascial space. Very contaminated wounds may be primarily closed and infection in the confined space facilitates extension into the perifascial space. Similarly, even open wounds with massive contamination evolve from local pyogenic infection into progressive and advancing NSTI into adjacent tissues.

The hallmark of diagnosis is recognition of the clinical setting where polymicrobial contamination is present and the patient is at risk of infection. The advance of polymicrobial NSTI is usually associated with a slower progression into adjacent tissues than with monomicrobial NSTI, but is nevertheless associated with necrotic fascia and muscle. The infection will commonly burrow along the fascial plane and may leave viable skin and some subcutaneous fat above the deeper necrosis. Definition of the pathogenic bacteria requires deep tissue specimens for culture and sensitivity. Gram-negatives and anaerobes are associated with major colonic contamination of the soft tissues, staphylococci and anaerobic streptococci with Meleney's synergistic gangrene, resistant hospital-associated gram-negatives and methicillin resistant staphylococci with decubitus ulcers and diabetic fetid feet, and a varying array of gram-positive, gram-negative, and anaerobic bacteria in Fournier's gangrene. Polymicrobial NSTIs may have four or more pathogens cultured at the site of infection.[34,35]

Treatment is complete debridement of dead tissue. Because the polymicrobial necrosis has a more indolent character than the spectacular infections of the monomicrobial infections, the tendency is to cheat on the debridement, which only results in more dead tissue 24–48 hours later. In the abdominal wall, the debridement cannot be abbreviated in the interest of concerns about reconstruction. Debridement must be complete and repeated debridement sessions are essential for success.

Antibiotics need to cover all of the participating pathogens (Table 14.1). Cultures are essential and initial therapy without culture guidance requires a very broad spectrum of coverage. Gram-positive, gram-negative, and anaerobic coverage are important. The ubiquitous presence of methicillin resistant staphylococci requires inclusion of this pathogen in the initial treatment choices. Once sensitivity data are available, the antibiotic therapy needs to be de-escalated to cover only the identified pathogens.

OTHER PATHOGENS

Infection that gains access to the perifascial space can result in a necrotizing soft tissue infection. Accordingly, any bacterial pathogen is potentially a culprit. Salt water injuries of the feet are associated with *Vibrio* sp.[36] and freshwater recreational injuries with *Aeromonas hydrophilia*.[37] Severe *Bacillus cereus*[38] and even fungal soft tissue infections[39,40] can be seen depending upon the circumstances of injury to the soft tissues and the adequacy of host defenses in the patient. Common gram-negative bacteria can likewise be pathogens of NSTI.[41,42] It is important to emphasize that NSTI is a clinical diagnosis that is not specific to only a limited number of bacterial pathogens.

CONCLUSIONS AND FUTURE PROSPECTS

In summary, NSTI infections are surgical diseases with many common clinical appearances. Mortality rates for NSTIs are identified as greater than 50% for *C. perfringens*, 20–30% for streptococcal infections, 10–20% for staphylococci, and about 10% for polymicrobial infections.[43,44] The overall frequency of these infections appears to be increasing, and clinicians must have a high index of suspicion for these infections. An organization has been formed to increase public awareness of NSTIs.[45] The treatment is the same regardless of the bacteriology. Antibiotic therapy is initially broad in spectrum if there is uncertainty about the pathogen, and is then de-escalated after culture results are available. Repeated debridement of the infection site is virtually always necessary.

References

1. Fry DE. Skin, skin structure, and soft tissue Infections. In: Fry DE, ed. *Surgical Infections*. London: JP Medical Publishers; 2013. pp. 63–74 [Chapter 5].
2. Sultan HY, Boyle AA, Sheppard N. Necrotising fasciitis. *BMJ* 2012;**345**:e4274.
3. Lancerotto L, Tocco I, Salmaso R, Vindigni V, Bassetto F. Necrotizing fasciitis: classification, diagnosis, and management. *J Trauma* 2012;**72**:560–6.
4. Wong CH, Khin LW, Heng KS, Tan KC, Low CO. The LRINEC (Laboratory Risk Indicator for Necrotizing Fasciitis) score: a tool for distinguishing necrotizing fasciitis from other soft tissue infections. *Crit Care Med* 2004;**32**:1535–41.
5. Sarani B, Strong M, Pascual J, Schwab CW. Necrotizing fasciitis: current concepts and review of the literature. *J Am Coll Surg* 2009;**208**:279–88.
6. Wilson MP, Schneir AB. A case of necrotizing fasciitis with a LRINEC score of zero: clinical suspicion should trump scoring systems. *J Emerg Med* 2013;**44**:928–31.
7. Swain RA, Hatcher JC, Azadian BS, Soni N, De Souza B. A five-year review of necrotising fasciitis in a tertiary referral unit. *Ann R Coll Surg Engl* 2013;**95**:57–60.
8. Brosnahan AJ, Schlievert PM. Gram-positive bacterial superantigen outside-in signaling causes toxic shock syndrome. *FEBS J* 2011;**278**:4649–67.
9. Lappin E, Ferguson AJ. Gram-positive toxic shock syndromes. *Lancet Infect Dis* 2009;**9**:281–90.
10. Mills MK, Faraklas I, Davis C, Stoddard GJ, Saffle J. Outcomes from treatment of necrotizing soft-tissue infections: results from the national surgical quality improvement program database. *Am J Surg* 2010;**200**:790–6.
11. Huang KF, Hung MH, Lin YS, Lu CL, Liu C, Chen CC, et al. Independent predictors of mortality for necrotizing fasciitis: a retrospective analysis in a single institution. *J Trauma* 2011;**71**:467–73.
12. Subhashraj K, Jayakumar N, Ravindran C. Cervical necrotizing fasciitis: an unusual sequel of odontogenic infection. *Med Oral Patol Oral Cir Bucal* 2008;**13**:E788–91.
13. Lanisnik B, Cizmarevic B. Necrotizing fasciitis of the head and neck: 34 cases of a single institution experience. *Eur Arch Otorhinolaryngol* 2010;**267**:415–21.
14. Bernal NP, Latenser BA, Born JM, Liao J. Trends in 393 necrotizing acute soft tissue infection patients 2000–2008. *Burns* 2012;**38**:252–60.
15. Kao LS, Lew DF, Arab SN, Todd SR, Awad SS, Carrick MM, et al. Local variations in the epidemiology, microbiology, and outcomes of necrotizing soft-tissue infections: a multicenter study. *Am J Surg* 2011;**202**:139–45.
16. Tsitsilonis S, Druschel C, Wichlas F, Haas NP, Schwabe P, Bail HJ, et al. Necrotizing fasciitis: is the bacterial spectrum changing? *Langenbecks Arch Surg* 2013;**398**:153–9.
17. Wong CH, Yam AK, Tan AB, Song C. Approach to debridement in necrotizing fasciitis. *Am J Surg* 2008;**196**:19–24.
18. Kaide CG, Khandelwal S. Hyperbaric oxygen: applications in infectious disease. *Emerg Med Clin North Am* 2008;**26**:571–95.
19. Fry DE. The story of hyperbaric oxygen continues. *Am J Surg* 2005;**189**:467–8.
20. Stevens DL, Aldape MJ, Bryant AE. Life-threatening clostridial infections. *Anaerobe* 2012;**18**:254–9.
21. Aldape MJ, Bryant AE, Stevens DL. *Clostridium sordellii* infection: epidemiology, clinical findings, and current perspectives on diagnosis and treatment. *Clin Infect Dis* 2006;**43**:1436–46.
22. Bisno AL, Stevens DL. Streptococcal infections of skin and soft tissues. *N Engl J Med* 1996;**334**:240–5.
23. Waldhausen JH, Holterman MJ, Sawin RS. Surgical implications of necrotizing fasciitis in children with chickenpox. *J Pediatr Surg* 1996;**31**:1138–41.
24. Sistla SC, Sankar G, Sistla S. Fatal necrotizing fasciitis following elective inguinal hernia repair. *Hernia* 2011;**15**:75–7.
25. Bharathan R, Hanson M. Diagnostic laparoscopy complicated by group A streptococcal necrotizing fasciitis. *J Minim Invasive Gynecol* 2010;**17**:121–3.
26. Stevens DL. Invasive group A streptococcus infections. *Clin Infect Dis* 1992;**14**:2–11.
27. Fraser JD, Proft T. The bacterial superantigen and superantigen-like proteins. *Immunol Rev* 2008;**225**:226–43.
28. Darenberg J, Ihendyane N, Sjölin J, Aufwerber E, Haidl S, Follin P, et al. StreptIg study group. intravenous immunoglobulin G therapy in streptococcal toxic shock syndrome: a European randomized, double-blind, placebo-controlled trial. *Clin Infect Dis* 2003;**37**:333–40.

29. Miller LG, Perdreau-Remington F, Rieg G, Mehdi S, Perlroth J, Bayer AS, et al. Necrotizing fasciitis caused by community-associated methicillin-resistant *Staphylococcus aureus* in Los Angeles. *N Engl J Med* 2005;**352**:1445–53.
30. Fry DE, Barie PS. The changing face of *Staphylococcus aureus*: a continuing surgical challenge. *Surg Infect* 2011;**12**:191–203.
31. Kim HJ, Kim DH, Ko DH. Coagulase-positive staphylococcal necrotizing fasciitis subsequent to shoulder sprain in a healthy woman. *Clin Orthop Surg* 2010;**2**:256–9.
32. Proctor RA. Challenges for a universal *Staphylococcus aureus* vaccine. *Clin Infect Dis* 2012;**54**:1179–86.
33. Altarac S, Katušin D, Crnica S, Papeš D, Rajković Z, Arslani N. Fournier's gangrene: etiology and outcome analysis of 41 patients. *Urol Int* 2012;**88**:289–93.
34. Elliott D, Kufera JA, Myers RA. The microbiology of necrotizing soft tissue infections. *Am J Surg* 2000;**179**:361–6.
35. Ustin JS, Malangoni MA. Necrotizing soft-tissue infections. *Crit Care Med* 2011;**39**:2156–62.
36. Hong GL, Lu CJ, Lu ZQ, Li MF, Qiu QM, Liang H, et al. Surgical treatment of 19 cases with vibrio necrotising fasciitis. *Burns* 2012;**38**:290–5.
37. Park SY, Nam HM, Park K, Park SD. Aeromonas hydrophila sepsis mimicking vibrio vulnificus infection. *Ann Dermatol* 2011;**23**:S25–9.
38. Hutchens A, Gupte A, McAuliffe PF, Schain D, Soldevila-Pico C, Knapik JA, et al. Bacillus cereus necrotizing fasciitis in a patient with end-stage liver disease. *Surg Infect* 2010;**11**:469–74.
39. Harada AS, Lau W. Successful treatment and limb salvage of mucor necrotizing fasciitis after kidney transplantation with posaconazole. *Hawaii Med J* 2007;**66**:68–71.
40. Neblett Fanfair R, Benedict K, Bos J, Bennett SD, Lo YC, Adebanjo T, et al. Necrotizing cutaneous mucormycosis after a tornado in Joplin, Missouri, in 2011. *N Engl J Med* 2012;**367**:2214–25.
41. Reisman JS, Weinberg A, Ponte C, Kradin R. Monomicrobial pseudomonas necrotizing fasciitis: a case of infection by two strains and a review of 37 cases in the literature. *Scand J Infect Dis* 2012;**44**:216–21.
42. Shaked H, Samra Z, Paul M, Madar-Shapiro L, Cohen J, Pitlik S, et al. Unusual "flesh-eating" strains of *Escherichia coli*. *J Clin Microbiol* 2012;**50**:4008–11.
43. Das DK, Baker MG, Venugopal K. Increasing incidence of necrotizing fasciitis in New Zealand: a nationwide study over the period 1990 to 2006. *J Infect* 2011;**63**:429–33.
44. Tunovic E, Gawaziuk J, Bzura T, Embil J, Esmail A, Logsetty S. Necrotizing fasciitis: a six-year experience. *J Burn Care Res* 2012;**33**:93–100.
45. *The national necrotizing fasciitis foundation*. <http://www.nnff.org/> [accessed 7.05.13].

CHAPTER

15

Significance of Surgical Intervention in the Management of Diabetic Foot Infections

Shalbha Tiwari[*,†], *Daliparthy Devi Pratyush*[*], *Sanjeev Kumar Gupta*[‡] and *Surya Kumar Singh*[*]

[*]Department of Endocrinology and Metabolism, Institute of Medical Sciences, Banaras Hindu University, Varanasi, India, [†]Chellaram Diabetes Institute, Pune, India, [‡]Department of General Surgery, Institute of Medical Sciences, Banaras Hindu University, Varanasi, India

INTRODUCTION

Diabetes is one of the major causes of morbidity and death worldwide. Over the past three decades, the incidence of diabetes worldwide has doubled.[1] The World Health Organization (WHO) predicts that developing countries will bear the brunt of this epidemic in the 21st century. According to the latest International Diabetes Federation (IDF) estimate of the global burden of diabetes, 382 million people have diabetes in 2013, and this is expected to rise to 592 million by 2035.[2] Foot ulcers are common in diabetic patients, with a prevalence as high as 25%.[3] Around 7 to 20% of patients with foot ulcers subsequently require an amputation. Foot ulceration is the precursor to approximately 85% of lower extremity amputations in people with diabetes. It is therefore not surprising that even in the United States diabetes is the leading cause of non-traumatic lower extremity amputation.[4]

Foot infections are major predictors of future lower extremity amputations in diabetic patients. Infection is frequent (40–80%) and costly.[5] The increased incidence of foot infection in subjects with diabetes mellitus is a result of impaired or altered innate immunity. The driving forces in the non-healing or recurrent diabetic foot infection seem to be predominantly related to underlying neuropathy, resulting in repeated unrecognized trauma, structural abnormalities and changes in inflammatory response. Tissue under-nutrition as

Microbiology for Surgical Infections
DOI: http://dx.doi.org/10.1016/B978-0-12-411629-0.00015-5

a result of macro-vascular disease and microvascular dysfunction[6] may also affect basic healing mechanisms. In this chapter, we focus on the epidemiology, pathogenetic mechanism, infection diagnosis and microbiological profile of diabetic foot infection and its management, with emphasis on appropriate antibiotic selection and surgical intervention.

DEFINITION OF DIABETIC FOOT

Diabetic foot describes the foot of a diabetic patient that has a potential risk of pathologic consequences, including infection, ulceration, and destruction of deep tissues associated with neurologic abnormalities, various degrees of peripheral arterial disease, and metabolic complications of diabetes in the lower limb (from the World Health Organization definition). The term "diabetic foot wound" refers to a variety of conditions like cellulitis, osteomyelitis or ulcers.[7] The diabetic foot and its sequelae account for billions of dollars in direct medical expenditure, as well as lengthy hospital stays and periods of disability.[4,8] The most characteristic lesion of the diabetic foot is a mal perforans ulceration which consequently is one of the major risk factors in amputation.[9]

EPIDEMIOLOGY

Studies report that 15% of all diabetic patients develop a foot ulcer at some point in their lifetime and around 28% of them may require some form of amputation.[10,11] Foot infections account for 20% of annual hospitalizations of diabetic patients.[12] The annual, population-based incidence of foot ulcers among people with diabetes ranges from less than 1% to 3.6% among people with type 1 diabetes mellitus (T1DM) and type 2 diabetes mellitus (T2DM), respectively.[13] Several other population-based studies indicate a 0.5% to 3% annual cumulative incidence of diabetic foot ulcers. According to one large British study of neuropathic patients, the one year incidence of foot ulcer was 7%.[14] The prevalence of foot ulcers reported for a variety of population ranges from 2% to 25%.[15] Foot ulceration is the precursor to approximately 85% of lower extremity amputations in persons with diabetes.

A previous report estimated that the risk of hospitalization and lower extremity amputation was respectively 56 and 155 times greater for diabetic people who had a foot infection than for those without.[16] Most infections involved soft tissue, but 20% of patients with foot infection had bone culture-proven osteomyelitis. In the Eurodiale study, 58% of diabetic patients consecutively attending one of 14 foot clinics in ten European countries with a new foot ulcer had a wound that was clinically infected.[17] A population-based study in southern India indicates that the prevalence of peripheral vascular disease (PVD) is 3.2% in the general population and 6.3% in the diabetic population.[18]

PATHOGENESIS OF DIABETIC FOOT ULCER

Certain foot deformities, past history of lower limb complications, reduced skin oxygenation and foot perfusion, poor vision, greater body mass and both sensory and autonomic

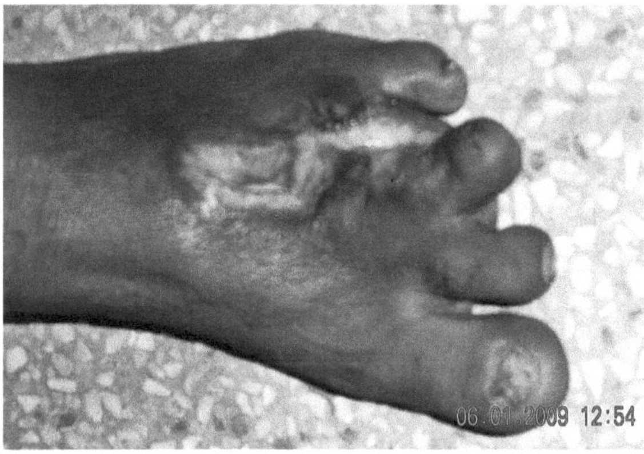

FIGURE 15.1 An ulcer in dorsum of the foot with deformity.

neuropathy independently influence foot ulcer risk, thereby providing support for a multifactorial etiology for diabetic foot ulceration.[19]

Many further factors in the causation of diabetic foot ulceration have been elucidated by numerous observational studies, such as peripheral sensory neuropathy, peripheral vascular disease (PVD), structural foot deformity, prolonged and elevated pressures due to compression, trauma, improperly fitted shoes, callus, history of prior ulcers/amputations, limited joint mobility, uncontrolled hyperglycemia, duration of diabetes, blindness/partial sight, chronic renal disease, older age etc.

Peripheral sensory neuropathy in the absence of perceived trauma is the primary factor leading to diabetic foot ulcerations.[14] Approximately 45–60% of all diabetic ulcerations are purely neuropathic, while up to 45% have neuropathic and ischemic components.[20] Other forms of neuropathy may also play a role in foot ulceration. Motor neuropathy can lead to foot deformities such as foot drop, equinus, hammer toes, and prominent plantar metatarsal heads.[21] Autonomic neuropathy may commonly result in dry skin with cracking and fissuring, thus creating a portal of entry for bacteria.[22] The alterations can subsequently be implicated in the pathogenesis of ulceration. Figure 15.1 represents an ulcer in dorsum with deformity.

PVD is defined as disease of any blood vessel that is not part of the heart or brain. It is most commonly reported in the lower extremities, and hence is also termed Lower Extremity Arterial Disease (LEAD). In a study of T2DM subjects, 4% had PVD, these included 15.1% with gangrene and 17.6% who had undergone amputations.[18] Newly detected type 2 diabetics have 11% prevalence of PVD. Its revalence was 36% in patients with clinical proteinuria.[23] Attempts to resolve any infection will be impaired due to lack of oxygenation and difficulty in delivering antibiotics to the site of infection. Early recognition and aggressive treatment of lower extremity ischemia is therefore vital for lower limb salvage. Trauma to the foot in the presence of peripheral sensory neuropathy is an important component cause of ulcerations. Shoe-related trauma has been identified as a frequent precursor to foot ulceration.[20] Other factors often associated with heightened risk of ulceration include: nephropathy, poor diabetes control, blindness, advanced age, and poor nutrition.[24]

TABLE 15.1 University of Texas Wound Classification System

	Grade 0	Grade 1	Grade 2	Grade 3
Stage A	Preulcer or postulcer lesion No skin break	Superficial ulcer	Deep ulcer to tendon/ capsule	Wound penetrating bone/ joint
Stage B	+Infection	+Infection	+Infection	+Infection
Stage C	+Ischemia	+Ischemia	+Ischemia	+Ischemia
Stage D	+Infection and Ischemia	+Infection and Ischemia	+Infection and Ischemia	+Infection and Ischemia

CLASSIFICATION OF DIABETIC FOOT

The most commonly used classification system was Wagner's grade, which has a limitation in that it cannot explain the infection status of the diabetic foot. More recently, the University of Texas (UT) classification addresses ulcer depth and includes the presence of infection and ischemia, and is now used to classify diabetic foot (Table 15.1).

Diabetic foot has also been classified as:

1. Non-limb threatening: The wound may be uninfected and lack purulence or any manifestation of infection, with no significant ischemia or may be mildly infected – presence of ≥2 manifestations of inflammation (purulence, erythema, pain, warmth, induration) but any erythema/cellulitis extends ≤2 cm around the lesion and infection is limited to the skin or superficial subcutaneous tissue; no other local complications or systemic illness or significant ischemia.
2. Limb threatening: Infection in a patient who is systemically well and metabolically stable but who has one of the following manifestations: cellulitis ≥2 cm, lymphangitic streaking spread beneath the superficial fascia, deep tissue abscess, gangrene, involvement of muscle, tendon bone or joint, ischemia. Figure 15.2 shows a limb threatening infection with gangrene. Figures 15.3A, B and C show an infected wound with cellulitis and partial gangrene before and after surgical intervention.
3. Life threatening: Infection in a patient with systemic toxicity and metabolic instability as fever/chills, tachycardia >100, hypotension <100 mm Hg, confusion, leukocytosis, acidosis, severe hyperglycemia >500 mg/dl, azotemia.

DIAGNOSIS OF DIABETIC FOOT INFECTION

As all chronic wounds are colonized by microorganisms, the diagnosis of diabetic foot infection (DFI) should not be based on the microbiological analysis of a wound culture but on clinical findings.[25,26] The International Working Group on the Diabetic Foot (IWGDF)

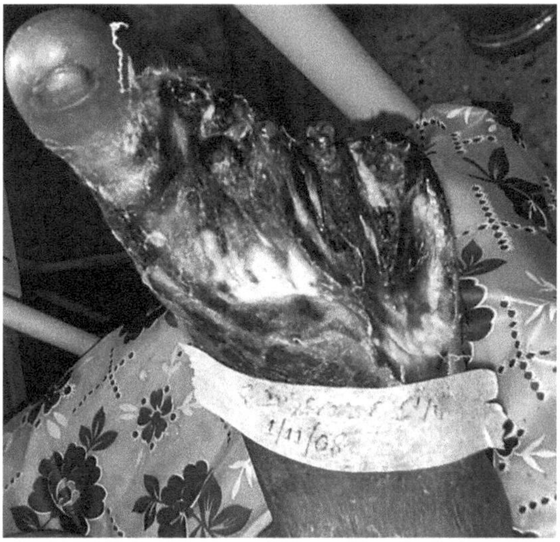

FIGURE 15.2 Limb threatening infection with gangrene.

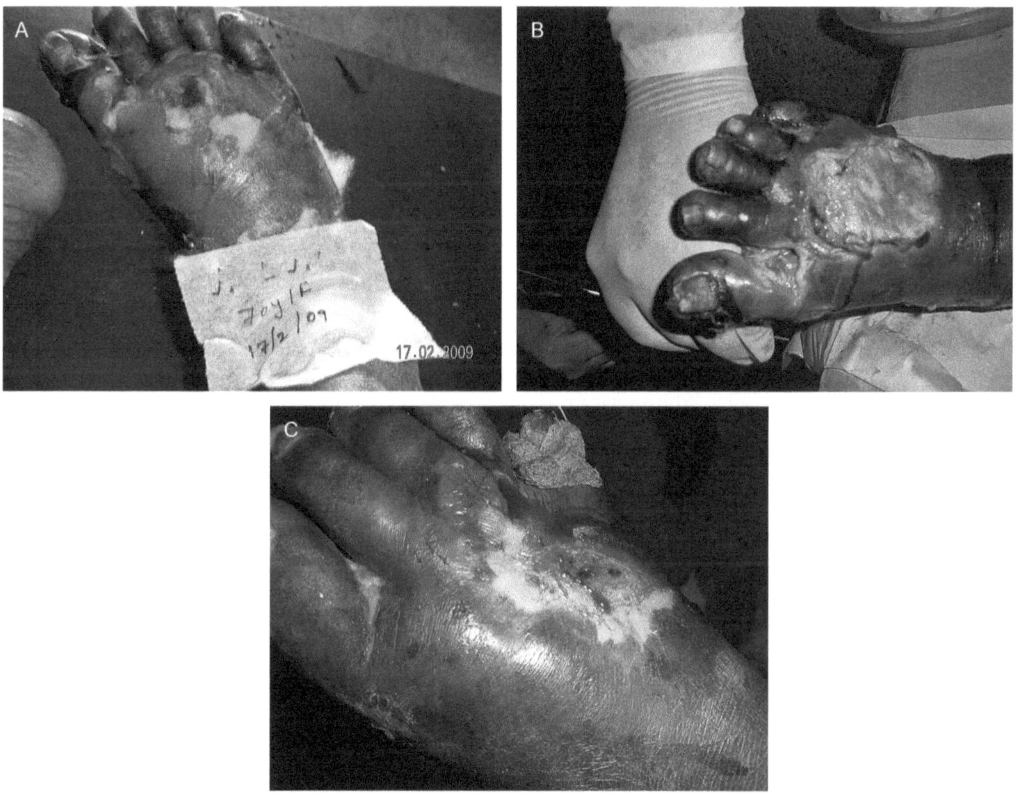

FIGURE 15.3 A, B and C **Infected wound with cellulitis and partial gangrene before and after surgical intervention.**

and the Infectious Diseases Society of America (IDSA) has developed clinical criteria for recognizing and classifying the severity of DFI.[27] Recently Lipsky et al.[28] developed a 10 item DFI wound score incorporating semi-quantitative grading of both wound measurements and various infection parameters.

The measurement of circulating markers of inflammation is usually regarded as lacking sensitivity and specificity to reliably identify infection: the sedimentation rate and leukocyte count are normal in approximately 50% of cases, even in severe deep infections. In contrast, measurement of some other inflammatory markers might be of value in discriminating between infected and non-infected diabetic foot ulcers (DFUs).[29] They could help in the more rational use of antibiotic agents.

The presence of bacteria in tissue is not sufficient evidence of infection. It only reflects colonization, which becomes clinically significant if their number reaches a critical value which host defense mechanisms fail to contain. This critical value or "critical colonization" is suggested to be a bacterial concentration of 10^5 colony forming units per gram of tissue.[30,31]

16S rDNA sequencing has been used to characterize and identify the bacteria in non-healing diabetic foot wounds.[32] In another study, enterobacterial repetitive intergenic consensus-polymerase chain reaction (ERIC-PCR) has been shown to be a sensitive detection technique for characterizing diversity among bacterial isolates at the genetic level in DFI.[33] Studies of the bacterial diversity in chronic wound pathogenic biofilms using different detection methods, such as partial ribosomal amplification and pyrosequencing (PRAPS), full ribosomal amplification, cloning and Sanger sequencing (FRACS) and partial ribosomal amplification, density gradient gel electrophoresis (DGGE) and Sanger sequencing (PRADS) have also been reported.[34]

Identification of the virulence of bacteria might also be of value in the future routine diagnosis of infection. New technologies such as DNA micro-array and multiplex real-time PCR help to analyze both the virulence and resistance potential of microorganisms. The method of miniaturized genotyping can rapidly and reliably detect the presence of genes encoding for various virulence and antibiotic resistance factors.[35] Using this method, both virulence and resistance genes were found to be present far more frequntly in clinically infected diabetic foot wounds than in uninfected wounds.[36] In another study, multiplex PCR assays were used to evaluate 31 of the most prevalent virulence-associated genes in *Staphylococcus aureus* strains isolated from DFUs.[37] Analysis of virulence potential (by genotyping methods) would be of value for the diagnosis of diabetic foot wounds in routine clinical practice. Table 15.2 summarizes the tools for diagnosing infections in diabetic foot.

MICROBIOLOGY OF DIABETIC FOOT INFECTION

In India, the observed prevalence of infection was 6% to 11%, and the prevalence of amputation was 3% in patients with type 2 diabetes in a 2005 study by Vishwanathan *et al*.[38] Infections in diabetic foot are mainly caused by aerobic bacteria,[39–42] but the impact of anaerobes has also been reported.[40,41,43] Reports are also available of fungal infections in diabetic foot.[44–46] A polymicrobial etiology of infection in DFU was reported in several studies.[33,39,40,42] In contrast, some other studies have also reported a higher prevalence of mono-microbial infection in diabetic foot.[43,47]

TABLE 15.2 Infection in Diabetic Foot – Diagnosis

• Clinical findings	• 16S rDNA sequencing
• Microbiological analysis of wound culture	• ERIC-PCR[33]
• Quantification of the bacterial burden	• Partial Ribosomal Amplification And Pyrosequencing (PRAPS),
• Measurement of circulating markers of inflammation	• Full Ribosomal Amplification, Cloning And Sanger Sequencing (FRACS) and
• DNA micro-array and multiplex real-time PCR offer a unique opportunity to analyze both the virulence and resistance potential of microorganisms.	• Partial Ribosomal Amplification, Density Gradient Gel Electrophoresis (DGGE) and Sanger sequencing (PRADS)

In one study,[44] it was noticed that *S. aureus* and *Pseudomonas aeruginosa* were the most common causes of diabetic foot infections. In a North Indian study of DFUs, it was found that gram-negative aerobes were the most frequently isolated (51.4%), followed by gram-positive aerobes and anaerobes. 72% of patients were positive for multidrug-resistant organisms (MDROs). Extended spectrum beta lactamase (ESBL) production and methicillin resistance was noted in 44.7 and 56% of bacterial isolates, respectively. MDRO-positive status was associated with presence of neuropathy, osteomyelitis, and ulcer size >4 cm^2 but not with duration of hospital stay. MDRO patients had poor glycemic control and needed to be surgically treated.[40]

The importance of individual virulence potential was recently illustrated by the Sidestep study: a favorable clinical response to ertapenem was noticed in patients in whom *Enterococcus* spp. and *P. aeruginosa* were isolated, despite the ertapenem resistance of the latter isolates.[48] The various organisms isolated from infected wounds do not have identical pathogenetic importance and rather than counting bacterial load, evaluating the intrinsic virulence potential of isolated bacteria to identify their real pathogenicity seems promising. Well-known virulent bacteria implied in DFI are *S. aureus*, β-hemolytic streptococci, Enterobacteriaceae, and anaerobes. However, bacteria of low virulence, including well described commensal flora such as coagulase-negative *Staphylococcus*, *Corynebacterium* spp. and *Propionibacterium* spp. together with *Enterococcus* spp., *Stenotrophomonas maltophilia* and *Pseudomonas* spp. can also be pathogenic.

The prevalence rate of MDROs dramatically reduced by half from 35.2% to 16.3%, and the prevalence of Methicillin-resistant *Staphylococcus aureus* (MRSA) reduced to one third from 52.2% to 18.9% between 2003 and 2007.[49] In contrast there was a less dramatic decline in the prevalence of pathogens of lower virulence, which reduced from 40.1% to 16.4%.[50]

MANAGEMENT OF DIABETIC FOOT INFECTION

The proper and effective management of diabetic foot wounds requires:

- Adequate glycemic control
- Wound care

- Surgical debridement
- Pressure offloading
- Maintaining adequate blood supply
- Evaluation of infection status clinically and microbiologically
- Selection of appropriate antibiotic treatment regimen

The majority of diabetic foot wounds are chronic in nature involving deep tissue complicated by infections, although some wounds are superficial, and do not extend into the subcutaneous fascia. Delayed healing could be a result of persistent infection, and may ultimately lead to systemic involvement in patients with diabetic foot. Superficial bullae, petechiae or ecchymoses, soft tissue crepitus, rapid spread of infection and tissue gas on X-rays[21,51] are commonly observed. Antimicrobial therapy alone is inadequate for the management of chronic wound infections, and therefore surgical interventions play a significant role in their effective treatment. A stepwise approach to their management requires a multidisciplinary team with special emphasis on the appropriate use of surgical intervention. This includes the following key elements: incision and drainage, wound investigation, debridement, wound irrigation and lavage, and definitive wound closure.[52]

ANTIBIOTIC SELECTION

The appropriate selection of antibiotics is crucial for managing diabetic foot infections. Antibiotic treatment is based on evidence provided by reports on bacteriological culture and sensitivity from different centers all over the world. Commonly prescribed conventional antibiotics include penicillins, cephalosporins, carbapenems, quinolones, aminoglycosides, macrolides, tetracyclines, nitrofurantoin, metronidazole, clindamycin, vancomycin, teicoplanin, co-trimoxazole, fusidic acid, etc. In the **penicillins**: penicillin V/G, ampicillin, amoxycillin, cloxacillin, ticarcillin, piperacillin; **cephalosporins**: 1st generation e.g., cefazolin, cefalexin (*Keflex*), 2nd generation e.g., cefuroxime (*Zinacef, Zinnat*), 3rd generation e.g., ceftriaxone (*Rocephin*), cefotaxime (*Claforan*), ceftazidime (*Fortum*), cefoperozone (*Cefobid*), ceftibuten (*Cedax*), 4th generation e.g., cefepime (*Maxipime*); **carbapenems**: imipenem, meropenem; **monobactam**: aztreonam; **macrolides**: erythromycin, clarithromycin (*Klacid*), azithromycin (*Zithromax*); **fluoroquinolones** (FQ): ofloxacin, levofloxacin (*Cravit*), ciprofloxacin (*Ciproxin*); **aminoglycosides**: gentamicin, amikacin; **tetracyclines**: doxycyline (*Vibramycin*), minocycline; **glycopeptides**: vancomycin, teicoplanin are commonly used. Recently, new antibiotics like linezolid, ertapenem, moxifloxacin have also been prescribed for the control of infection in diabetic foot.

The antibiotics commonly used against gram-positive infections are cloxacillin, amoxycyclin with β-lactamase inhibitors, cefazolin, cephalexin, and clindamycin. During polymicrobial infection; i.e., where gram-positive cocci, gram negative bacilli and anaerobes are present, broad spectrum antibiotics are used, such as β-lactam + β-lactamase inhibitors + amikacin, 3rd generation cephalosporin + clindamycin, ciprofloxacin + clindamycin, ciprofloxacin + linezolid. A combination of carbapenems ± vancomycin is used if life threatening infection is present.

A study reviewed patients admitted to a tertiary care teaching center over a six month period to determine the percentage of empirically treated soft tissue infections susceptible

TABLE 15.3 Percentage of Empirically Treated Soft Tissue Infections Susceptible to the Prescribed Antibiotic Therapy

Antibiotic	Frequency of Use (%)
Cefazolin (IV)	47
Ciprofloxacin (IV)	21
Cefotaxime (IV)	20
Clindamycin (IV)	16
Cloxacillin (IV)	11
Metronidazole (IV)	8
Ciprofloxacin (PO)	7
Clindamycin (PO)	7
Metronidazole (PO)	4
Vancomycin (IV)	4
Cephelexin (PO)	4
Moxifloxacin (PO)	4
Piperacillin-tazobactam (IV)	3
Penicillin G (IV)	1
Ampicillin (IV)	1
Cloxacillin (PO)	1
Levofloxacin (IV)	1

IV Intravenous; PO By mouth.

to the prescribed antibiotic therapy.[53] The data of this study is shown in Table 15.3. It was concluded that broad spectrum antibiotic therapy should be reserved for patients who do not improve or have culture results that suggest additional antibiotics.

In another study, piperacillin + tazobactum showed the highest sensitivity against isolated bacteria from diabetic foot infection.[33] Fincke et al. found large variations in the antibiotic regimens used to treat patients with diabetic foot infections.[54] Tiwari et al. suggested that piperacillin/tazobactam should be the treatment of choice on an empirical basis prior to a definitive bacteriological study.[47]

SURGICAL INTERVENTION

Incision and Drainage

Central, medial and lateral are three major plantar spaces which are at risk of ulceration and subsequent infection. A plantar incision could drain infection from these spaces but

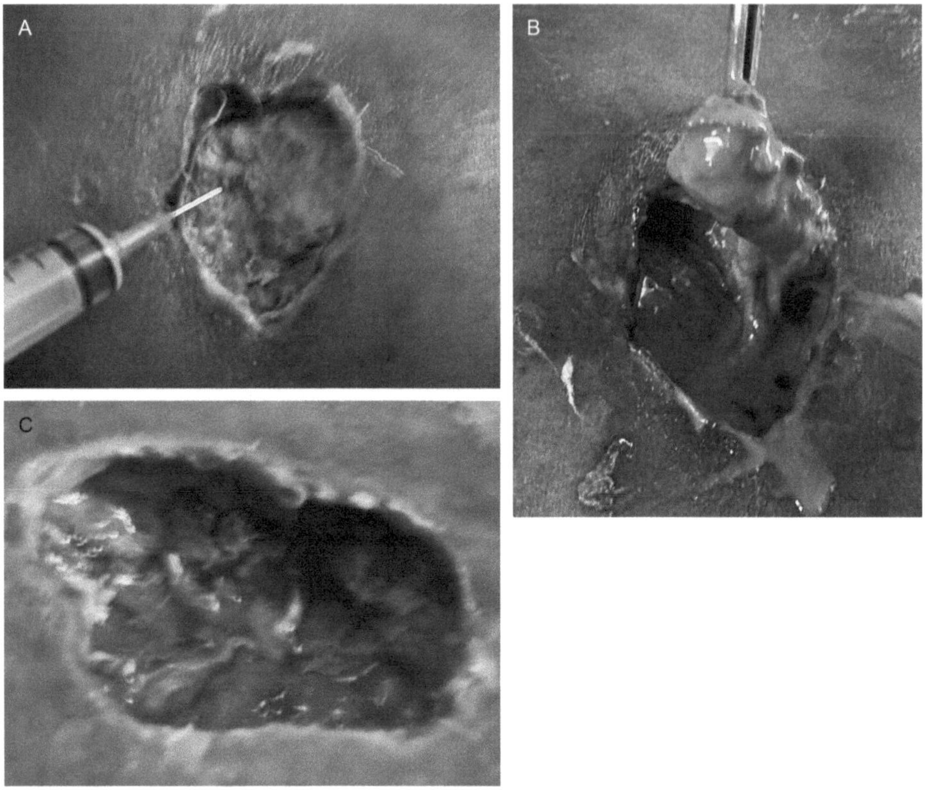

FIGURE 15.4 A, B and C **Sequential surgical intervention procedure (incision and drainage and removal of dead tissue).**

results in an unsightly scar. However, careful dissection and handling can prevent this, or alternatively a medial approach of incision has also been recommended.[55]

Debridement

Dead and devitalized tissue not only delays wound healing, but also predisposes to infection and interferes with adequate assessment of wound. Debridement is the process by which these dead and necrotic tissues and calluses are removed. Scalpel, forceps and/or similar tools are used for this purpose, and after removal of non-viable tissue, healthy tissues are exposed. This exposure accelerates the rate of wound healing, reduces the risk of infection and shortens the wound healing time. Debridement may be associated with excessive bleeding and may result in a larger wound area, but the benefits far outweigh the disadvantages and debridement remains an integral part of the care of diabetic foot ulcers. Repeated debridements are required for appropriate management in most cases. Where bleeding is a matter of concern, alternate wet and dry dressings, various topical enzymes or moisture-retentive dressings are used. Figure 15.4A, B and C shows points in

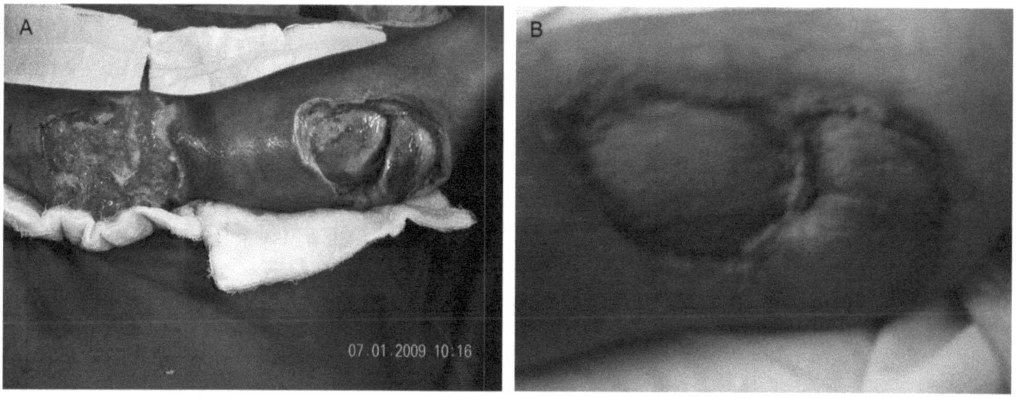

FIGURE 15.5 A and B **Pre and post wound closure by skin grafting.**

a surgical intervention procedure where incision and drainage were performed, followed by removal of dead tissue from the wound area. After debridement, wounds should be thoroughly investigated to determine their depth, the presence of infection or any other foreign materials, abscesses, sinus tract or exposed bone etc. Probe to bone test is used to identify osteomyelitis or joint infection.

Wound Lavage

Colonizing bacteria present at the margins of a wound could progress to infection. Proper cleansing of the wound should therefore be performed to lower the bacterial load. Aggressive rinsing of the wound can help to eliminate necrotic tissue and blood clots which are the barriers to natural immunity of the host cells. Irrigation with saline alone[56] or in combination with an appropriate antibiotic[57] is now widely used to prevent postoperative wound infection. Alternatively, wound irrigation with tap water has also been shown to be as effective as using saline irrigation.[58] Thus, it is recommended that diabetic foot wounds are irrigated, preferably with normal saline, to make treatment more effective.

Wound Closure

Wound closure is the final step of surgical intervention. There are two major types of wound closure: primary and secondary. In primary closure, the skin is closed at the end of the surgery, whereas in the secondary closure the wound is left open at the end of surgery and heals by granulation and contraction. A further type is delayed primary closure, in which wounds are closed a few days after the first amputation and infection drainage. The choice of closure type depends entirely on the degree of drainage, the existing amount of tissue for wound closure and infection status at the site of wound closure. Topical negative pressure therapy, full- and split-thickness grafts, and local and distant flaps also are useful adjuncts for closing deep, complex wounds following debridement. Figure 15.5A and B shows pre- and post-wound closure by skin grafting.

Revascularization

Arterial perfusion is required for antibiotic delivery and the wound healing process. Ischemia and infection are the two significant factors that impair healing in patients with diabetic foot ulcers. Leg claudication, hair loss, poor capillary refill, absence of palpable foot pulses, skin atrophy and nail cornification are characteristic of ischemic wounds.[58] A suspected ischemic non-healing ulcer should be evaluated for its suitability for a revascularization procedure such as percutaneous transluminal angioplasty or a surgical bypass graft.[59] Revascularization procedures not only reduce pain but also accelerate healing, improve wellbeing and decrease amputation rates at all levels.[60] Thus, revascularization is recommended for the treatment of diabetic foot ulcers.[61]

Amputation

Amputation should be used sparingly and for very limited indications. The indications for amputation include gangrene, severe soft tissue infection, arterial occlusion, extensive osteomyelistis and/or a non-healing ulcer.[8] Risk factors for a patient with diabetes to require an amputation include lower extremity ischemia, peripheral neuropathy, elevated glycated hemoglobin levels, a history of foot ulcers and retinopathy. The goal of amputation must be the most distal amputation that would lead to the healing of the diabetic foot wound. Figure 15.6A and B shows an infected wound with localized gangrene, where partial amputation was performed and the wound healed. Figure 15.7A and B shows partial amputation of a second gangrenous toe.

CONCLUSIONS AND FUTURE PROSPECTS

Diabetic foot is a feared complication that accounts for frequent hospital admission and major use of health care system resources. The prevalence of the diabetic foot complication is variable in different populations, and this relates to the existence of basic diabetes education and access to specialized health care facilities. In the developing world, the prevalence data and intervention outcomes are limited and therefore it is speculated to be worse than in the developed world.

The pathogenesis of diabetic foot is multi-factorial, including neuropathy, vasculopathy, impaired host defense, poor tissue reparative processes and altered anatomy of the foot. Infection of foot ulcer is the most important confounder of wound healing, which needs to be contained and eradicated. Such measures are cost effective and have potential for preventing amputation.

Infection is the most significant risk factor for the non-healing diabetic foot. The pattern of infection is not consistent. Studies based on microbial culture and isolation showed a high prevalence of mono-microbial infection with gram-negative infections predominating. Contrary to this, molecular studies support a polymicrobial etiology of diabetic foot wounds. Combination antibiotic therapy such as piperacillin plus tazobactam is suggested for non-healing diabetic foot wounds with infection.

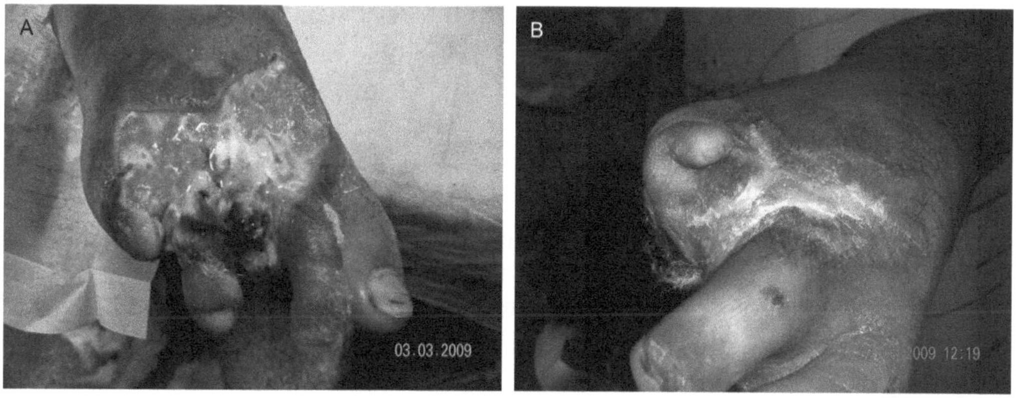

FIGURE 15.6 A and B **Infected wound with localized gangrene where partial amputation was performed and wound was healed.**

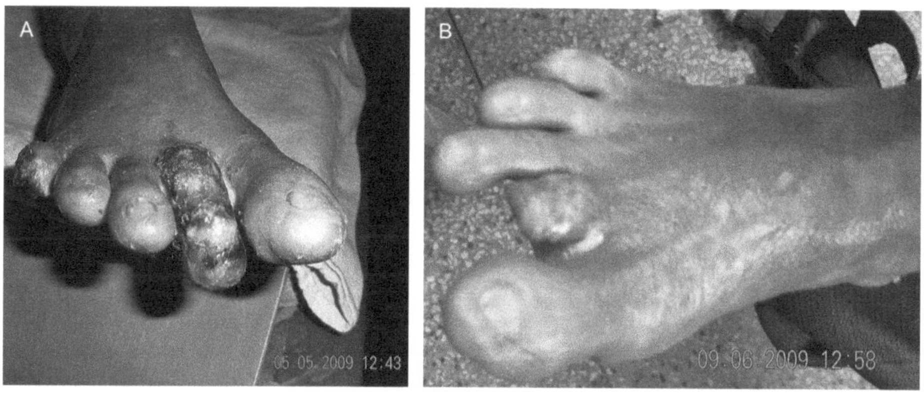

FIGURE 15.7 A and B **Partial amputation of second gangrenous toe followed by healing.**

Surgical debridement is an important tool to help to control the spread of infection by improving blood supply and reducing the microbial load at the wound site. Revascularization is the definitive means of restorating most of the abnormalities associated with hostile wound for healing. Timed intervention with a combination of glycemic management, use of antibiotics conforming to a bacteriogram, in conjunction with surgery, can prevent limb- and life-threatening situations in diabetes.

Careful evaluation of clinical and laboratory findings helps to determine associated metabolic risk, complexity of wound and severity of infection as well as deciding the timing of surgical interventions. Surgical intervention is a proven method for managing diabetic foot infection, but the decision regarding frequency, extent and type of debridement is subjective and depends on the surgeon's experience. Hence, no definite recommendation relating to surgical intervention can be proposed for the management of diabetic foot infection in clinics. Innovations in the form of newer techniques such as vacuum-assisted closure devices, laser surgery or debriding tools, including biological agents, will be interesting developments which could minimize the dependency on surgeon's experience and may rationalize intervention in diabetic foot infection.

ACKNOWLEDGMENT

We would like to thank Balram Gupta (PhD scholar) and Dr. J Swain (SR III), Department of Endocrinology and Metabolism, and Dr. Pavan Yadav (JR III), Department of Tuberculosis and Respiratory Diseases, Institute of Medical Sciences, Banaras Hindu University, Varanasi for their support in manuscript drafting and editing.

References

1. Danaei G, Finucane MM, Lu Y, Singh GM, Cowan MJ, Paciorek CJ, et al. National, regional, and global trends in fasting plasma glucose and diabetes prevalence since 1980: systematic analysis of health examination surveys and epidemiological studies with 370 country-years and 2.7 million participants. *Lancet* 2011;**378**:31–40.
2. Whiting DR, Guariguata L, Weil C, Shaw J. IDF Diabetes Atlas: global estimates of the prevalence of diabetes for 2011 and 2030. Available from http://www.idf.org/sites/default/files/EN_6E_Atlas_Full_0.pdf.
3. Singh N, Armstrong DG, Lipsky BA. Preventing foot ulcers in patients with diabetes. *J Am Med Assoc* 2005;**293**:217–28.
4. Frykberg RG, Armstrong DG, Giurini J, Edwards A, Kravette M, Kravitz S. Diabetic foot disorders: a clinical practice guideline. American college of foot and ankle surgeons. *J Foot Ankle Surg* 2000;**39**:1–60.
5. Prompers L, Huijberts M, Schaper N, Apelqvist J, Bakker K, Edmonds M, et al. Resource utilisation and costs associated with the treatment of diabetic foot ulcers. Prospective data from the Eurodiale Study. *Diabetologia* 2008;**51**:1826–34.
6. Casqueiro J, Casqueiro J, Alves C. Infections in patients with diabetes mellitus: a review of pathogenesis. *Indian J Endocr Metab* 2012;**16**:27–36.
7. Kahn R. American diabetes association, Consensus development conference on diabetic foot wound care. Boston, Massachusetts. *Diabetes Care* 1999;**22**:1354–60.
8. Reiber GE, Boyko EJ, Smith DG. Lower extremity foot ulcers and amputations in diabetes. In: Harris MI, Cowie C, Stern MP, eds. *Diabetes in America*. 2nd ed. Rockville: NIH Publication; 1995. pp. 409–27. No. 95–1468.
9. Frykberg RG. Diabetic foot ulcers: pathogenesis and management. *Am Fam Physician* 2002;**66**:1655–62.
10. Reiber GE, Lipsky BA, Gibbons GW. The burden of diabetic foot ulcers. *Am J Surg* 1998;**176**:5S–10S.
11. Boulton AJ, Vileikyte L, Ragnarson-Tennvall G, Apelqvist J. The global burden of diabetic foot disease. *Lancet* 2005;**366**:1719–24.
12. Lavin ME, O'Neal LW. *The diabetic foot*. St. Louis: CW Mosby Co.; 1988. pp. 203–5.
13. Joseph W, Master L, Reiber GE. Epidemiology and economic impact of foot ulcers. In: Boulton AJM, Cavanagh Peter R, Rayman Gerry, eds. *The foot in diabetes*. 4th ed. JohnWiley & Sons, Ltd.; 2006.
14. Abbott CA, Vileikyte L, Williamson S, Carrington AL, Boulton AJ. Multicenter study of the incidence of and predictive risk factors for diabetic neuropathic foot ulceration. *Diabetes Care* 1998;**21**:1071–5.
15. Frykberg RG, Zgonis T, Armstrong DG, Driver VR, Giurini JM, Kravitz SR, et al. Diabetic foot disorders: a clinical practice guideline (2006 revision). *J Foot Ankle Surg* 2006;**45**:S1–66.
16. Lavery LA, Armstrong DG, Wunderlich RP, Mohler MJ, Wendel CS, Lipsky BA. Risk factors for foot infections in individuals with diabetes. *Diabetes Care* 2006;**29**:1288–93.
17. Prompers L, Huijberts M, Schaper N, Apelqvist J, Bakker K, Edmonds M, et al. Resource utilization and costs associated with the treatment of diabetic foot ulcers. Prospective data from the Eurodiale Study. *Diabetologia* 2008;**51**:1826–34.
18. Premalatha G, Shanthirani S, Deepa R. Prevalance and risk factors of peripheral vascular disease in a selected South Indian population. The chennai urban population study (CUPS). *Diabetes Care* 2000;**23**:1295–300.

19. Boyko EJ, Ahroni JH, Stensel V, Forsberg RC, Davignon DR, Smith DG. A prospective study of risk factors for diabetic foot ulcer. the Seattle diabetic foot study. *Diabetes Care* 1999;**22**:1036–42.
20. Reiber GE, Vileikyte L, Boyko EJ, del Aguila M, Smith DG, Lavery LA, et al. Causal pathways for incident lower-extremity ulcers in patients with diabetes from two settings. *Diabetes Care* 1999;**22**:157–62.
21. Boulton AJ, Meneses P, Ennis WJ. Diabetic foot ulcers. A framework for prevention and care. *Wound Rep Reg* 1999;**7**:7–17.
22. Shaw JE, Boulton AJM. The pathogenesis of diabetic foot problems. An overview. *Diabetes* 1997;**46**:S58–61.
23. Singh SK, Kumar R. Peripheral vascular disease and diabetic nephropathy. American Diabetes Association 2004; Abstract # 25LB.
24. Boulton AJM. The pathway to ulceration. In: Boulton AJM, Connor H, Cavanagh PR, eds. *The foot in diabetes*. 3rd ed. Chichester, UK: John Wiley and Sons; 2000. pp. 19–31.
25. Jeffcoate WJ, Lipsky BA, Berendt AR, Cavanagh PR, Bus SA, Peters EJ, et al. Unresolved issues in the management of ulcers of the foot in diabetes. *Diabet Med* 2008;**25**:1380–9.
26. Lipsky BA. Infectious problems of the foot in diabetic patients. In: Bowker JH, Pfeifer MA, eds. *Levin and O'Neal's* The diabetic foot. 7th ed. Philadelphia: Mosby Elsevier; 2008. pp. 305–18.
27. Lipsky BA. A report from the international consensus on diagnosing and treating the infected diabetic foot. *Diabetes Metab Res Rev* 2004;**20**:S68–77.
28. Lipsky BA, Polis AB, Lantz KC, Norquist JM, Abramson MA. The value of a wound score for diabetic foot infections in predicting treatment outcome: a prospective analysis from the SIDESTEP trial. *Wound Repair Regen* 2009;**17**:671–7.
29. Jeandrot A, Richard JL, Combescure C, Jourdan N, Finge S, Rodier M, et al. Serum procalcitonin and C-reactive protein concentrations to distinguish mildly infected from non-infected diabetic foot ulcers: a pilot study. *Diabetologia* 2008;**51**:347–52.
30. Bowler PG, Duerden BI, Armstrong DG. Wound microbiology and associated approaches to wound management. *Clin Microbiol Rev* 2001;**14**:244–69.
31. Williams DT, Hilton JR, Harding KG. Diagnosing foot infection in diabetes. *Clin Infect Dis* 2004;**39**:S83–6.
32. Redkar R, Kalns J, Butler W, Krock L, McCleskey F, Salmen A, et al. Identification of bacteria from a non-healing diabetic foot wound by 16S rDNA sequencing. *Mol Cell Probes* 2000;**14**:163–9.
33. Singh SK, Gupta K, Tiwari S, Shahi SK, Kumar S, Kumar A, et al. Detecting aerobic bacterial diversity in patients with diabetic foot wounds using ERIC-PCR: a preliminary communication. *Int J Low Extrem Wounds* 2009;**8**:203–8.
34. Dowd SE, Sun Y, Secor PR, Rhoads DD, Wolcott BM, James GA, et al. Survey of bacterial diversity in chronic wounds using pyrosequencing, DGGE, and full ribosome shotgun sequencing. *BMC Microbiol* 2008;**8**:43.
35. Monecke S, Ehricht R. Rapid genotyping of methicillin resistant *Staphylococcus aureus* (MRSA) isolates using miniaturized oligonucleotide arrays. *Clin Microbiol Infect* 2005;**11**:825–33.
36. Sotto A, Richard JL, Jourdan N, Combescure C, Bouziges. N, Lavigne JP. Miniaturized oligonucleotide arrays: a new tool for discriminating colonization from infection due to *Staphylococcus aureus* in diabetic foot ulcers. *Diabetes Care* 2007;**30**:2051–6.
37. Sotto A, Lina G, Richard JL, Combescure C, Bourg G, Vidal L, et al. Virulence potential of *Staphylococcus aureus* strains isolated from diabetic foot ulcers: a new paradigm. *Diabetes Care* 2008;**31**:2318–24.
38. Vishwanathan V, Thomas N, Tandon N, Asirvatham A, Rajasekar S, Ramachandran A, et al. Profile of diabetic foot complications and its associated complications – a multicentric study from india. *J Assoc Physicians India* 2005;**53**:933–6.
39. Anandi C, Aaguraja D, Natarajan V, Ramanathan M, Subramaniam CS, Thulasiram M, et al. Bacteriology of diabetic foot lesions. *Ind J Med Microbiol* 2004;**2**:175–8.
40. Gadepalli R, Dhawan B, Kapil A. A clinico-microbiological study of diabetic foot ulcers in an Indian tertiary care hospital. *Diab Care* 2006;**29**:1727–32.
41. Citron DM, Goldstein EJC, Merriam CV, Lipsky BA, Abramson MA. Bacteriology of moderate-to-severe diabetic foot infections and *in vitro* activity of antimicrobial agents. *J Clin Microbiol* 2007;**45**:2819–28.
42. Ramakant P, Verma AK, Misra R, Prasad KN, Chand G, Mishra A, et al. Changing microbiological profile of pathogenic bacteria in diabetic foot infections: time for a rethink on which empirical therapy to choose? *Diabetologia* 2011;**54**:58–64.

43. Colayco CAS, Mendoza MT, Alejandria MM, Ang CF. Microbiologic and clinical profile of anaerobic diabetic foot infections. *Phil J Microbiol Infect Dis* 2002;**31**:151–60.
44. Abdulrazak A, Bitar ZI. Bacteriological study of diabetic foot infections. *J Diab Compl* 2005;**19**:138–41.
45. Raja NS. Anaerobic culture of diabetic foot infections: organisms and antimicrobial susceptibilities. *J Microbiol Immunol Infect* 2007;**40**:39–44.
46. Bansal E, Garg A, Bhatia S, Attri AK, Chander J. Spectrum of microbial in diabetic foot ulcers. *Indian J Pathol Microbiol* 2008;**51**:204–8.
47. Tiwari S, Pratyush DD, Dwivedi A, Gupta SK, Rai M, Singh SK. Microbiological and clinical characteristics of diabetic foot infections in Northern India. *J Infect Dev Ctries* 2012;**6**:329–32.
48. Lipsky BA, Armstrong DG, Citron DM, Tice AD, Morgenstern DE, Abramson MA. Ertapenem versus piperacillin/tazobactam for diabetic foot infections (SIDESTEP): prospective, randomised, controlled, double-blinded, multicenter trial. *Lancet* 2005;**366**:1695–703.
49. Sotto A, Richard JL, Combescure C, Jourdan N, Schuldiner S, Bouziges N, Lavigne JP. Beneficial effects of implementing guidelines on microbiology and costs of infected diabetic foot ulcers. *Diabetologia* 2010;**53**:2249–55.
50. Richard JL, Sotto A, Lavigne JP. New insights in diabetic foot infection. *World J Diab* 2011;**2**:24–32.
51. Eneroth M, Larsson J, Apelqvist J. Deep foot infections in patients with diabetes and foot ulcer: an entity with different characteristics, treatments, and prognosis. *J Diabetes Complications* 2000;**13**:254–63.
52. Fisher TK, Scimeca CL, Bharara M, Mills JL, Armstrong DG. A step-wise approach for surgical management of diabetic foot infections. *J Vasc Surg* 2010;**52**:72S–5S.
53. Shortt R, Thoma A. Empirical antibiotics use in soft tissue infections. *Can J Plast Surg* 2008;**16**:201–4.
54. Fincke BG, Miller DR, Christiansen CL, Turpin RS. Variation in antibiotic treatment for diabetic patients with serious foot infections: A retrospective observational study. *BMC Health Serv Res* 2010;**10**:193.
55. Grodinsky M. A study of the fascial spaces of the foot and their bearing on infections. *Surg Gynecol Obstet* 1929;**49**:739–51.
56. Badia JM, Torres JM, Tur C, Sitges-Serra A. Saline wound irrigation reduces the postoperative infection rate in guinea pigs. *J Surg Res* 1996;**63**:457–9.
57. Howell JM, Stair TO, Howell AW, et al. The effect of scrubbing and irrigation with normal saline, povidone iodine, and cefazolin on wound bacterial counts in a guinea pig model. *Am J Emerg Med* 1993;**11**:134–8.
58. Bansal BC, Wiebe RA, Perkins SD, Abramo TJ. Tap water for irrigation of lacerations. *Am J Emerg Med* 2002;**29**:469–72.
59. Calhoun JH, Overgaard KA, Stevens CM, Dowling JPF, Mader JT. Diabetic foot ulcers and infections: current concepts. *Adv Skin Wound Care* 2002;**15**:31–45.
60. Gibbons GW, Burgess AM, Guadagnoli E, et al. Return to well-being and function after infrainguinal revascularization. *J Vasc Surg* 1995;**21**:35–45.
61. Armstrong DG, Lipsky BA. Diabetic foot infections: stepwise medical and surgical management. *Int Wound J* 2004;**1**:123–32.

SECTION V

SKELETAL INFECTIONS

CHAPTER

16

Spinal Epidural Abscesses

Andreas K. Demetriades and Florian Ringel†*

*Department of Neurosurgery, Western General Hospital, Edinburgh, UK, †Department of Neurosurgery, Klinikum rechts der Isar, Technische Universität München, München, Germany

INTRODUCTION

Spinal epidural abscess (SEA) is an uncommon condition, which, if left untreated, can lead to irreversible neurological injury, paralysis, and even death. It is a condition whose diagnosis requires a high index of suspicion and whose treatment is a medical emergency. It may present to a variety of specialties, hence the need for all doctors to be aware of its importance.

In SEA there is an accumulation of purulent fluid material in the epidural space of the spinal canal. Although originally recognized postmortem in 1769 by Giovanni Batista Morgagni, it remained largely lethal until the advent of surgical decompression by means of a laminectomy and the introduction of antimicrobial treatment in the 1930s. In the current era, the diagnostic and therapeutic capabilities are much improved, yet SEA remains associated with a high morbidity and mortality.

EPIDEMIOLOGY

The incidence of SEA varies between 2 and 25 per 100,000 admissions and is believed to have steadily increased over the last few decades.[1,2] This increase is partly due to the availability of magnetic resonance imaging (MRI); the increasing aging population; the increasing number of patients undergoing invasive spinal or epidural procedures, such as for anesthesia or pain control; an increasing population of intravenous drug abusers; and an increasing prevalence of people with compromised immunity, such as diabetes and renal disease.

Microbiology for Surgical Infections
DOI: http://dx.doi.org/10.1016/B978-0-12-411629-0.00016-7

The disease may affect all age groups; however, 70% of those affected are between the ages of 31–70. The male:female ratio is approximately 2:1.[1]

PATHOPHYSIOLOGY OF INFECTION

Bacteria can gain access into the spinal canal by three mechanisms:

i. *Direct extension from infected contiguous tissue*: infection near the spine can spread from an adjacent infected focus. Such a mechanism is observed in SEA in relation to a psoas abscess; vertebral body osteomyelitis; discitis; retropharyngeal abscess; paravertebral abscess; or infected dermal sinus tract.
ii. *Direct inoculation*: This is often an iatrogenic condition, occurring after invasive techniques. Such techniques may be either therapeutic or diagnostic and include spinal surgery, spinal or epidural anesthesia, or epidural or facet joint or nerve root injections for pain management. The percentage of SEA that arise due to this group collectively has been quoted to be as high as 14–22%,[3,4] although the risk for SEA after epidural catheterization *per se* seems to be much smaller (1:1000 or less).[5]
iii. *Hematogenous spread*: infection can reach the spinal canal from non-contiguous foci in the skin; oral cavity (e.g., dental abscess); abdominal viscera; infective endocarditis; lung infection; genitourinary infection; or infected urinary catheter.

ANATOMICAL CONSIDERATIONS

The majority of SEA occur in the thoracolumbar spine (cervical 15%; thoracic 50%; lumbar 35%). The majority (80%) are dorsally rather than ventrally located. This relates partly to the fact that the thoracolumbar spine is much longer than the cervical spine, but also due to certain anatomical predispositions. The epidural space in the thoracolumbar spine is larger. Since the dura are adherent ventrally to the vertebral bodies from the foramen magnum to L1, this is only a potential rather than a true epidural space. As a result, most SEAs are posterior and often below L1. An actual epidural space exists posterolaterally all along the spinal canal, but this space is smaller in the cervical region and larger in the sacral region. The adipose tissue as well as the arteries and veins which are present in the posterior epidural space provide an avenue for hematogenous spread. Finally, the vertical sheath nature of the epidural space allows the vertical spread of SEA from the level of origin to multiple levels.

PATHOPHYSIOLOGY OF NEUROLOGICAL DEFICIT

This is not yet fully understood, as the pyogenic process of inflammation advances and the SEA spreads by 3–5 spinal cord segments on average, neurological compromise can arise from a combination of mechanisms:

i. *Direct mechanical compression* of the cord or cauda equina as the abscess expands in volume. The improvement after surgical decompression renders support to this argument. On the contrary, however, autopsy reports have shown an absence of compression in some patients, suggesting that more factors are at play.[4]

ii. *Thrombophlebitis and thrombosis* of epidural veins causing venous ischemia.[6]

iii. *Arterial ischemia* due to the compromised and compressed arterial blood supply. The ischemia hypothesis may be supported by the speed of symptom onset.[6]

iv. *Vasculitis* due to inflammatory mediators and toxins released by bacterial pathogens. This is suggested by the often severe neurological deficit in the presence of SEA that is small in volume, although experimental studies have failed to prove this.[7]

WHEN SHOULD SEA BE SUSPECTED?

Presenting Symptoms

Part of the challenge in reaching a timely diagnosis is the fact that often the initial symptoms are non-specific, such as malaise and pyrexia, and also variable. The speed of onset is also variable, with hematogenous spread progressing over days, whereas spread from osteomyelitis or discitis progresses over weeks or even months.

Symptoms include localized spinal pain; chills or fever; paravertebral muscle stiffness or spasms; radicular pain and paraesthesias; muscular weakness and sensory loss; sphincter dysfunction; and paraparesis, paraplegia or quadriplegia. Back pain, which can be severe and constant, is the commonest symptom affecting 80–90%, followed by fever in 50–60%, and neurological deficit (30%); these making up the "classical triad" of SEA symptoms.[8] This classic triad, however, is only seen in 8–37% of cases.[6,12]

Over time and if untreated, there is a sequence of neurological symptom progression as described by Heusner[9]:

Stage I: Back pain, fever, tenderness on palpation.
Stage II: Radicular pain, hyper-reflexia, nuchal rigidity.
Stage III: Sensory loss and motor weakness. Bladder and bowel dysfunction.
Stage IV: Paralysis.

Heusner's stages are unpredictable at best, as is the time for each, or taken to progress through stages. What is clear is that by the time that paralysis has manifested, it might be too late to reverse, hence the need for urgent spinal decompression.

Clinical Signs on Examination

Pyrexia is present in two thirds of cases whereas localized spinal tenderness in only one third. Neurological compromise can be found in 71%.[10] This compromise can vary from a single nerve root to a poly-radiculopathy, depending on the vertical spread of the abscess. In advanced cases there might be a transverse cord syndrome with motor, sensory and reflex levels. Incomplete or complete spinal cord syndromes will manifest paraparesis or paraplegia and sphincter disturbance. Reflexes may vary from hyperreflexia, with clonus and extensor plantars, to areflexia due to spinal shock and a transient inhibition of spinal reflexes.

Radiculopathy has been observed in 12–47% of case series. An associated weakness was present in 26–60%.[6] Paraplegia or tetraplegia can be the presenting sign in up to 25% of cases.[11] Nuchal rigidity may be present in cervical SEA.

Diagnostic Delay

The difficulties in diagnosing SEA can irreversibly affect outcome. Difficulties arise due to the non-specific nature of symptoms in the early stages of the disease; the frequently absent pyrexia or chills, and the absence of the "classical triad" in the majority of cases. A high index of suspicion is paramount. The impact of diagnostic delay on outcome was addressed in a study by Davis et al., who showed delays in diagnosis in 75% of SEA patients. Of these, there was neurological deterioration in 57%, and 45% had persistent motor weakness as opposed to 13% of those diagnosed in a timely fashion.[12]

Risk Factors

The commonest risk factor is diabetes mellitus. Others include alcoholism; intravenous drug abuse; immune compromise; spinal trauma, which might be at a different spinal location; spinal surgery; renal disease; pregnancy; spinal or epidural anesthesia or injections; tattooing; and acupuncture. Immune compromise includes malignancy, HIV or other infection, bacteremia; cirrhosis, and chronic steroid administration.[1,8,10] Davis et al. aptly highlighted that, given the variability of clinical presentation and the absence of the "classical triad", the consideration of risk factors becomes very crucial; one or more risk factors were present in 98% of those with SEA.[12] This again supports the need for a high index of suspicion and risk stratification based on symptoms and risk factor assessment.

DIFFERENTIAL DIAGNOSIS

The etiology of back pain and neurological complaints can be extensive, and it is the presence of fever that should alert the clinician to the possibility of SEA. The differential diagnosis should consider discogenic radiculopathy; spondylosis; transverse myelitis; degenerative disc disease; metastatic tumors; discitis; vertebral osteomyelitis; meningitis; psoas abscess; retropharyngeal abscess; and back pain.

INVESTIGATIONS

Laboratory

As leucocytosis is only seen in 60–80% of patients, the *white blood cell count (WBC)* is not very reliable as neither the total WBC nor the neutrophil count is sensitive enough to make the diagnosis.[6]

The *erythrocyte sedimentation rate (ESR)* is more sensitive and is elevated in 94% of cases.[1] It is slow to return to normal, however, and as such may not be very useful in monitoring the response to treatment, whereas C-reactive proteins (CRP) return to normal sooner.[6]

Blood cultures are used to try and identify the causative organism and the origin of the infection. They can be negative in up to 40% of cases.[3,12]

Screening includes *urinary analysis and culture* in all cases of SEA. In suspected endocarditis, an *echocardiogram* is indicated, such as in the intravenous drug abusers. In a suspected dental abscess an *orthopantomogram and dental/maxillofacial review* is indicated.

Lumbar punctures, previously performed, are NOT indicated, so as to prevent spread of the infection into the intradural space.

Radiology

In suspected cases of SEA it is essential to image the spine. Barring any contra-indications, such as a pacemaker, MRI is the investigation of choice. It is sensitive and specific from the early stages of the infection and it assesses its location and extent, as well as relation to adjacent structures, e.g., osteomyelitis, discitis or psoas abscess. A whole spine MRI is required both pre and post contrast administration to show the extent of the collection, which can involve the whole spine, and to identify potential skip lesions. The MRI scan should be performed as an emergency and should there be no such facility, the patient must be urgently transferred to an appropriately equipped center.

SEA demonstrates low T1 and high T2 signal on MRI. They exhibit two types of enhancement; homogeneous enhancement when there is inflammatory tissue without purulent fluid, or, peripheral enhancement with a necrotic core. Linear enhancement may also be seen, due to the extension of the inflammation into the dura. Engorgement of the epidural veins can also be seen just above and below the SEA in the sagittal plane.[13]

A computer tomography (CT) scan with contrast may be acceptable if an MRI is contra-indicated. CT myelography is cumbersome and unnecessary, as in the absence of MRI the quality of modern CT scanners is very good. Plain radiographs, though they might identify bony erosion due to discitis or osteomyelitis, are not diagnostic of SEA. They can however point out relevant information such as the presence of an indwelling foreign body or spinal deformity, which will be relevant during the surgical treatment.

MICROBIOLOGY

Most infective agents are bacterial but mycobacterial, parasitic and fungal abscesses also occur. The commonest pathogens are *Staphylococcus aureus* (50–93%), Streptococci (18%), and several types of gram-negative bacilli (13%).[5] In less developed countries, and in immunocompromised populations, tuberculous abscesses or *Candida albicans* occur more frequently.[5] In a meta-analysis of 915 patients, Reihsaus identified the following causative organisms:

- Gram-positive cocci, eg *Staphylococcus aureus* and *epidermidis*, coagulase-negative staphylococci

- *Streptococcus pneumonia*, *viridans*, *pyogenes*, and *milleri*
- Gram-negative rods, eg *Escherichia coli*, *Enterobacter*, *Salmonella*, *Klebsiella*; *Bacteroides*; *Pseudomonas aeruginosa*
- Mycobacteria
- Mixed bacteria
- Fungi, e.g., *Aspergillus fumigatus*
- Parasites, e.g., *Echinococcus*[1,5]

CLINICAL MANAGEMENT

The most important factor in getting a prompt diagnosis is a high index of suspicion. Symptoms and signs of spinal tenderness and neurological involvement with fever should alert to the possibility of SEA. The same suspicion should be aroused with symptoms of spinal pain and neurological involvement without pyrexia but in the presence of risk factors. The diagnostic tests should include inflammatory markers (WBC, CRP, ESR), blood cultures, urinalysis and culture, in the first instance. An urgent MRI should be obtained and, if positive, an urgent consultation with the local/regional neurosurgical team should be sought for possible transfer and treatment.

The goals of treatment are eradication of the infection, preservation of neurological function, pain control and maintenance of spinal stability. The treatment options are largely three:

i. *Surgical debridement and targeted antimicrobial therapy*

   This is the treatment of choice for the vast majority of patients. By means of a laminectomy, the spinal canal is accessed and the abscess removed and washed out. Samples are sent for cultures and sensitivities. The presence of spinal instability due to any associated deformity, such as with osteomyelitis, can be addressed with stabilization if necessary. The patient remains on broad-spectrum antibiotics until the culture results, after which the prolonged specific treatment is coordinated between the surgical and infectious diseases teams. The progress is monitored with inflammatory markers and serial MRI scans.

ii. *Aspiration biopsy and targeted antimicrobial therapy*

   Occasionally a smaller procedure is undertaken either for reasons of medical comorbidity or due to a small volume of SEA, which would not benefit from open surgery. The antibiotic treatment is followed as above.

iii. *Empirical antimicrobial therapy*

   Non-operative treatment is reserved for cases where the patient is a poor surgical candidate, either due to medical comorbidities or futile surgical treatment, such as with established paralysis of over 48–72 hours duration.[6]

Timing of Antibiotic Treatment

The choice of antibiotic treatment is an important consideration. If the patient is stable and surgery can take place expeditiously, systemic antibiotics can be withheld until after intraoperative culture samples can be obtained. This increases the diagnostic accuracy of the cultures, which will allow specific antibiotic treatment after the pathogen's sensitivities are identified. If, on the other hand, surgery will not happen expeditiously, or the patient is septic, or there is neurological compromise, then empiric broad-spectrum antibiotic therapy should commence. This should take into account the most likely pathogens (staphylococcus and streptococcus), the need for gram-negative cover, and the patient's specific history. After culture results a switch to appropriate antibiotics is mandatory, and follow-up as described above.

Comparative Effectiveness of Surgical vs. Non-Operative Treatment

There have been reports of recovery with non-operative treatment of patients with no or early neurological compromise. A retrospective review of SEA has shown that prolonged use of parenteral antibiotics alone or combined with percutaneous needle drainage yielded clinical outcomes at least comparable with antibiotics plus surgical intervention; there were no significant differences in the rates of complete recovery, and this was unrelated to abscess size, age or neurological deficit at presentation.[14] Other reports, where diagnosis was made by culture, also reported successful outcomes with medical treatment, but drew attention to the retrospective and selective nature of the reports.[15] On the other hand, a very recent retrospective review showed a 65% improvement with surgical therapy and identified age as a prognostic factor.[16] Overall, these highlight the difficulties of retrospective analysis and the limitations of selection bias and reporting bias in a rare condition.

It is generally accepted that where there is no neurological deficit, medical therapy after a positive culture is acceptable; and that if there is neurological deficit, the chances of recovery are enhanced by early surgical debridement followed by targeted antibiotic therapy.[1,12,13,16]

Antibiotics

While definitive therapy will depend on the culture and sensitivities, empirical broad-spectrum therapy should be commenced after consultation with a microbiologist and/or infectious disease specialist. Such empirical treatment, until switching to specific treatment, must be able to cover against staphylococci, streptococci, and gram-negative bacilli. Such a regime may be vancomycin plus metronidazole plus cefotaxime or ceftriaxone or ceftazidime. This way vancomycin covers against methicillin resistant *Staphylococcus aureus* (MRSA) as well as covering penicillin allergy. If later the culture shows methicillin sensitive *S. aureus* (MSSA), vancomycin can be changed to oxacillin or nafcillin. An alternative for MRSA is linezolid.[5]

The duration of treatment with antibiotics is usually 6–8 weeks or until the abscess has completely resolved on serial MRI scanning. Such scanning can start at four weeks post-treatment initiation, or earlier if clinically indicated.

PROGNOSIS

The presence of paralysis for more than 24 hours usually signifies a permanent event, and delay to diagnosis directly affects outcome. 45% of those with delayed diagnosis will have persistant motor deficit, while permanent paraplegia has been observed at up to 22%.[12,17] The mortality of SEA, either from sepsis or prolonged immobilization, is 5%, markedly improved from 81% in the 1920s.[5,17]

Patient age, duration of symptoms, and degree of neural compression are three prognostic factors identified.[18] Any future accumulation of data will be subject to the same constraints as existing reports due to the small numbers of people suffering from the condition and its presentation as an emergency.

CONCLUSION AND FUTURE PROSPECTS

Spinal epidural abscess can present to a variety of clinicians and its prompt diagnosis will determine the outcome. Due to the often non-specific mode of presentation, a high index of suspicion is required in people with back pain and who are immunocompromised; back pain and pyrexia; who have a recent history of a spinal invasive procedure and show signs of an infection. These cases may present with or without a neurological deficit. Investigations should be pursued as an emergency. Laboratory investigations include serology (WBC, CRP, ESR) and cultures (blood and urine). Urgent MRI is the imaging investigation of choice. Upon diagnosis of SEA, an urgent referral to the neurosurgical unit should follow. The mainstay of treatment is surgical debridement and long term antibiotics, though selected cases may be treated non-operatively. The multidisciplinary input of the referring, neurosurgical, microbiological and radiological teams is essential for prompt and effective management which will directly affect outcome.

Future prospects are limited by the rarity of the condition and its presentation as an emergency, which preclude the feasibility of trial randomization. The most important future factor will continue to be an increased awareness and familiarity and high index of suspicion. This has been and will continue to be relevant with an increasing aging population; with the improved survival of immunocompromised patients, such as those with malignancy, diabetes, HIV; and an ever-rising number of spinal procedures.

References

1. Reihsaus E, Waldbaur H, Seeling W. Spinal epidural abscess: a meta-analysis of 915 patients. *Neurosurg Rev* 2000;**23**:175–204.
2. Johnson KG. Spinal epidural abscess. *Crit Care Nurs Clin North Am* 2013;**25**:389–97.
3. Danner RL, Hartman BJ. Update on spinal epidural abscess: 35 cases and review of the literature. Rev Infect Dis 1987;**9**:265–74.
4. Hlavin ML, Kaminski HJ, Ross JS, Ganz E. Spinal epidural abscess: a ten-year perspective. *Neurosurgery* 1990;**27**:177–84.
5. Grewal S, Hocking G, Wildsmith JA. Epidural abscesses. *Br J Anaesth* 2006;**96**:292–302.
6. Tompkins M, Panuncialman I, Lucas P, Palumbo M. Spinal epidural abscess. *J Emerg Med* 2010;**39**:384–90.
7. Feldenzer JA, McKeever PE, Schaberg DR, Campbell JA, Hoff JT. Experimental spinal epidural abscess: a pathophysiological model in the rabbit. *Neurosurgery* 1987;**20**:859–67.
8. Darouiche RO, Hamill RJ, Greenberg SB, Weathers SW, Musher DM. Bacterial spinal epidural abscess. Review of 43 cases and literature survey. *Medicine (Baltimore)* 1992;**71**:369–85.
9. Heusner AP. Nontuberculous spinal epidural infections. *N Engl J Med* 1948;**239**:845–54.

10. Rigamonti D, Liem L, Sampath P, Knoller N, Namaguchi Y, Schreibman DL, et al. Spinal epidural abscess: contemporary trends in etiology, evaluation, and management. *Surg Neurol* 1999;**52**:189–96.
11. Soehle M, Wallenfang T. Spinal epidural abscesses: clinical manifestations, prognostic factors, and outcomes. *Neurosurgery* 2002;**51**:79–85.
12. Davis DP, Wold RM, Patel RJ, Tran AJ, Tokhi RN, Chan TC, Vilke GM. The clinical presentation and impact of diagnostic delays on emergency department patients with spinal epidural abscess. *J Emerg Med* 2004;**26**:285–91.
13. Avilucea FR, Patel AA. Epidural infection: Is it really an abscess? *Surg Neurol Int* 2012;**3**:S370–6.
14. Siddiq F, Chowfin A, Tight R, Sahmoun AE, Smego Jr RA. Medical vs surgical management of spinal epidural abscess. *Arch Intern Med* 2004;**164**:2409–12.
15. Wheeler D, Keiser P, Rigamonti D, Keay S. Medical management of spinal epidural abscesses: case report and review. *Clin Infect Dis* 1992;**15**:22–7.
16. Connor Jr DE, Chittiboina P, Caldito G, Nanda A. Comparison of operative and nonoperative management of spinal epidural abscess: a retrospective review of clinical and laboratory predictors of neurological outcome. *J Neurosurg Spine* 2013;**19**:119–27.
17. Darouiche RO. Spinal epidural abscess. *N Engl J Med* 2006;**355**:2012–20.
18. Khanna RK, Malik GM, Rock JP, Rosenblum ML. Spinal epidural abscess: evaluation of factors influencing outcome. *Neurosurgery* 1996;**39**:958–64.

SECTION VI

SURGICAL INFECTIONS OF NON-BACTERIAL ORIGIN

CHAPTER

17

State of the Art in the Laboratory Methods for the Diagnosis of Invasive Fungal Diseases

Guillermo Quindós, Elena Eraso*, Guillermo Ezpeleta*[†], *Javier Pemán*** and Ferran Sanchez-Reus*[‡]

*UFI 11/25, Departamento de Inmunología, Microbiología y Parasitología, Universidad del País Vasco/Euskal Herriko Unibertsitatea (UPV/EHU), Bilbao, Spain, [†]Departamento de Medicina Preventiva y Salud Pública, Facultad de Medicina y Odontología, Universidad del País Vasco/Euskal Herriko Unibertsitatea (UPV/EHU), Bilbao, Spain, **Servicio de Microbiología, Hospital Universitario y Politécnico La Fe, Valencia, Spain, [‡]Servicio de Microbiología, Hospital de la Santa Creu i Sant Pau, Barcelona, Spain

INTRODUCTION

Invasive fungal diseases (IFDs) possess a great clinical relevance because of their growing incidence, morbidity and mortality in oncohematological and surgical patients. Immunosuppression and breakdown of anatomical barriers are major risk factors and most at-risk patients are in surgery services, AIDS clinics, intensive care units (ICU), or transplantation and oncology units. Invasive candidiasis (IC) and to a lesser extent pulmonary and disseminated aspergillosis, pulmonary pneumocystosis and meningeal cryptococcosis are the most frequent IFDs. Organ transplant recipients form an important group of at-risk patients, with the type of transplanted organ causing major differences in etiology; thus, aspergillosis and other filamentous fungi infections are common in bone marrow transplants. However, IFDs are dominated by *Candida* in the first month after solid organ transplantation, in a manner usually related to surgical issues and nosocomial risk factors. Thus, anastomotic leaks, early graft failure, reoperation, and central venous catheter-associated candidemia are common. In the absence of complications of transplant, IFD by

Microbiology for Surgical Infections
DOI: http://dx.doi.org/10.1016/B978-0-12-411629-0.00017-9

filamentous fungi are uncommon. After the first month, intense immunosuppression becomes manifest and IFD by filamentous fungi, such as aspergillosis, mucormycosis, fusariosis or scedosporiosis, are relevant. *Pneumocystis jirovecii* pneumonia is also common during this period among patients not receiving prophylaxis. The peak incidence of aspergillosis is between one and four months after transplantation. The development of IFD during this time interval often follows evidence of active disease with an immunomodulating virus, such as cytomegalovirus, Epstein-Barr, or hepatitis B and C viruses.[1–3]

Invasive candidiasis is a leading cause of mortality, and candidemia is the most common clinical presentation. *Candida albicans* remains the predominant etiology, accounting for nearly 50% of all cases. However, there has been an epidemiological shift in the last decades, and a number of species of *Candida* other than *C. albicans* have emerged as important causes of severe candidemia. Moreover, these non-*C. albicans* species can exhibit resistance to fluconazole and other antifungal agents. *Candida parapsilosis* has been associated with candidemia in neonates and young adults. This species usually has an exogenous origin and contaminates medical devices, causing central venous catheter-associated infections. *Candida glabrata*, *Candida tropicalis* and *Candida krusei* have been isolated from older patients (>65 years) with important risk factors, such as major abdominal surgery, solid tumors and hematologic malignancies, transplants, and/or prolonged treatment with corticoids. *C. parapsilosis* predominates in Australia, Latin America and Mediterranean countries of Africa, Asia and Europe. In contrast, *C. glabrata* has an important etiological role in the USA and Central and Northern Europe. An important and worrying issue is that mortality due to invasive candidiasis remains unacceptably high.[3]

Aspergillus is the etiological agent responsible for a variety of infections and conditions referred to as aspergillosis. These fungi are ubiquitous and can be found everywhere in the human environment. Their manifestations include a spectrum of diseases from allergic responses, colonization and invasive infection. The main organ of entrance is the lung and pneumonia is usually the first manifestation. Invasive aspergillosis is a significant cause of morbidity and mortality in high-risk patients, including patients undergoing hematopoietic stem cell or solid organ transplantation and those using corticosteroids and other immunosuppressive therapies. *Aspergillus fumigatus* is the predominant species, but others such as *Aspergillus flavus*, *Aspergillus niger* or *Aspergillus terreus* are important in some clinical settings. As in the case of candidiasis, aspergillosis caused by non-*A. fumigatus* species is a major challenge for the clinician, as diagnosis and treatment are more complicated. Cryptococcosis is an invasive fungal infection caused by the encapsulated yeasts *Cryptococcus neoformans* and *Cryptococcus gattii*. These fungi are found in soil and vegetation, often associated with bird droppings. Infection involves the lungs and the central nervous system. A higher incidence of cryptococcosis is observed in immunodeficient patients, especially among patients with AIDS or organ transplant recipients, but cryptococcosis can occur in immunocompetent patients, mainly due to *C. gattii*.[4] *P. jirovecii* is an important cause of pneumonia in immunodeficient hosts, especially those with hematologic malignancies, in receipt of organ transplant, those with AIDS, and those receiving potent immunosuppressive drugs. The incidence of IFD caused by unusual fungi, such as the Mucorales, *Trichosporon*, *Magnusiomyces* (formerly *Saprochaete* or *Blastoschizomyces*), *Fusarium* or *Scedosporium*, is very low.[1–4]

In most cases, clinical suspicion of IFD begins after persistent fever despite antibacterial therapy. However, fever has very low specificity and radiological findings are neither specific nor sensitive. Laboratory diagnosis of IFD is a serious challenge because conventional methods, such as direct microscopy, histology or culture of clinical specimens, are not sensitive or specific enough, and the results are often available too late to be clinically useful. This delay habitually enables the IFD to progress to the point where antifungal therapy becomes ineffective, thus there is a close relationship between early diagnosis, prompt treatment, and lower mortality.[5,6] The prompt initiation of antifungal therapy is critical for improving the IFD outcome, and in many cases this has led to empirical treatment.[1] Early mycological diagnosis is the cornerstone of a prompt and appropriate treatment, and for improving the survival of patients. Diagnostic methods are in continual evolution, and more rapid and sensitive methods have been developed, such as the detection of biomarkers or surrogate markers, e.g., 1,3-β-D-glucan (BG), glucuronoxylomannan (GXM), galactomannan (GM), mannan (MN), antimannan and anti-*Candida* germtube antibodies (CAGTA), and fungal DNA detection by polymerase chain reaction (PCR). However, many of these approaches are unfamiliar and they are generally not available in most clinical settings.[7–11]

TRADITIONAL MICROBIOLOGICAL DIAGNOSIS

Etiological diagnosis combines clinical, radiological, microbiological and histological findings. Microbiologically essential steps include the identification of the causal fungus by direct observation and culture from normally sterile sites (such as blood, cerebrospinal fluid [CSF], pleural or peritoneal fluids, or deep tissue biopsy) combined with clinical signs of infection at sites of fungal isolation and the absence of other potential pathogens. Conventional diagnostic procedures depend on the possibility of obtaining specimens from infected deep tissues, but these methods are slow, present poor sensitivity or specificity and do not clearly differentiate fungal colonization from invasion. Thus, a histological diagnosis is the only method that confirms tissue invasion, but the isolation of the etiological agent in culture gives essential additional information.

Microscopic examination of clinical specimens can show characteristic fungal structures and it is a reliable diagnostic procedure. However, histological examination and biopsy culture require an aggressive approach that often precludes their use in many patients. In pulmonary mycoses (aspergillosis, cryptococcosis or pneumocystosis), bronchoalveolar lavage (BAL) is an optimal specimen, but the bronchoscopy procedure depends on the clinical status of patient and it is made after radiological detection of pulmonary infiltrates in patients with persistent fever, causing a diagnostic delay. India ink examination of CSF is a very useful and rapid diagnostic test for cryptococcal meningitis. Tissue sections and body fluids stained with calcofluor white (for fungal chitin), Gomori-Grocott methenamine silver, or periodic acid-Schiff, facilitate the observation of fungal elements, tissue alterations and the inflammatory responses against the invasion. For capsulated fungi, such as *Cryptococcus*, mucicarmine and alcian blue stains identify the capsular material surrounding the fungi, and Fontana-Mason stain (for melanin) can help in the detection of criptococci and dematiaceous fungi. However, fungal structures and organic lesions can be difficult to interpret, and the fungal burden may be too low for visualization. For these reasons, specific antibodies marked with fluorescein, DNA probes for hybridization, or

PCR with specific primers can help because they have the advantage of detecting low numbers of fungi in tissues.[9–14]

Culture of adequate clinical specimens is essential to establish or confirm the diagnosis of IFD. The isolation of fungi in culture gives invaluable information on the main characteristics of the fungus (such as identity, virulence factors, genotype and antifungal susceptibility). However, contamination, sample volume, optimum sampling frequency, and the difficulty of distinguishing colonization from infection limits current utility. Cultures of blood and other normally sterile specimens provide useful information, so increasing the likelihood of diagnosing an IFD from the number of positive cultures. The incubation period must be at least 5–7 days for yeasts, but two or more weeks could be necessary if filamentous fungi are involved. The use of an automated validated blood culture system is recommended (BACTEC, Becton Dikinson, USA, and BacT/Alert, bioMérieux, France). Mycological media give the highest performance, but their routine usage is unusual. However, blood cultures are only positive in 50–75% of autopsy-proven IC and their usefulness is even worse in other IFDs. Moreover, the sensitivity of blood cultures decreases in neutropenic patients and in patients under antifungal prophylaxis. This can be related to the intermittent presence of fungi in blood or low concentration of fungal cells. Hence it can be seen that negative blood cultures do not rule out an IFD. Lysis-centrifugation or culture of the buffy coat could improve the yield of fungemia detection in invasive infections by unusual fungi. Sequencing of fungal DNA from positive blood culture bottles can facilitate early diagnosis of fungemia.[10,15–18]

Most fungi grow in current bacteriological media such as blood agar, but specific media like Sabouraud dextrose agar with cloramphenicol can facilitate the isolation. *Candida*, *Trichosporon* and other yeasts grow in white to yellow mucoid or dry colonies. *Rhodotorula* and *Sporobolomyces* colonies usually are red or orange to salmon. *Aspergillus* and other filamentous fungi develop a rich spectrum of colonies with different colors and textures. CHROMagar *Candida* (CHROMagar company, France), ChromID *Candida* (bioMérieux), and chromogenic agar for *Candida* (Laboratorios Conda, Spain) are specific chromogenic agars used for primary culture and presumptive identification of *C. albicans*, *Candida dubliniensis*, *C. glabrata*, *C. krusei*, and *C. tropicalis*, based on colony colors. These chromogenic media are also helpful for distinguishing polymicrobial infections. *C. neoformans* and *C. gattii* develop brown to black colonies when cultured on birdseed agar or caffeic acid synthetic medium. Culture in canavanine-glycine-bromothymol blue agar separates both species of *Cryptococcus*, as the medium turns blue because of glycin degradation by *C. gattii*. Special media or Sabouraud agar supplemented with drops of sterile olive oil are required for the isolation of *Malassezia*.[10,19,20]

Morphological and biochemical characteristics allow the identification of many fungi. The germ tube test in serum is rapid, but nonspecific for *C. albicans* identification. Direct microscopic identification is based on the observation of fungal structures, like septate hyphae, pseudohyphae, arthroconidia, blastoconidia, or appresoria. *C. neoformans* and other cryptococci are budding, encapsulated yeast cells that can be visualized in CSF using India ink. *Trichosporon* develops cylindrical arthroconidia in addition to septate hyphae and blastoconidia. Cells of *Rhodotorula* are sub-globose, oval to elongate, with or without small capsules and sometimes rudimentary hyphae but not the ballistoconidia that are characteristic of *Sporobolomyces*. Morphology is even more important for filamentous fungi differentiation, as the presence of septa in regular hyphae, such as in *Fusarium*, *Aspergillus* or *Scedosporium*, differentiate them from Mucorales, with non-septate, wide and irregular hyphae. Moreover, in many fungi the hyphae are transparent, (hyalinum), such as *Aspergillus*, in contrast to others, such as *Alternaria* which have brown, melanised hyphae and are known as dematiaceous fungi. Finally, observation of different reproductive structures (conidia) is very useful for distinguishing fungal genera.[7,10–12]

Biochemical, immunological and molecular methods are needed for the identification of many yeast-like fungi. There are several commercial systems based on carbohydrate assimilation, such as the API ID 32C system (bioMérieux), Auxacolor 2 (Bio-Rad, USA), Vitek Systems (bioMérieux) or Baxter Microscan (Baxter Microscan, USA), that facilitate the identification of most pathogenic yeasts in 24–48 hours.[20–22] Rapid immunological agglutination tests for *C. albicans*, *C. dubliniensis* and *C. krusei* (Bichro-latex albicans, Bichro-dubli or krusei color [Fumouze Diagnostics, France]) give a reliable identification within 5–10 min. A rapid trehalose test (Fumouze Diagnostics) for *C. glabrata* and urea hydrolysis by *Cryptococcus* and *Trichosporon* allow a presumptive identification within a few hours.[7,12,23–25]

DNA amplification, DNA sequencing or the use of DNA hybridization probes are useful for identifying uncommon fungi. Internal transcriber spacers (ITS1 and ITS2), the D1/D2 domain and the intergenic spacer region (IGS1) of the ribosomal DNA remain the most commonly sequenced because they contain species-specific sequences.[10,26–28] Peptide nucleic acid fluorescence *in situ* hybridization (PNA-FISH) (Yeast Traffic Light, AdvanDx, USA) distinguishes among different *Candida* species from positive blood culture bottles, being rapid (2 h), highly specific, and sensitive. The assay uses PNA probes which target *Candida*-specific rRNA.[29] Some DNA-microarray systems based on the arrayed-primer extension (APEX) technique have been developed for the rapid identification of pathogenic fungi, but information on their usefulness is limited.[30] Matrix-assisted laser desorption ionization-time of flight mass spectrometry (MALDI-TOF MS) is a suitable method for a fast and secure identification of *Candida* isolates, and can detect *Candida* directly in blood culture bottles in two hours.[31–33] However, identification performance is lower for other yeasts, including *Cryptococcus*, *Trichosporon* and *Geotrichum*, and for filamentous fungi, due to the low representation of these genera in the databases, or because their cell walls are rich in external sugars, making protein extraction more difficult.[34,35] Other techniques have been developed for rapidly identifying fungal isolates, usually in combination with DNA amplification by PCR, such as electro spray ionization mass spectrometry (ESI MS).[36,37]

The emergence of species with reduced susceptibility or resistance to current antifungal agents, such as *C. glabrata*, *C. krusei*, *A. terreus* or *Scedosporium prolificans*, makes identification very important, as the antifungal susceptibility of clinical isolates could be inferred from their identity.[7–12] However, *in vitro* antifungal susceptibility testing has an important role and there are standardized tests (CLSI [Clinical and Laboratory Standards Institute] and EUCAST [European Committee on Antimicrobial Susceptibility Testing] methods), and several commercialized kits, such as Sensititre Yeast One (TREK Diagnostics, USA), ATB Fungus or VITEK 2 (bioMérieux), than can facilitate the study of clinical isolates of yeasts. Moreover, Etest strips (bioMérieux) are very useful for studying the *in vitro* susceptibility of either yeast-like or filamentous fungi.[8,18,22,38]

ALTERNATIVE METHODS FOR DIAGNOSIS: NON-CULTURE BASED METHODS

The main advantage of molecular methods for detecting fungal biomarkers is their speed, as they do not need 24–48 hours or even weeks for fungal isolation. There are two complementary approaches; the detection of panfungal surrogate markers, like BG and common sequences of fungal DNA, or the detection of markers of specific IFDs, such as GM (aspergillosis), GXM (cryptococcosis), mannan and antimannan antibodies, CAGTA (IC) or fungal specific DNA sequences that facilitate a more reliable diagnosis. Rapid elimination and/or low production of some biomarkers are important disadvantages that could be avoided by analyzing serial samples.[8–12,39] These techniques combined with risk stratification and conventional laboratory methods permit an early diagnosis of IFD, and the implementation of pre-emptive therapeutic strategies are especially useful for patients with neutropenia. Circulating fungal markers are an early indicator of opportunistic mycoses, and identifying them provides a tool for evaluating the response to treatment. GXM, GM and BG assays in combination with radiologic procedures, such as computed tomography scan (CT-scan) or positron emission tomography (PET) are part of the diagnostic strategy in many hospitals. PCR and other approaches to the detection of fungal DNA in blood or BAL specimens are promising tools, and some commercialized kits are available. These include LightCycler SeptiFast Test MGRADE (Roche Molecular Diagnostics, USA), MycAssay Aspergillus and MycAssay Pneumocystis (Myconostica, UK), Magicplex Sepsis (Seegene, South Korea), Luminex xTAG Fungal ASR (Luminex Molecular Diagnostics, USA), and VYOO/LOOXTER (SIRS-Lab, Germany).[40–42]

1,3-β-D-Glucan (BG) Detection

BG is a cell wall component of most fungi, with the exception of *Cryptococcus* and the Mucorales, which is not present in bacteria, viruses and mammalian cells. BG detection in blood and human tissues is a panfungal marker. Serum BG concentrations show a constant rise before clinical and microbiological evidence of IFD, then decrease, and eventually become negative if patients respond to antifungal therapy. Conversely, patients not responding to antifungal treatment do not show a decrease or show a continuous rise in serum BG. There are several techniques for the detection of BG, with Fungitell being the most used test in Europe and the USA (Table 17.1).[1,43–45] Several studies have confirmed the high specificity and good positive (PPV) and negative predictive values (NPV) of BG assays for the diagnosis of a wide range of IFDs, such as those caused by *Aspergillus*, *Candida*, *Fusarium*, *Trichosporon*, *Saccharomyces* or *Pneumocystis*. BG detection twice a week is very useful for ruling out IFD.[46–55] Initial studies suggest that BG assays may be combined with either measurement of GM, PCR or detection of antibodies to *Candida*-specific antigens to improve sensitivity and PPV.[1,10,43,55]

There are potential drawbacks associated with BG detection for IFD diagnosis. Although the ability of BG testing to detect a wide range of fungi is beneficial, this can also be a disadvantage when a more specific diagnosis is needed.[1,56] Furthermore, administration of immunoglobulin or albumin preparations contaminated with fungal components,

TABLE 17.1 Evidence and Strength of Diagnostic Tests Recommendations for Invasive Mycoses

Indications and Diagnostic Limits	1–3-β-D-Glucan (BG) Detection	Galactomannan (GM) Detection
Quality of evidence and strength of recommendations	Early diagnosis of invasive fungal disease (Panfungal biomarker) in adults: invasive candidiasis and invasive aspergillosis (AI-BII), and pneumocystosis (BII).	Early diagnosis of invasive aspergillosis in: Hematologic patients with malignancies (AI-AII-BII). HIV-infected patients and other immunodeficient patients (BII) Patients in medical and surgical ICUs (BII).
Methods and suggested use	Fungitell (Associates of Cape Cod, USA). Wako colorimetric assay (Wako Pure Chemical Ind., Japan). Fungitec-G (Seikagaku, Japan). Maruha colorimetric assay (Maruha-Nichiro Foods, Japan). Detection each 3–4 days in serum.	Platelia Aspergillus EIA – EBA-2 monoclonal antibody- (Bio-Rad Laboratories, France). Detection each 3–4 days in serum (and in bronchoalveolar lavage or cerebrospinal fluid if available).
Diagnostic value	Mycological criterion of probable invasive mycoses according to EORTC/MSG and the ECIL-3. Included in the IDSA guidelines for aspergillosis and candidiasis. Recommended by ESCMID for the diagnosis of candidaemia and invasive candidiasis. Detection of BG previous to positivity in other diagnostic tests. Positive: >60–80 pg/ml (Fungitell) or 7 pg/ml (Wako).	Mycological criterion of probable invasive aspergillosis according to EORTC/MSG and the ECIL-3. Included in the IDSA guidelines for aspergillosis. Detection of GM previous to positivity in other diagnostic tests. Combination with CT-scan and other image-based methods improves its diagnostic usefulness. Positive: >1 (in bronchoalveolar lavage), >0.7 (in a single serum test) or >0.5 (in two consecutive serum tests or in a single cerebrospinal fluid test)
Prognostic value	BG decrease is considered as good prognosis.	Decrease of GM is considered as good prognosis. GM index >1 in serum after the initiation of treatment or without decrease after 1 week of treatment points to a therapeutic failure.
False positives	Contamination of laboratory ware. Gram-positive (*Streptococcus pneumoniae*) and Gram-negative (*Alcaligenes* and *Pseudomonas aeruginosa*) bacteremia. Contact with fat and sponge or gauze containing glucan during surgery. Hemodialysis (glucose acetate filters and membranes). Intravenous albumin, immunoglobulins, coagulation factors, plasmatic proteins. Antineoplastic (lentinan and K polysaccharide) and antimicrobial agents (amoxicillin-clavulanic acid, cefepine, meropenem, or piperacillin-tazobactam)*.	Potential cross reactivity with other fungi: *Acremonium, Alternaria, Botrytis, Cladosporium, Fusarium, Geotrichum, Magnusiomyces, Paecilomyces, Penicillium, Rhodotroula, Trichosporon, Wangiella*, etc. Gastro-intestinal colonization by *Bifidobacterium* species (neonates and children). Immunoglobulins, coagulation factors, plasmatic proteins, parenteral food solutions (Plasma-Lyte, Baxter, USA), soya rich diets, cyclophosphamide, and some antimicrobial agents of fungal origin

(Continued)

TABLE 17.1 (Continued)

Indications and Diagnostic Limits	1–3-β-D-Glucan (BG) Detection	Galactomannan (GM) Detection
		(amoxicillin-clavulanic acid, cefepine, meropenem, or piperacillin-tazobactam). Chronic Graft versus Host disease.
False negatives	Hiperpigmented serum (↑ bilirubin and/or triglycerides). Antifungal prophylaxis or empiric treatment. Intravenous azithromycin or pentamidine.	Osteomyelitis, tracheobronchitis and other focalized aspergillosis. Some *Aspergillus* species, such as *A. fumigatus*, produce less GM than others. Antifungal prophylaxis or empiric treatment.

EORTC/MSG; European Organization for Research and Treatment of Cancer/Invasive Fungal Infections Cooperative Group and the National Institute of Allergy and Infectious Diseases Mycoses Study Group. ECIL-3: 3rd European Conference on Infections in Leukemia. IDSA: Infectious Diseases Society of America. ESCMID: European Society of Clinical Microbiology & Infectious Diseases. **Strength of recommendation:** A = Good evidence to support a recommendation for or against use. B = Moderate evidence to support a recommendation for or against use. C = Poor evidence to support a recommendation. **Quality of evidence:** I = Evidence from ≥1 properly randomized, controlled trial. II = Evidence from ≥1 well-designed clinical trial, without randomization; from cohort or case-controlled analytic studies (preferably from >1 center); from multiple time-series; or from dramatic results from uncontrolled experiments. III = Evidence from opinions of respected authorities, based on clinical experience, descriptive studies, or reports of expert committees.
**This problem with some antimicrobial agents has been solved for most marketed products.*

hemodialysis with cellulose membranes, bacteremias, surgical use of gauze containing glucan, and certain antibiotics can also lead to false positive results (Table 17.1). Two recent meta-analyses have reported sensitivity and specificity values of 50–90% and 70–100%, respectively for proven or probable IFD.[45,57] BG detection showed a greater utility in patients without hematological diseases such as surgical or medical ICU patients suffering from IC. Estimated PPV and NPV for an IFD prevalence of 10% in patients with hematological malignancies were 84% and 95%, respectively.[45] The interpretation of BG among pediatric patients is more difficult because of false positives and limited data.[58] Normal mean BG values in children (68 pg/ml) are in general higher than those from adults (48 pg/ml) probably due to *Candida* colonization.[59–61] However, BG is higher in children with proven IFD than in uninfected comparators, and most patients show positive results before CT-scan findings or fungal isolation in most patients.[62–64]

Candida Antigens and Anti-*Candida* Antibody Detection

The diagnosis of IFD through detection of antibodies or antigens in serum or other bodily fluids has been disappointing to date. Immunocompromised individuals with IC often produce low antibody titers, which can lead to false negative test results, while elevated antibody levels due to *Candida* colonization can result in false positives. However, Platelia Candida Antigen Plus and Antibody Plus (Bio-Rad) for detecting mannan and anti-mannan antibodies and *Candida albicans* IFA IgG (Vircell, Spain) for detecting CAGTA have showed promising results (Table 17.2).[55,65] Mannan, a major component of *Candida*

TABLE 17.2 Recommendations of Serological Tests for Invasive Candidiasis

Indications and Diagnostic Limits	Mannan and Anti-Mannan Antibody Detection	Anti-Germ Tube Antibodies (CAGTA) Detection
Quality of evidence and strength of recommendations	Early diagnosis of invasive candidiasis (AII-BII).	Early diagnosis of invasive candidiasis (AII-BII).
Methods and suggested use	Enzyme immune assay (Platelia Candida Antigen Plus and Antibody Plus, Bio-Rad Lab., France). Detection each 3–4 days.	Indirect immune-fluorescence (*Candida albicans* IFA IgG, Vircell, Spain). Detection each 4–7 days.
Diagnostic value	Mycological criterion of probable invasive candidiasis according to ESCMID and EORTC/MSG definitions. Detection of mannan + antimannan antibodies previous to positivity in other tests.	Detection of CAGTA previous to positivity in other tests. The combination of BG and CAGTA increases the diagnosis of invasive candidiasis. Positive: >1/160 Detection of CAGTA and later decrease of titers are considered sign of good prognosis.
Prognostic value	Decrease of titers is considered as good prognosis.	Decrease of titers is considered as good prognosis.
False positives	Allergic aspergillosis. *Saccharomyces cerevisiae* fungemia.	Allergic aspergillosis.
False negatives	Some immunodeficient patients.	Some immunodeficient patients.

EORTC/MSG: European Organization for Research and Treatment of Cancer/Mycoses Study Group of the National Institute of Allergy and Infectious Diseases. ESCMID: European Society of Clinical Microbiology & Infectious Diseases. **Strength of recommendation:** A = Good evidence to support a recommendation for or against use. B = Moderate evidence to support a recommendation for or against use. C = Poor evidence to support a recommendation. **Quality of evidence:** I = Evidence from ≥ 1 properly randomized, controlled trial. II = Evidence from ≥ 1 well-designed clinical trial, without randomization; from cohort or case-controlled analytic studies (preferably from >1 center); from multiple time-series; or from dramatic results from uncontrolled experiments. III = Evidence from opinions of respected authorities, based on clinical experience, descriptive studies, or reports of expert committees.

cell wall, and antimannan antibodies are detected in serum from patients with IC. Detectable concentrations of mannan are of short duration and serial determinations are necessary. Antimannan antibodies are detected many times after mannan disappearance. Thus, the combined detection of mannan and antimannan antibodies is recommended for the diagnosis of candidemia in adults and neonates as it can be positive prior to blood cultures, with sensitivity, specificity, and NPV rates around 70–85%.[18,52,66,67] Arendrup et al. reported even higher sensitivity (100%) but lower specificity (74%), with 80% and 100% PPV and NPV, respectively, for candidemia employing lower cut-offs than those suggested by the manufacturers.[16] There are also encouraging results in sensitivity (88%), specificity (95%), PPV (78%) and NPV (98%) for CAGTA detection in hematological patients with IC.[68] Moreover, CAGTA detection can be a complement of blood cultures for diagnostic and therapeutic monitoring of ICU patients with IC.[65]

Galactomannan (GM) Detection

GM is a component of the *Aspergillus* cell wall that can be detected by enzyme immune assay (EIA) (Platelia Aspergillus, Bio-Rad) in the tissues and fluids of patients with invasive aspergillosis (IA). Different cut-offs have been proposed for serum (0.5–0.7 ng/ml), urine and BAL (1 ng/ml), and pericardial fluid and CSF (0.5 ng/ml) but there is a variable performance in BAL and urine, and excellent GM detection but limited data are available in CSF. GM detection assay is well standardized, requires small amounts of specimen for analysis (300 μl) and it is fast (~4 hours). The test is useful for the early diagnosis for IA, particularly if at-risk patients are serially screened, with an overall sensitivity and specificity of 65–75% and 80–95%, respectively. However, sensitivity is higher among patients with hematologic malignancies (70%) and bone marrow transplant recipients (80%) than in solid organ transplant recipients (22%). The sensitivity is better in IA caused by *Aspergillus* species other than *A. fumigatus*. Of interest, declining GM levels have been observed in patients who responded to antifungal therapy and rising concentrations in those with fatal outcomes.[43,69]

GM detection also has different characteristics among children with cancer or bone marrow transplant, with GM sensitivity and specificity values of 32% and 98%, respectively, and PPV and NPV of 70% and 92%, respectively.[58] The meta-analysis of Pfeiffer et al. revealed that GM detection sensitivity in children was 76% compared to 73% in adults, while specificity reached 86% in children vs. 90% in adults.[70] GM detection could give false positives in neonates and other patients without IA as it is absorbed through altered gastro-intestinal mucosa of patients receiving cytotoxic chemotherapy. A lipoteichoic acid antigen of bifidobacteria, present in high concentrations in the gastro-intestinal microbiota of neonates, the use of betalactam antibiotics, other IFD (geotrichosis, penicilliosis, histoplasmosis, or blastomycosis), or use of plasmalyte in BAL fluid or other infusion solutions could increase false positive results (Table 17.1). False negatives have been reported in <10% of patients with documented IA, probably associated with variability in GM production prior or concomitant antifungal therapy, the presence of interfering anti-GM antibodies, or a low level of angioinvasion in very localized infections.[44,71,72] Recently, a novel lateral flow device that uses a monoclonal antibody that targets an extracellular glycoprotein of *Aspergillus*, has been developed for IA diagnosis. This device removes most technical requirements and reduces processing time (<30 min); preliminary clinical results with serum, BAL and urine are very promising.[73,74]

Pazos et al. observed that the combination of BG and GM detection was more effective than either test alone in diagnosing IA.[43] Interestingly, both tests were positive in the same patients with IA. The kinetics of both markers were very similar but BG tended to become positive earlier than GM. Discrepancies were observed with false positive results, since patients without IA but positive for BG detection were negative by GM detection and patients with false positive results for GM were negative by BG detection. Consideration of the results obtained for both markers in combination showed an improvement over the diagnostic efficacy of each individual test. The sensitivity, specificity, PPV and NPV were 87.5%, 100%, 100%, and 96.3%, respectively. The combination of BG and GM detection was very useful in confirming the existence of IA (both markers were positive in patients with IA). BG tended to become positive earlier than GM.

This precocity and the fact that BG is a panfungal marker should lead to initiation of the diagnostic workup of IFD when a positive result is detected and should enable prompt antifungal treatment with a broad-spectrum antifungal drug.

Glucuroxylomannan (GXM) Detection

Detection of GXM in CSF is widely accepted as a screening tool for the diagnosis of cryptococcal meningitis, due to its high sensitivity (93 to 100%). Any of the commercially available EIA or latex agglutination methods can be used to perform the test, using serum and CSF. Serial monitoring of GXM titer in CSF may correlate with clinical response to treatment.[75,76] A new, rapid (<10 min), commercial method using a lateral flow device (IMMY Diagnostics, USA) has become available for detection of GXM that can be used to test serum, CSF or urine and should improve the early diagnosis of cryptococcal meningitis.[77–79] False positive results have been reported during invasive *Trichosporon* infections, presence of rheumatoid factor, and *Capnocytophaga canimorsus* or *Stomatococcus mucilaginosus* bacteremias.[75,80] Some authors consider that the combined detection of GXM and BG is very useful for the early diagnosis of invasive trichosporonosis.[80]

DNA Detection

Nucleic acid-based diagnostic techniques are the fastest growing segment of fungal diagnostics. Detection of fungal DNA in blood or BAL specimens is promising for IFD caused by fungal pathogens that grow very slowly, or are difficult to isolate in culture media.[44] PCR-based assays have a low threshold of detection and as few as 1–5 cells/ml can be detected. For this reason, these tests have increased sensitivity, especially in cases of IC, such as hepatosplenic candidiasis, for which blood culture results are negative. However, PCR-based assays measure DNA not only from viable cells but also from dead cells or even free DNA. Moreover, PCR-based assays can detect multiple species in a single sample of blood and may have a role in monitoring the effectiveness of antifungal treatment of IC.[69,81] Most studies report similar results in children and adults, with sensitivity close to 100%, specificity ranges from 60% to 90%, NPV between 88% and 100%, but extremely variable PPV, indicating that they are more useful in excluding rather than establishing the diagnosis of IC. A recent meta-analysis of 54 studies using PCR-based methods to diagnose IC from blood samples showed good sensitivity and specificity for rapid diagnosis of IC. Sensitivity and specificity were very good for patients with proven candidemia, but the sensitivity was lower for patients with probable or possible IC.[82]

PCR-based assays of blood can detect *Aspergillus* DNA 14 days before the clinical diagnosis of IA, preceding positive GM (range, 4–63 days) or positive CT findings (range 1–84 days). PCR studies for IA show variability in PPV, but NPV can reach 100%, highlighting that PCR-based assays are currently more useful in excluding rather than confirming the diagnosis of IA.[69] A meta-analysis of the use of PCR from blood, serum, or plasma samples reported that the sensitivity and specificity for two consecutive positive samples were 75% and 87%, respectively.[83] Similar meta-analyses evaluating PCR on BAL revealed superior sensitivity and specificity values (90%) for the diagnosis of IA.[84,85] Regarding other clinical samples, in a small study that included six patients with cerebral aspergillosis, a PCR-based assay of CSF demonstrated 100% sensitivity.[86] Despite a low sensitivity rate, positive results for both GM and *Aspergillus* DNA in BAL make a pulmonary aspergillosis highly likely. However, if either positive BAL GM or positive BAL PCR results define test positivity, the sensitivity increases and the specificity decreases to values around 83%, which may be more convenient for aiding clinical diagnosis of IA.[87,88]

Data concerning DNA detection of *Aspergillus* in pediatric patients are limited. One recent multicenter study has evaluated a nested PCR assay of blood, CSF and BAL samples from 71 pediatric and adolescent patients with suspected IA. The sensitivity and specificity were high (80% and 81%, respectively), and PPV and NPV were 40% and 96%, respectively.[86]

One advantage of PCR is that a negative result nearly always correlates with absence of disease. However, PCR has a number of drawbacks, which include the lack of standardization and reliable validation of methods and results obtained among laboratories. Moreover, the difficulty of interpreting the findings may result in inconsistencies and unreliable results in both adult and pediatric populations. To overcome these problems and to optimize PCR findings, a real-time quantitative PCR assay based on minimum information for the publication of real-time quantitative PCR experiments (MIQE) guidelines have been proposed.[89] There are a limited number of commercialized kits avaiable, such as LightCycler SeptiFast Test MGRADE and MycAssay Aspergillus, but performance studies are underway for many others.[40–42] For instance, MycAssay™ Aspergillus has been tested on tissues from patients with IFD. The observed sensitivity was 82–90%, the specificity 64–79%, and PPV and NPV were 69–88% compared to microscopy and culture.[41] LightCycler SeptiFast detects a number of bacteria and some fungi, including *A. fumigatus* and five species of *Candida*: *C. albicans*, *C. glabrata*, *C. krusei*, *C. parapsilosis*, and *C. tropicalis*.[40] Fernandez-Cruz et al. have recently assessed the value of persistence of bacterial and fungal DNA in blood for predicting complicated bloodstream infections. Patients with a positive SeptiFast result between days 3 and 7 after a positive blood culture had a higher risk of developing complicated bloodstream infections.[42]

CONCLUSION: PROPOSED CLINICAL USES AND LIMITATIONS OF FUNGAL BIOMARKERS DETECTION

Management of IFD presents a therapeutic and diagnostic challenge because prompt initiation of targeted antifungal treatment is life saving. An ideal test for diagnosing an IFD should have good performance characteristics, give a high level of certainty with respect to the presence or absence of disease, and should become positive at an early stage of infection. However, culture lacks sensitivity, histology requires invasive procedures, imaging lacks specificity, and fungal biomarker detection requires higher standardization. There is no single test available with all the required features, and combining information from different laboratory tests and clinical procedures is required.

Non-culture methods based on biomarker detection allow a faster and more sensitive diagnosis than traditional microbiological methods. Their usefulness increases when samples from patients are tested regularly. BG and GM detection are useful tests for the diagnosis of IA in high-risk patients with hematological malignancies. In addition, a combination of GM and PCR-based testing could also help identify cases of IA with poor prognosis. One possible diagnostic approach would be to screen high risk patients for the presence of biomarkers during the periods of highest risk. This selection is very important, because the incidence of IFD, especially when it is very low, greatly influences the PPV and NPV of diagnostic methods. For instance, when the incidence of IA is 0.5%, as in kidney transplant recipients, only one out of 10 patients with a positive GM and one out of 70 with a positive PCR will actually suffer from IA. Although some questions remain to be solved, such as the kinetics and release of biomarkers, assay performances in different groups of patients, or the effects of antifungal treatment, the broader use of these assays could help to better understand the pathogenesis of IFD and improve their diagnosis, treatment and prognosis. Moreover, currently most tests lack standardization, and a future consensus will be necessary for facilitating large, prospective, clinical trials focused on the integration of these techniques into clinical decision-making. In conclusion, the implementation of non-culture based methods, alone or in combination with other microbiological tests, into the clinical setting could improve therapeutic approaches and outcome, by treatment monitoring, and personalizing the therapy.

References

1. Ostrosky-Zeichner L. Invasive mycoses: diagnostic challenges. *Am J Med* 2012;**125**:S14–24.
2. Kauffman CA, Pappas PG, Sobel JD, Dismukes WE, eds. *Essentials of Clinical Mycology*. 2nd ed. New York: *Springer Science + Business Media*; 2011.
3. Quindós G. Epidemiology of candidaemia and invasive candidiasis. A changing face. *Rev Iberoam Micol* 2013. Available from: http://dx.doi.org/10.1016/j.riam.2013.10.001.
4. Gullo FP, Rossi SA, Sardi JCO, Teodoro VL, Mendes-Giannini MJ, Fusco-Almeida AM. Cryptococcosis: epidemiology, fungal resistance, and new alternatives for treatment. *Eur J Clin Microbiol Infect Dis* 2013;**32**:1377–91.
5. Garey KW, Rege M, Pai MP, Mingo DE, Suda KJ, Turpin RS, et al. Time to initiation of fluconazole therapy impacts mortality in patients with candidemia: a multi-institutional study. *Clin Infect Dis* 2006;**43**:25–31.
6. Grim SA, Berger K, Teng C, Gupta S, Layden JE, Janda WM, et al. Timing of susceptibility-based antifungal drug administration in patients with *Candida* bloodstream infection: correlation with outcomes. *J Antimicrob Chemother* 2012;**67**:707–14.
7. Ayats J, Martín-Mazuelos E, Pemán J, Quindós G, Sánchez F, García-Rodríguez J, Grupo de Estudio de Micología Médica de la SEIMC (GEMICOMED), et al. Spanish Society of Clinical Microbiology and Infectious Diseases (SEIMC) guidelines for the diagnosis of invasive fungal infections. 2010 update. *Enferm Infecc Microbiol Clin* 2011;**29**: 39.e1–15
8. Cuenca-Estrella M, Bassetti M, Lass-Flörl C, Rácil Z, Richardson M, Rogers TR. Detection and investigation of invasive mould disease. *J Antimicrob Chemother* 2011;**66**:i15–24.
9. Arendrup MC, Bille J, Dannaoui E, et al. ECIL-3 classical diagnostic procedures for the diagnosis of invasive fungal diseases in patients with leukemia. *Bone Marrow Transplant* 2012;**47**:1030–45.
10. Quindós G, Eraso E, López-Soria LM, Ezpeleta G. Enfermedad fúngica invasora: ¿Diagnóstico micológico convencional o molecular? *Enferm Infecc Microbiol Clin* 2012;**30**:560–71.
11. Posteraro B, Torelli R, De Carolis E, Posteraro P, Sanguinetti M. Update on the laboratory diagnosis of invasive fungal infections. *Mediterr J Hematol Infect Dis* 2011;**3**:e2011002.
12. Colombo AL, Cortés JA, Zurita J, Guzman-Blanco M, Alvarado Matute T, de Queiroz Telles F, et al. Recommendations for the diagnosis of candidemia in Latin America. *Rev Iberoam Micol* 2013;**30**:150–7.
13. Guarner J, Brandt ME. Histopathologic diagnosis of fungal infections in the 21st century. *Clin Microbiol Rev* 2011;**24**:247–80.

14. Rickerts V, Khot PD, Myerson D, Ko DL, Lambrecht E, Fredricks DN. Comparison of quantitative real time PCR with sequencing and ribosomal RNA-FISH for the identification of fungi in formalin fixed, paraffin-embedded tissue specimens. *BMC Infect Dis* 2011;**11**:202.
15. Horvath LL, George BJ, Murray CK, Harrison LS, Hospenthal DR. Direct comparison of the BACTEC 9240 and BacT/ALERT 3D automated blood culture systems for *Candida* growth detection. *J Clin Microbiol* 2004;**42**:115–8.
16. Arendrup MC, Bergmann OJ, Larsson L, Nielsen HV, Jarløv JO, Christensson B. Detection of candidaemia in patients with and without underlying haematological disease. *Clin Microbiol Infect* 2010;**16**:855–62.
17. Cateau E, Cognee AS, Tran TC, Vallade E, Garcia M, Belaz S, et al. Impact of yeast-bacteria coinfection on the detection of *Candida* sp. in an automated blood culture system. *Diagn Microbiol Infect Dis* 2012;**72**:328–31.
18. Cuenca-Estrella M, Verweij PE, Arendrup MC, Arikan-Akdagli S, Bille J, Donnelly JP, et al. ESCMID Fungal Infection Study Group. ESCMID* guideline for the diagnosis and management of *Candida* diseases 2012: diagnostic procedures. *Clin Microbiol Infect* 2012;**18**:9–18.
19. Eraso E, Moragues MD, Villar-Vidal M, Sahand IH, González-Gómez N, Pontón J, et al. Evaluation of the new chromogenic medium Candida ID 2 for isolation and identification of *Candida albicans* and other medically important *Candida* species. *J Clin Microbiol* 2006;**44**:3340–5.
20. McTaggart L, Richardson SE, Seah C, Hoang L, Fothergill A, Zhang SX. Rapid identification of *Cryptococcus neoformans* var. *grubii*, *C. neoformans* var. *neoformans*, and *C. gattii* by use of rapid biochemical tests, differential media, and DNA Sequencing. *J Clin Microbiol* 2011;**49**:2522–7.
21. Baires-Varguez L, Cruz-García A, Villa-Tanaka L, Sánchez-García S, Gaitán-Cepeda LA, Sánchez-Vargas LO, et al. Comparison of a randomly amplified polymorphic DNA (RAPD) analysis and ATB ID 32C system for identification of clinical isolates of different *Candida* species. *Rev Iberoam Micol* 2007;**24**:148–51.
22. Cejudo MA, Gallego AG, Cantón-Lacasa E, Aller AI, Romero A, García JP, et al. Evaluation of the VITEK 2 system to test the susceptibility of *Candida* spp., *Trichosporon asahii* and *Cryptococcus neoformans* to amphotericin B, flucytosine, fluconazole and voriconazole: a comparison with the M27-A3 reference method. *Med Mycol* 2010;**48**:710–9.
23. Quindós G, San Millán R, Robert R, Bernard C, Ponton J. Evaluation of bichro-latex albicans, a new method for rapid identification of *Candida albicans*. *J Clin Microbiol* 1997;**35**:1263–5.
24. Freydiere AM, Perry JD, Faure O, Willinger B, Tortorano AM, Nicholson A, et al. Routine use of a commercial test, GLABRATA RTT, for rapid identification of *Candida glabrata* in six laboratories. *J Clin Microbiol* 2004;**42**:4870–2.
25. Sahand IH, Moragues MD, Robert R, Quindós G, Pontón J. Evaluation of Bichro-Dubli Fumouze to distinguish *Candida dubliniensis* from *Candida albicans*. *Diagn Microbiol Infect Dis* 2006;**55**:165–7.
26. Chagas-Neto TC, Chaves GM, Melo AS, Colombo AL. Bloodstream infections due to *Trichosporon* spp.: species distribution, *Trichosporon asahii* genotypes determined on the basis of ribosomal DNA intergenic spacer 1 sequencing, and antifungal susceptibility testing. *J Clin Microbiol* 2009;**47**:1074–81.
27. Landlinger C, Preuner S, Willinger B, Haberpursch B, Racil Z, Mayer J, et al. Species-specific identification of a wide range of clinically relevant fungal pathogens by use of Luminex xMAP technology. *J Clin Microbiol* 2009;**47**:1063–73.
28. Xiao M, Guo LN, Kong F, Wang H, Sorrell TC, Li RY, et al. Practical identification of eight medically important *Trichosporon* species by reverse line blot hybridization (RLB) assay and rolling circle amplification (RCA). *Med Mycol* 2013;**51**:300–8.
29. Hall L, Le Febre KM, Deml SM, Wohlfiel SL, Wengenack NL. Evaluation of the Yeast Traffic Light PNA FISH probes for identification of *Candida* species from positive blood cultures. *J Clin Microbiol* 2012;**50**:1446–8.
30. Tavanti A, Landi S, Senesi S. APEX DNA microarray for the identification of pathogenic fungi. *Methods Mol Biol* 2013;**968**:63–70.
31. Marklein G, Josten M, Klanke U, Müller E, Horré R, Maier T, et al. Matrix-assisted laser desorption ionization-time of flight mass spectrometry for fast and reliable identification of clinical yeast isolates. *J Clin Microbiol* 2009;**47**:2912–7.
32. Sendid B, Ducoroy P, Francois N, Lucchi G, Spinali S, Vagner O, et al. Evaluation of MALDI-TOF mass spectrometry for the identification of medically-important yeasts in the clinical laboratories of Dijon and Lille hospitals. *Med Mycol* 2013;**51**:25–32.

33. Mancini N, De Carolis E, Infurnari L, Vella A, Clementi N, Vaccaro L, et al. Comparative evaluation of the Bruker biotyper and Vitek MS Matrix-assisted laser desorption ionization-time of flight (MALDI-TOF) mass spectrometry systems for identification of yeasts of medical importance. *J Clin Microbiol* 2013;**51**:2453–7.
34. Firacative C, Trilles L, Meyer W. MALDI-TOF MS enables the rapid identification of the major molecular types within the *Cryptococcus neoformans/C. gattii* species complex. *PLoS ONE* 2012;**7**:e37566.
35. Normand AC, Cassagne C, Ranque S, L'ollivier C, Fourquet P, Roesems S, et al. Assessment of various parameters to improve MALDI-TOF MS reference spectra libraries constructed for the routine identification of filamentous fungi. *BMC Microbiol* 2013;**13**:76.
36. Massire C, Buelow DR, Zhang SX, Lovari R, Matthews HE, Toleno DM, et al. PCR followed by electrospray ionization mass spectrometry for broad-range identification of fungal pathogens. *J Clin Microbiol* 2013;**51**:959–66.
37. Shin JH, Ranken R, Sefers SE, Lovari R, Quinn CD, Meng S, et al. Detection, identification, and distribution of fungi in bronchoalveolar lavage specimens by use of multilocus PCR coupled with electrospray ionization/mass spectrometry. *J Clin Microbiol* 2013;**51**:136–41.
38. Eraso E, Ruesga M, Villar-Vidal M, Carrillo-Muñoz AJ, Espinel-Ingroff A, Quindós G. Comparative evaluation of ATB Fungus 2 and Sensititre YeastOne panels for testing *in vitro Candida* antifungal susceptibility. *Rev Iberoam Micol* 2008;**25**:3–6.
39. Ha JF, Italiano CM, Heath CH, Shih S, Rea S, Wood FM. Candidemia and invasive candidiasis: a review of the literature for the burns surgeon. *Burns* 2011;**37**:181–95.
40. von Lilienfeld-Toal M, Lehmann LE, Raadts AD, Hahn-Ast C, Orlopp KS, Marklein G, et al. Utility of a commercially available multiplex real-time PCR assay to detect bacterial and fungal pathogens in febrile neutropenia. *J Clin Microbiol* 2009;**47**:2405–10.
41. Lass-Flörl C, Follett SA, Moody A, Denning DW. Detection of *Aspergillus* in lung and other tissue samples using the MycAssay Aspergillus real-time PCR kit. *Can J Microbiol* 2011;**57**:765–8.
42. Fernández-Cruz A, Marín M, Kestler M, Alcalá L, Rodriguez-Créixems M, Bouza E. The value of adding Septifast(R) to a score for predicting complicated bloodstream infections caused by Gram-positive bacteria or *Candida* species. *J Clin Microbiol* 2013;**51**:1130–6.
43. Pazos C, Pontón J, Del Palacio A. Contribution of (1,3)-β-D-glucan chromogenic assay to diagnosis and therapeutic monitoring of invasive aspergillosis in neutropenic adult patients: a comparison with serial screening for circulating galactomannan. *J Clin Microbiol* 2005;**43**:299–305.
44. Quindós G. New microbiological techniques for the diagnosis of invasive mycoses caused by filamentous fungi. *Clin Microbiol Infect* 2006;**12**:40–52.
45. Lamoth F, Cruciani M, Mengoli C, Castagnola E, Lortholary O, Richardson M, et al. Third European Conference on Infections in Leukemia (ECIL-3). β-Glucan antigenemia assay for the diagnosis of invasive fungal infections in patients with haematological malignancies: a systematic review and meta-analysis of cohort studies from the Third European Conference on Infections in Leukemia (ECIL-3). *Clin Infect Dis* 2012;**54**:633–43.
46. Obayashi T, Yoshida M, Mori T, Goto H, Yasuoka A, Iwasaki H, et al. Plasma (1→3)-beta-D-glucan measurement in diagnosis of invasive deep mycosis and fungal febrile episodes. *Lancet* 1995;**345**:17–20.
47. Odabasi Z, Mattiuzzi G, Estey E, Kantarjian H, Saeki F, Ridge RJ, et al. Beta-D-glucan as a diagnostic adjunct for invasive fungal infections: validation, cutoff development, and performance in patients with acute myelogenous leukemia and myelodysplastic syndrome. *Clin Infect Dis* 2004;**39**:199–205.
48. Koo S, Bryar JM, Page JH, Baden LR, Marty FM. Diagnostic performance of the (1,3)-beta-D-glucan assay for invasive fungal disease. *Clin Infect Dis* 2009;**49**:1650–9.
49. Cuétara MS, Alhambra A, Moragues MD, González-Elorza E, Pontón J, del Palacio. Detection of (1,3)-beta-D-glucan as an adjunct to diagnosis in a mixed population with uncommon proven invasive fungal diseases or with an unusual clinical presentation. *Clin Vaccine Immunol* 2009;**16**:423–6.
50. Pappas PG, Kauffman CA, Andes D, Benjamin Jr DK, Calandra TF, Edwards Jr JE, et al. Infectious Diseases Society of America. Clinical practice guidelines for the management of candidiasis: 2009 update by the Infectious Diseases Society of America. *Clin Infect Dis* 2009;**48**:503–35.
51. Jaijakul S, Vazquez JA, Swanson RN, Ostrosky-Zeichner L. (1,3)-β-D-glucan as a prognostic marker of treatment response in invasive candidiasis. *Clin Infect Dis* 2012;**55**:521–6.
52. Marchetti O, Lamoth F, Mikulska M, Viscoli C, Verweij P, Bretagne S, European Conference on Infections in Leukemia (ECIL) Laboratory Working Groups. ECIL recommendations for the use of biological markers for

the diagnosis of invasive fungal diseases in leukemic patients and hematopoietic SCT recipients. *Bone Marrow Transplant* 2012;**47**:846–54.

53. Nakase K, Suzuki K, Kyo T, Kohara T, Sugawara Y, Katayama N. Is elevation of the serum beta-d-glucan level a paradoxical sign for *Trichosporon* fungemia in patients with hematologic disorders? *Int J Infect Dis* 2012;**16**: e2–4.
54. Kushima H, Ishii H, Komiya K, Tokimatsu I, Kadota J. Prognostic significance of serum β-d-glucan levels in 78 patients with *Trichosporon* fungemia. *Int J Infect Dis* 2013;**17**:e134–5.
55. Pazos C, Moragues MD, Quindós G, Pontón J, del Palacio A. Diagnostic potential of (1,3)-beta-D-glucan and anti-*Candida albicans* germ tube antibodies for the diagnosis and therapeutic monitoring of invasive candidiasis in neutropenic adult patients. *Rev Iberoam Micol* 2006;**23**:209–15.
56. Theel ES, Doern CD. Point-Counterpoint: B-D-glucan is an important test in the diagnosis of invasive fungal infections. *J Clin Microbiol* 2013;**51**:3478–83.
57. Karageorgopoulos DE, Vouloumanou EK, Ntziora F, Michalopoulos A, Rafailidis PI, Falagas ME. Beta-D-glucan assay for the diagnosis of invasive fungal infections: a meta-analysis. *Clin Infect Dis* 2011;**52**:750–70.
58. Roilides E, Pana ZD. Application of diagnostic markers to invasive aspergillosis in children. *Ann N Y Acad Sci* 2012;**1272**:1–8.
59. Smith PB, Benjamin Jr DK, Alexander BD, Johnson MD, Finkelman MA, Steinbach WJ. Quantification of 1,3-beta-D-glucan levels in children: preliminary data for diagnostic use of the beta-glucan assay in a pediatric setting. *Clin Vaccine Immunol* 2007;**14**:924–5.
60. Mokaddas E, Burhamah MH, Khan ZU, Ahmad S. Levels of (1→3)-β-D-glucan, *Candida* mannan and *Candida* DNA in serum samples of pediatric cancer patients colonized with *Candida* species. *BMC Infect Dis* 2010;**10**:292.
61. Montagna MT, Coretti C, Lovero G, De Giglio O, Montagna O, Laforgia N, et al. Diagnostic performance of 1→3-β-D-Glucan in neonatal and pediatric patients with candidemia. *Int J Mol Sci* 2011;**12**:5871–7.
62. Mularoni A, Furfaro E, Faraci M, Franceschi A, Mezzano P, Bandettini R, et al. High Levels of beta-D-glucan in immunocompromised children with proven invasive fungal disease. *Clin Vaccine Immunol* 2010;**17**:882–3.
63. Oz Y, Kiraz N. Diagnostic methods for fungal infections in pediatric patients: microbiological, serological and molecular methods. *Expert Rev Anti Infect Ther* 2011;**9**:289–98.
64. Goudjil S, Kongolo G, Dusol L, Imestouren F, Cornu M, Leke A, et al. (1–3)-β-D-glucan levels in candidiasis infections in the critically ill neonate. *J Matern Fetal Neonatal Med* 2013;**26**:44–8.
65. Zaragoza R, Pemán J, Quindós G, Iruretagoyena JR, Cuétara MS, Ramírez P, et al. *Candida albicans* Germ Tube Antibody Detection in Critically Ill Patients (CAGTAUCI). Clinical significance of the detection of *Candida albicans* germ tube-specific antibodies in critically ill patients. *Clin Microbiol Infect* 2009;**15**:592–5.
66. Oliveri S, Trovato L, Betta P, Romeo MG, Nicoletti G. Experience with the Platelia Candida ELISA for the diagnosis of invasive candidosis in neonatal patients. *Clin Microbiol Infect* 2008;**14**:391–3.
67. Mikulska M, Calandra T, Sanguinetti M, Poulain D, Viscoli C. Third European Conference on Infections in Leukemia Group. The use of mannan antigen and anti-mannan antibodies in the diagnosis of invasive candidiasis: recommendations from the Third European Conference on Infections in Leukemia. *Crit Care* 2010;**14**: R222.
68. García Ruiz JC, Arilla MC, Regúlez P, Moragues MD, Quindós G, Pontón J. Detection of antibodies to *Candida albicans* germ tubes for the diagnosis and therapeutic monitoring of invasive candidiasis in patients with hematologic malignancies. *J Clin Microbiol* 1997;**35**:3284–7.
69. Kourkoumpetis TK, Fuchs BB, Coleman JJ, Desalermos A, Mylonakis E. Polymerase chain reaction-based assays for the diagnosis of invasive fungal infections. *Clin Infect Dis* 2012;**54**:1322–31.
70. Pfeiffer CD, Fine JP, Safdar N. Diagnosis of invasive aspergillosis using a galactomannan assay: a meta-analysis. *Clin Infect Dis* 2006;**42**:1417–27.
71. Marr KA, Balajee SA, McLaughlin L, Tabouret M, Bentsen C, Walsh TJ. Detection of galactomannan antigenemia by enzyme immunoassay for the diagnosis of invasive aspergillosis: variables that affect performance. *J Infect Dis* 2004;**190**:641–9.
72. Bonini A, Capatti C, Parmeggiani M, Gugliotta L, Micozzi A, Gentile G, et al. Galactomannan detection in *Geotrichum capitatum* invasive infections: report of 2 new cases and review of diagnostic options. *Diagn Microbiol Infect Dis* 2008;**62**:450–2.

73. Thornton CR. Development of an immunochromatographic lateral-flow device for rapid serodiagnosis of invasive aspergillosis. *Clin Vaccine Immunol* 2008;**15**:1095–105.
74. White PL, Parr C, Thornton C, Barnes RA. Evaluation of real-time PCR, galactomannan enzyme-linked immunosorbent assay (ELISA), and a novel lateral-flow device for diagnosis of invasive aspergillosis. *J Clin Microbiol* 2013;**51**:1510–6.
75. Babady NE, Bestrom JE, Jespersen DJ, Jones MF, Beito EM, Binnicker MJ, et al. Evaluation of three commercial latex agglutination kits and a commercial enzyme immunoassay for the detection of cryptococcal antigen. *Med Mycol* 2009;**47**:336–8.
76. Jarvis JN, Percival A, Bauman S, Pelfrey J, Meintjes G, Williams GN, et al. Evaluation of a novel point-of-care cryptococcal antigen test on serum, plasma, and urine from patients with HIV-associated cryptococcal meningitis. *Clin Infect Dis* 2011;**53**:1019–23.
77. Lindsley MD, Mekha N, Baggett HC, Surinthong Y, Autthateinchai R, Sawatwong P, et al. Evaluation of a newly developed lateral flow immunoassay for the diagnosis of cryptococcosis. *Clin Infect Dis* 2011;**53**:321–5.
78. McMullan BJ, Halliday C, Sorrell TC, Judd D, Sleiman S, Marriott D, et al. Clinical utility of the cryptococcal antigen lateral flow assay in a diagnostic mycology laboratory. *PLoS ONE* 2012;**7**:e49541.
79. Hansen J, Slechta ES, Gates-Hollingsworth MA, Neary B, Barker AP, Bauman S, et al. Large-scale evaluation of the immuno-mycologics lateral flow and enzyme-linked immunoassays for detection of cryptococcal antigen in serum and cerebrospinal fluid. *Clin Vaccine Immunol* 2012;**20**:52–5.
80. Liao Y, Hartmann T, Ao J, Yang RY. Serum glucuronoxylomannan may be more appropriate for the diagnosis and therapeutic monitoring of *Trichosporon* fungemia than serum beta-d-glucan. *Int J Infect Dis* 2012;**16**:e638.
81. Sugawara Y, Nakase K, Nakamura A, Ohishi K, Sugimoto Y, Fujieda A, et al. Clinical utility of a panfungal polymerase chain reaction assay for invasive fungal diseases in patients with haematologic disorders. *Eur J Haematol* 2013;**90**:331–9.
82. Avni T, Leibovici L, Paul M. PCR diagnosis of invasive candidiasis: systematic review and meta-analysis. *J Clin Microbiol* 2011;**49**:665–70.
83. Mengoli C, Cruciani M, Barnes RA, Loeffler J, Donnelly JP. Use of PCR for diagnosis of invasive aspergillosis: systematic review and meta-analysis. *Lancet Infect Dis* 2009;**9**:89–96.
84. Tuon FF. A systematic literature review on the diagnosis of invasive aspergillosis using polymerase chain reaction (PCR) from bronchoalveolar lavage clinical samples. *Rev Iberoam Micol* 2007;**24**:89–94.
85. Zou M, Tang L, Zhao S, Chen L, Chen P, Huang Z, et al. Systematic review and meta-analysis of detecting galactomannan in bronchoalveolar lavage fluid for diagnosing invasive aspergillosis. *PLoS One* 2012;**7**:e43347.
86. Hummel M, Spiess B, Kentouche K, Niggemann S, Böhm C, Reuter S, et al. Detection of *Aspergillus* DNA in cerebrospinal fluid from patients with cerebral aspergillosis by a nested PCR assay. *J Clin Microbiol* 2006;**44**:3989–93.
87. Reinwald M, Spiess B, Heinz WJ, Vehreschild JJ, Lass-Flörl C, Kiehl M, et al. Diagnosing pulmonary aspergillosis in patients with haematological malignancies: a multicenter prospective evaluation of an *Aspergillus* PCR assay and a galactomannan ELISA in bronchoalveolar lavage samples. *Eur J Haematol* 2012;**89**:120–7.
88. Hummel M, Spiess B, Roder J, von Komorowski G, Dürken M, Kentouche K, et al. Detection of *Aspergillus* DNA by a nested PCR assay is able to improve the diagnosis of invasive aspergillosis in pediatric patients. *J Med Microbiol* 2009;**58**:1291–7.
89. Johnson GL, Bibby DF, Wong S, Agrawal SG, Bustin SA. A MIQE-compliant real-time PCR assay for *Aspergillus* detection. *PLoS One* 2012;**7**:e40022.

Index

Note: Page numbers followed by "*f*", and "*t*" refers to figures and tables respectively.

Q

R

FSC
www.fsc.org
MIX
Papier | Fördert
gute Waldnutzung
FSC® C083411

Kolibri 360 GmbH
Ferdinand-Jühlke-Straße 7
99095 Erfurt, Deutschland
produktsicherheit@kolibri360.de